THE FRONTIERS COLLECTION

THE FRONTIERS COLLECTION

The books in this collection are devoted to challenging and open problems at the forefront of modern physics and related disciplines, including philosophical debates. In contrast to typical research monographs, however, they strive to present their topics in a manner accessible also to scientifically literate non-specialists wishing to gain insight into the deeper implications and fascinating questions involved. Taken as a whole, the series reflects the need for a fundamental and interdisciplinary approach to modern science. It is intended to encourage scientists in all areas to ponder over important and perhaps controversial issues beyond their own speciality. Extending from quantum physics and relativity to entropy, time and consciousness - the Frontiers Collection will inspire readers to push back the frontiers of their own knowledge.

Information and Its Role in Nature
By J. G. Roederer

Relativity and the Nature of Spacetime
By V. Petkov

Quo Vadis Quantum Mechanics?
Edited by A. C. Elitzur, S. Dolev, N. Kolenda

Life - As a Matter of Fat
The Emerging Science of Lipidomics
By O. G. Mouritsen

Quantum-Classical Analogies
By D. Dragoman and M. Dragoman

Knowledge and the World
Challenges Beyond the Science Wars
Edited by M. Carrier, J. Roggenhofer, G. Küppers, P. Blanchard

Quantum-Classical Correspondence
By A. O. Bolivar

Mind, Matter and Quantum Mechanics
By H. Stapp

Quantum Mechanics and Gravity
By M. Sachs

Extreme Events in Nature and Society
Edited by S. Albeverio, V. Jentsch, H. Kantz

S. Albeverio V. Jentsch H. Kantz (Eds.)

EXTREME EVENTS IN NATURE AND SOCIETY

With 115 Figures, 7 in Color

Springer

Prof. Dr. Sergio Albeverio
Universität Bonn
Interdisziplinäres Zentrum
für Komplexe Systeme
und Institut
für Angewandte Mathematik
Wegelerstraße 6
53115 Bonn, Germany
e-mail: albeverio@uni-bonn.de

Dr. Volker Jentsch
Universität Bonn
Interdisziplinäres Zentrum
für Komplexe Systeme
und Institut
für Angewandte Mathematik
Wegelerstraße 6
53115 Bonn, Germany
e-mail: jentsch@uni-bonn.de

Prof. Dr. Holger Kantz
Max-Planck-Institut
für Physik komplexer Systeme
Nöthnitzer Straße 38
01187 Dresden, Germany
e-mail: kantz@mpiks-dresden.mpg.de

Series Editors:

Cover figure: Detail from the image "Continuum Mechanics Models of DNA Molecules" by P. Furrer, J.H. Maddocks, R.S. Manning, R.C. Paffenroth and O. Gonzalesz

Library of Congress Control Number: 2005937085

ISSN 1612-3018
ISBN-10 3-540-28610-1 Springer Berlin Heidelberg New York
ISBN-13 978-3-540-28610-3 Springer Berlin Heidelberg New York

Springer is a part of Springer Science+Business Media
springer.com

Typesetting by Stephen Lyle using a Springer TEX macro package
Final processing by LE-TEX Jelonek, Schmidt & Vöckler GbR, Leipzig
Cover design by KünkelLopka, Werbeagentur GmbH, Heidelberg

Printed on acid-free paper SPIN: 11308621 57/3141/YL - 5 4 3 2 1 0

Preface

Somebody once remarked on how unjust it is that chaos has always had such a bad press. Whenever there is a traffic jam in the morning, when the children don't keep their things in order, when politics is turning crazy, it is always the fault of chaos. And yet, if there was no chaos, things would be pretty boring. Nothing unexpected would ever happen, and we could predict that the same dull things would happen today as they did yesterday. That's if we could predict anything at all – without chaos it would be quite likely that our thoughts would be trapped in some limited cycle and our brains would be quite useless!

The same also applies to extremes. Usually, when one thinks of extremes, negative connotations come to mind. Extremely hot weather is as unpleasant as extremely cold weather, and if its rains like crazy it is just as bad as when it is extremely dry. Extreme stock market fluctuations often result in large financial losses; earthquakes and floods can kill thousands of people, and global terrorism is strongly linked to political extremism.

But now try to imagine a world without extremes. Putting grand events like the Big Bang or the extinction of the dinosaurs – without which we humans would not exist – to one side, consider a world with constant lukewarm weather, where no-one ever fell in love, where there was never any deviation from the average. One can argue that even catastrophes have their positive sides, since they force us to look beyond our comfortable, well trodden paths. Although instinctively we would like to minimize their effects, that fact that we have to deal with them often leads to progress. Without extremes, there would be no shake-ups leading to novel situations and opportunities. And the Olympic Games would not be much fun either!

The present collection of articles, all written by well known experts in their fields, demonstrates these two aspects of extremes perfectly. On the one hand, we have to cope with their unpleasant sides, by predicting them as much as possible and by minimizing their effects. Most of the articles are therefore written from the point of view of the engineer or applied scientist who has to deal with this. But despite of the diversity of extreme phenomena – ranging from economic and geologic disasters via the breaking of steel to extreme neural bursts in epileptic seizures – the authors manage to show that there is a common underlying conceptual frame that links them. Indeed, as well as being linked by these concepts, various mathematical tools can be applied to most problems involving extremes. Therefore, this book demonstrates (without overstressing the point – just by providing the facts) that there is an emerging unifying and truly interdisciplinary science of extreme events.

Finally, the authors would not be good scientists if the fascinating and exciting aspects of the science of extremes did not permeate through every page. This another positive aspect of extremes: that they have led to this fascinating book, which is a real pleasure to read and which is sure to stimulate much further research.

Jülich, June 2005 *Peter Grassberger*

Acknowledgement

We, the editors, would like thank the authors for providing the articles. We also thank Springer Verlag for his continuous help in editing this book. We are thankful to Hurshid Kadirov who produced its TEX version.

Bonn, June 2005 *Sergio Albeverio*
Volker Jentsch
Holger Kantz

Contents

List of Contributors

Sergio Albeverio
Interdisciplinary Center
of Complex Systems (IZKS)
Universität Bonn
Bonn, Germany
albeverio@uni-bonn.de

Eduardo G. Altmann
Max Planck Institute
for the Physics of Complex Systems
Dresden, Germany
edugalt@mpipks-dresden.mpg.de

Hendrik Ammoser
Institute
for Transport and Economics
Dresden University of Technology
Dresden, Germany
ammoser@vwi.tu-dresden.de

Philippe Blanchard
Faculty of Physics and BiBoS
Universität Bielefeld
Bielefeld, Germany
blanchard
@physik.uni-bielefeld.de

Zuzana Chladná
International Institute
for Applied Systems Analysis
Laxenburg, Austria
and
Department of Applied
Mathematics and Statistics
Comenius University
Bratislava, Slovakia
chladna@iiasa.ac.at

Petra Friederichs
Meteorologisches Institut
Universität Bonn
Bonn, Germany
pfried@uni-bonn.de

Santo Fortunato
Institut für Physik
Universität Bielefeld
Bielefeld, Germany
fortunat@physik.
uni-bielefeld.de

Peter Grassberger
Forschungszentrum Jülich
Jülich, Germany
p.grassberger@fz-juelich.de

Matz Haaks
Helmholtz Institut für
Strahlen- und Kernphysik
Universität Bonn
Bonn, Germany
haaks@iskp.uni-bonn.de

Sarah Hallerberg
Max Planck Institute
for the Physics of Complex Systems
Dresden, Germany
sarah@mpipks-dresden.mpg.de

Dirk Helbing
Institute for Transport
and Economics
Dresden University of Technology
Dresden, Germany
helbing1@vwi.tu-dresden.de

Eric J. Heller
Department of Physics,
Chemistry and Chemical Biology
Harvard University
Cambridge, MA 02138 USA,
and
Wissenschaftskolleg zu Berlin
Berlin, Germany
heller@physics.harvard.edu

Andreas Hense
Meteorologisches Institut
Universität Bonn
Bonn, Germany
ahense@uni-bonn.de

Jürgen Herget
Geographisches Institut
Universität Bonn
Bonn, Germany
herget@giub.uni-bonn.de

Detlef Holstein
Max Planck Institute
for the Physics of Complex Systems
Dresden, Germany
holstein@mpipks-dresden.mpg.de

Volker Jentsch
Interdisciplinary Center
of Complex Systems (IZKS)
Universität Bonn
Bonn, Germany
jentsch@uni-bonn.de

Holger Kantz
Max Planck Institute
for the Physics of Complex Systems
Dresden, Germany
kantz@mpipks-dresden.mpg.de

Tyll Krueger
Faculty of Physics and BiBoS
Universität Bielefeld
Bielefeld, Germany
tyll.krueger@freenet.de

Christian Kühnert
Institute for Transport
and Economics
Dresden University of Technology
Dresden, Germany
kuehnert@vwi.tu-dresden.de

Klaus Lehnertz
Department of Epileptology
Universität Bonn
Bonn, Germany
klaus.lehnertz@ukb.uni-bonn.de

Michael Lehning
WSL, Swiss Federal Institute
for Snow and Avalanche Research
SLF
Switzerland
lehning@slf.ch

Karl Maier
Helmholtz Institut
für Strahlen- und Kernphysik
Universität Bonn
Bonn, Germany
maier@iskp.uni-bonn.de

Elena Moltchanova
International Institute
for Applied Systems Analysis
Laxenburg, Austria
moltchan@iiasa.ac.at

Mihai Nadin
Instutute for Research
in Anticipatory Systems
University of Texas at Dallas
Richardson, TX, USA
nadin@utdallas.edu

Michael Obersteiner
International Institute
for Applied Systems Analysis
Laxenburg, Austria
and
Department of Economics
Institute for Advanced Studies
Vienna, Austria
oberstei@iiasa.ac.at

Vladimir Piterbarg
Department of Statistics
Department of Probability
Moscow State University
Moscow, Russia

Anja Riegert
Max Planck Institute
for the Physics of Complex Systems
Dresden, Germany
riegert@mpipks-dresden.mpg.de

Didier Sornette
Institute of Geophysics
and Planetary Physics
and
Department of Earth
and Space Sciences
University of California
Los Angeles, CA 90095, USA
and
Laboratoire de Physique
de la Matière Condensée
CNRS UMR 6622
Université
de Nice – Sophia Antipolis
Nice Cedex 2, France
sornette@moho.ess.ucla.edu

Dietrich Stauffer
Institut für Theoretische Physik
Universität zu Köln
Köln, Germany
stauffer@thp.uni-koeln.de

Christian Wilhelm
Office for Natural Hazards
Forestry Department
Canton of Grisons
Switzerland

1 Extreme Events: Magic, Mysteries, and Challenges

Volker Jentsch, Holger Kantz, and Sergio Albeverio

Summary. Extreme events (henceforth Xevents) occur in natural, technical and societal environments. They may be natural or anthropogenic in origin, or they can arise simply from "chance". They often entail loss of life and/or materials. They usually occur "by surprise" and therefore often only become the focus of scientific attention after their onset. Knowledge of Xevents is often rather fragmentary, and recorded experience is limited. Indeed, scientists do not really understand *what* causes extreme events, *how* they develop, and *when* and *where* they occur. In addition, we are rarely able to cope with their consequences, due to lack of anticipation and preparedness. All this has motivated us, the editors of this volume, to bring together specialists from a variety of fields of expertise, all of whom have a common background in mathematics and physics. We asked them to write their views about Xevents. The result is the present book of essays that will (hopefully) enable the reader to unlock the mysteries surrounding Xevents.

1.1 Why Study Xevents?

There is a long tradition of phenomenological studies of Xevents in human history. Let's focus on two examples. The first refers to the water levels of the Nile, which have been recorded for over 5000 years, providing a remarkable hydrological chronology of the lowest and highest water levels. The water level of the Nile has been discussed and analysed from ancient times in relation to religion, philosophy, and economics and human welfare: hunger when the water level sank to a minimum and disaster when the Nile was too high. Moderate flooding of the Nile delta, however, has been known to be beneficial to agriculture for many millennia.

The second example refers to earthquakes. Here, the chronology is not reflected in numbers, as in the case of the Nile, but mostly in oral or written history, poems, or newspaper articles. Records go back 3000 years, beginning with the Mt. Taishan earthquake in the Shandong Province of China. The Lisbon earthquake of November 1755 received much attention. It not only triggered earthquake research in Europe, but also served as the focus for various publications, ranging from Kant's essays about the causes of earthquakes to Voltaire's *Poème sur le désastre de Lisbonne.* What all of these reports have in common, however, is that they strongly convey the unpredictability and unimaginability of when and where the earth would tremble.

Today, Xevents attract both public and scientific interest, for various reasons. For example, we fear that natural Xevents could increase in frequency and intensity, possibly triggered by human activity. Furthermore, we are shaken by the sudden and abrupt collapse of structures, such as buildings, power plants, and traffic and transportation systems, which are almost always man-made and subject to further complications due to ignorance and/or negligence. It all boils down to the question of vulnerability – how can populations be protected from Xevents, especially in view of the global interdependencies of technology, economy, ecology, and society? Science can make a significant contribution in this respect, as it aims to understand the dynamics of Xevents (the processes occurring before, during and after the event); to predict the occurrence of an event and its impact; and to define the limits of prediction.

1.2 What are Xevents? A First Approach

Before we proceed any further with this discussion, we need to define exactly what we mean by an "extreme event". In the context of an extreme event, an "event" is something that happens within a limited space and time. Its occurrence can arise by chance or necessity or through a combination of both; through natural or human-made causes or a combination of both. The interpretation of "extreme" cannot be defined so easily. It encompasses a collection of attributes, such as rare, exceptional, catastrophic, surprising, and the like. An insurer would translate "rare" as "low-probability" and "catastrophic" as "of great consequence", the latter emphasising the event's potential for impact and change. Therefore, a hurricane is an Xevent only if it causes loss of life and material damage; it is considered to be an ordinary event if it hits uninhabited areas. An asteroid strike is an extreme event only when it strikes the earth and changes the course of evolution, which seems to have happened 65 million years ago. The degree of "extremeness" of a Xevent, an important consideration for insurance companies, politicians and journalists, may thus be intuitively expressed as the product of the change and the impact caused by the event divided by the frequency of occurrence.

From a science viewpoint, on the other hand, the impact aspect is not the most important. What stirs scientific passion are huge deviations in a series of measurements, the burst-like nature of extremes, their apparent uniqueness: in short, the unexplainable and unpredictable. This means that the occurance of an asteroid strike is an extreme event, regardless of its impact on human life; as are magnetic storms in the magnetosphere, even if there is no recordable impact on electronic devices on earth. However, society's need to cope with the consequences of Xevents is becoming more and more urgent and so we can no longer afford to leave all Xevent-related considerations solely to policy- or decision-makers. Powerful simulation tools may help and thus we will discuss them in this book.

Xevents can also be individual, such as a first love, the birth of a child, the death of a spouse, the awarding of a Nobel Prize, to mention only a few examples. Xevents can also be general, in that they affect people and the environment: societal disasters (pandemics such as influenza and AIDS); natural disasters (floods, droughts, cyclones); technical breakdown (power outages, material ruptures, explosions, chemical contaminations); or market turbulence (huge losses or gains in the stock market), to mention but a few. World wars are undoubtedly among the most extreme of extreme events. We remind the reader of Eric Hobsdawn's fulminate *Age of Extremes* [1], which describes and analyses the social catastrophes of the twentieth century, in particular the two World Wars and the revolutions that followed each war.

The connection between wars and Xevents raises the question of morality, which quickly dominates all other issues involved: should a specific Xevent be judged as positive or negative? The wars of the twentieth century (but not only these!) are rightly considered as human tragedies unmatched in scale and consequence. However, tragedies – on whatever scale – almost always contain the *seeds* of positive change. The World Wars ultimately led to the spread of democracy around the globe (especially in Germany), ending the era of colonialism. The meteorite that is believed to have struck the earth millions of years ago extinguished the dinosaurs and facilitated the evolution of mammals; the nuclear disaster in Chernobyl, which killed thousands of people, fostered the development and implementation of alternative sustainable forms of energy; while the unification of West Germany and East Germany – widely welcomed as a positive Xevent – also gave rise to high unemployment and social displacement and deprivation. Therefore, rating Xevents as positive or negative is purely subjective; there are always trade-offs between the risks from and the benefits of the event.

1.3 What are Xevents? A Second Approach

So it seems that there are many definitions of an Xevent. This indicates that the issue is multifaceted, intricate, complex, and subject to various interpretations, perceptions, assessments, and even emotions. For science, this is not a comfortable situation. From a scientific perspective, the aim must be to free Xevents from their apparent subjectivity so that a more objective definition can be obtained. Defining a quantity (mathematical, physical, or whatever), on the other hand, requires adequate knowledge of it. This is not yet available. All we can do at the moment is to characterise Xevents by their most important elements: their statistical and dynamic properties, possible commonalities and analogies, observations, mechanisms, predictability, prediction, and management. It may be helpful to present a few remarks on these topics prior to explaining the idea underlying this volume and its components.

1.3.1 Statistical Characterisation of Xevents

From a statistical perspective, Xevents occur in the tails of probability distributions that define the occurrence of events of a given size (in terms of energy, duration, and so on). In a Gaussian distribution, these tails (situated to the far left and right of the peak value) are exponentials. For many Xevents, the tails are "heavy": for instance (algebraic) power laws with some fixed power, $p(x) \propto x^{-\alpha}, \alpha > 0$. Power laws fall off much more slowly than exponential (Gaussian) distributions, indicating an enhanced probability of occurrence. We note in passing that power laws (not exponentials) possess scale invariance (corresponding to self-similarity, in terms of geometry), which is important for many natural phenomena (see below). This property can be expressed mathematically as $p(bx) = b^{-\alpha}p(x)$, meaning that the change of variable from x to bx results in a "scaling factor" independent of x, while the shape of p is conserved. So power laws represent "scale-free systems". A typical Gaussian distribution is that representing a the heights of a number of people, with a well-defined mean value and a relatively small variance. Typical power laws include the distribution of wealth (known as Pareto's law, with a fraction of people presumably several times wealthier than the reader) and the size distributions of earthquakes (Gutenberg-Richter), forest fires and avalanches, among other examples.

The statistics of Xevents is known as *extreme value statistics* (this form of statistics dates back to 1958, when E.J. Gumbel published his seminal book [2]. The aim is to obtain as much information as possible on their (unknown) distribution functions. Typical problems include finding the probability that the size of an event exceeds a given value, or the largest event that will occur in a given period of time, for a given location. The assumptions used in common theories are still quite limiting, dealing mostly with independently and identically distributed events, which is rarely the case in reality. However, if the distribution function of Xevents can be estimated with sufficient accuracy, all relevant quantities (including those mentioned above) can be evaluated.

1.3.2 Dynamic Characterisation of Xevents

From what has been said so far, it appears as though Xevents are generated randomly, as with throwing dice. This is a wrong assumption. They occur in systems with complex dynamics, usually far from equilibrium, where the system's variability (not its mean values) and collective effects (not its individual aspects) are dominant. Consider weather extremes. What we call weather is the state of the Earth's atmosphere in the region relevant to us, which is continually and dynamically evolving according to well-known equations of motion (such as the Navier–Stokes equations). Therefore, modern weather prediction performed by running numerical simulations of model equations, fed by observations (measurements) as initial conditions. In fact, all natural

Xevents are almost undoubtedly phenomena that occur as manifestations of the complex dynamics of a certain system. Hence, we search for dynamic mechanisms that allow a given system to make an excursion far from its normal state. Several such scenarios are known, among them the concepts called the theory of large deviations, self-organised criticality (SOC), deterministic chaos, and fully developed turbulence, to mention just a few. SOC, for instance, suggests that a system reacts to a sequence of perturbations by manoeuvring itself into a critical state (with no external tuning or organisation required) where huge fluctuations are the rule rather than the exception, and which cause power law distribution functions for the relevant observables. For instance, the aforementioned Gutenberg-Richter law of earthquake magnitude distribution can be reproduced by suitable SOC models. However, there is a wide range of potential dynamic scenarios for Xevents, some of them generating precursors, some of them requiring nonlinear positive feedback loops with evident instabilities. Hence, there is definitely no universal dynamic mechanism at work; but the number of potential mechanisms is small. Despite a huge body of knowledge about dynamics, accumulated mainly over the past three decades, Xevents have only rarely been the focus of such studies.

1.3.3 Shaping Evolution

During the evolution of the Earth's surface, state economies, and political structures, to name three examples, Xevents have obviously had significant roles to play: they shape the future courses of such systems. Indeed, the worst earthquakes in California, with a recurrence rate of about once every two centuries, account for a significant fraction of the region's total tectonic deformation; landscapes are changed by the "millennium" flood, which is more effective than the concerted action of all other eroding agents; the largest volcanic eruptions lead to major topographic changes and to severe climatic disruptions; financial crashes, which in an instant can cause the loss of trillions of dollars, loom and affect the psychological state of investors, society, and the world economy.

1.3.4 Commonalities, Analogies, Universality

Newton's law of gravity is universal, as it applies to any particle of matter. Could a similar statement hold true for Xevents? Certainly not – they are too complex and too diverse. So let us modify the question: is there some (simple or complicated) mechanism that produces similarities in behaviour between different Xevents? Or will the behavior depend crucially on the specifics of each system or classes of systems, provided that such classes exist and can be defined and identified? In other words, is there any kind of universality that expresses the common nature or essence that the members of a class (individual events) share with one another? Is this universality purely formal, or can it be imagined as being dynamic and developing where universality and individuality merge?

Physicists and mathematicians are used to thinking of universality. We remind the reader of bifurcations and phase transitions (where phenomena as diverse as ferromagnetism, superconductivity, and the spread of epidemics – percolation theory – enjoy a unified theoretical description in which details of the system become irrelevant). However, as already mentioned above, we cannot expect universality. The most we can hope to find is the existence of several universality classes. Things might be more complicated than this, however, since the different facets of Xevents (origin, impact, and phenomenology) allow us to search for and to discover commonalities on different levels of description. Therefore systems falling into the same universality class when considering physical aspects might appear very disparate on the sociological level.

If there are commonalities in cause, there are many more commonalties in effect. Indeed, Xevents entail casualties: deaths, heavy financial costs, environmental destruction, and undermining the fabric of society. These result from side effects or secondary events deriving from the primary Xevent, such as the disruption of communication networks, the contamination of water, and the breakdown of health support, energy supplies, and so on.

1.3.5 Prediction, Anticipation and Management

Xevents call for *prediction.* In a sense, we believe in "savoir pour prévoir", as stated by the French philosopher Auguste Comte. Prediction implies movement from the past through to the present towards the future: a cause or several causes may lead to an effect – the extreme event – to be predicted. In a wider sense, prediction may be complemented by a proactive dimension: how do we cope with a predicted event when we don't know when it is going to occur. This requires backward travel from the future – as a possibility – to the present. *Anticipation* describes this (nonreactive) perspective. Combining prediction and anticipation is a prerequisite to managing Xevents.

Predicting may be a dangerous activity, however. There are many reasons for this. An important one is that forecast models are constructed on mean quantities. However, a real-world complex system is often better viewed as a collection of "hot spots" rather than a reasonably homogeneous background contaminated by small-scale noise. This explains why classical methods often fail. We select four striking examples from a long list of drastic failures, which were severely underestimated or not predicted at all: the storms that struck Western Europe in December 1999; heavy and devastating precipitation in Northern Italy in autumn 2000; the terrorist attacks in Madrid and Beslan (Russia) in 2004; and most recently, the Asian apocalypse in which more than 200,000 people were killed and many more made homeless on 26 December 2004 by a tsunami.

Any quantity that is to be predicted must be predictable. This depends on the quality and availability of data, existence and type of precursors, and the amount of determinism involved, such as memory effects or some dependence (in time or space) between the observables. The signal-to-noise ratio is an important quantity in this context. If it is low, faithful predictions are impossible. Another important quantity is the time horizon for prediction. In general, we may distinguish between short-, medium- and long-term prediction. While this holds true for ordinary time series prediction, this concept must be revised as far as Xevents are concerned. Here our desire is to predict, among other things, the largest event that will occur in a given period of time, for a given location.

However, more important than predicting an event (or more correctly, the probability of occurrence) is the specification of confidence intervals indicating the upper and lower bounds of probability. As a matter of fact, rather than trying to predict Xevents, one may define the range of all possible Xevents, just to provide the information we so desperately seek.

When we discuss consequences, we should bring up management, which refers to mechanisms used to cope with the impacts of Xevents. The effectiveness of management depends on understanding, anticipation, preparedness, and response to these events. There are avoidable (usually human-made) and unavoidable (usually natural) catastrophes. If an Xevent is avoidable, then prevention is of great importance. If it is unavoidable, then management searches for mitigation and adaptation (mathematically, this means some kind of optimisation); this is known as a vulnerability reduction strategy.

1.3.6 Trends

Regular structures, especially trends, can be superimposed on the irregular behaviour of a time series. So it is useful to check possible trends in the extreme values of the time series and to evaluate the extent to which they depend on technology, behaviour, habits, and so on. Some people claim that, due to human interaction, climate extremes are becoming more extreme and temperature swings occur more often. Ultimately, Xevents may become so frequent that they are no longer extreme, but define a system's norm. This conclusion is certainly very speculative, but it seems to have many adherents among the worldwide climate community.

1.3.7 Building Models

As with other scientific problems, the modelling approach is the most effective and appropriate one to apply to Xevents. One may roughly differentiate between diagnostic models, which investigate *what has happened*, and prognostic models, which investigate *what will happen*. These (microscopic or macroscopic, general or specific) models are always some simplification of reality, which is usually reduced to some basic physical laws and formalised in mathematical terms. Computer simulations also play an important role. Simulations are used to answer question like "Suppose some relevant parameter is changed, what is the response of the system if all other factors are kept constant?" In order to obtain valid answers, the current state of the system must be properly reproduced by the simulation, which requires the optimisation of system parameters and functions. This is difficult to achieve in most cases, which sheds some doubt on simulations in general. Dynamic models contain nonlinear feedback, and the solutions to these are usually obtained by numerical methods. Statistical models are data driven; in their simplest version they try to fit a given set of data using various techniques. There are hybrids, coupling dynamic and statistical aspects, including deterministic and stochastic elements. Simulations are often based on cellular automata and network formalisms, connecting input and output in nonlinear ways. These models are calibrated by training the networks, so that the error between output and given test data is minimised.

1.3.8 Observations

The underlying reality of theories is data. In fact, observations constitute a firm base from which scientific reasoning can start and to which it must always return in order to test its validity. So without data – and this is a possible way of thinking – there is no theory, or at least no verifiable theory. In our context, this means a sequence of data forming a time series, in which a measurement point is associated with each time point (often equally spaced). Collecting data, organising data and drawing conclusions (statistical inference) is achieved through ordinary data handling. For Xevents, it is not the mean value that matters, but rather the deviations from this, in particular the greatest deviations. A distribution function of the extreme values is required. A special problem arises here which is sometimes called the "curse of few observations", meaning that Xevents tend to be rare and thus impede meaningful statistical inference. The lack of observations is, in many cases, overcome by using extreme value statistics (see above).

1.3.9 Risk

For an individual, "risk" means the probability of an undesired outcome, such as disease or death, resulting from bad habits or an unfavourable environment, among other possible causes. For finance, risk is the uncertainty that the actual return of an investment will be less than the expected one, due to inflation or fluctuating currency exchange rates, for example.

Sociologists go beyond the microscopic view of risk. Risk becomes a macroscopic or social phenomenon. Ulrich Beck received much attention when he created his "risk society" [3]. He observed a transition from "old" society – whose fate was determined by naturally occurring hazards along with socially induced hazards, such as wars – to a "modern" one, whose course is governed by risks, especially those driven by industry and new technology. According to Beck, risk, rather than social deprivation or inequality, is the constituent element of society, causing new conflicts and social formations. A "risk society" seems to entail an "insurance society", in which damage is compensated by money. For insurers and re-insurers, risk is a measure of uncertainty ranging between 0 (highly uncertain) and 1 (certainty), proportional to the product of probability of occurrence and damage. Risk can be insured if it is computable and identifiable. However, this is not always possible, since Xevents are – as has already been stated above – difficult to estimate, both in cause and in effect; in addition, they can be accompanied by enormous and highly correlated losses. Other forms of protection, such as prevention or precaution, must be developed in order to "tame" Xevents.

1.4 How the Book is Organised

1.4.1 Background

The papers presented in this volume draw largely upon complex systems research carried out by a number of eminent researchers who have published widely (we restrict ourselves to more recent publications, notably [4–14]). They are also influenced by recent workshops (including the Extreme Events Workshop held in Boulder, CO, USA, sponsored by the National Science Foundation,, which tackled the research agenda for the twenty-first century; for more information about this visit the website at http://www.isse.ucar.edu/extremes/index.html). The objective of the workshop was to reconsider research on Xevents in terms of a more unified perspective. Networks have also been formed that document natural hazards and discuss management possibilities [15, 16]. In addition, a collection of the various aspects of Xevents, entitled "The Science of Disasters", has been published [17].

1.4.2 Rationale

The essays in this volume were selected to represent some important speciality fields of Xevents that have been investigated in great depth over the last ten years. Often in books such as this one, one aspect – say floods – is explored in all of its glory. This is a well-proven procedure, but it is not our intention. We want to explore the individuality of Xevents, but at the same time, we want to demonstrate that Xevents do not vary as much as they might first appear, and that it is useful to compare and contrast these phenomena. It is our conviction that a collective view of Xevents in diverse systems will allow us to gain insights into what may lie behind them. This is the physicist's approach: to try to extract general features and laws that apply to many phenomena, not just one. The differences and the commonalities, the specific and the universal, are reviewed on a more general level above. It is up to the reader to deepen their understanding of these statements with the help of the essays presented in this book.

Another reason to bring together these essays on Xevents is a very pragmatic one. We strongly believe that the cooperation of scientists belonging to different fields yet aiming at the same goal (describing and explaining Xevents) yields a better result than the collective efforts of experts working independently and separately. Admittedly, bringing together scientists is a difficult endeavour, and one that is seldom successful in practice due to differences in language, background, and goals. The reader will soon realise this when comparing the essays written by theoreticians and researchers, specialists and generalists, philosophers and practitioners. Some write more technically, others prefer a review-like style. All authors discuss some of their own as-yet unpublished research.

Another aspect, probably the most important, is the methodological aspect. Can the methods we apply to finance also be applied to, say, epilepsy? Method transfer has been shown to be successful in many cases and contexts. Xevents is undoubtedly another field where method transfer is of the utmost importance.

The topics were selected to meet the criteria mentioned above. The opening essay deals with our theme as a whole, describing and discussing the various aspects, be they philosophical or oriented towards information science. The closing essay, although more technical, provides a broad outlook on how to cope with Xevents. In-between, three articles shed light on the fundamental mathematical and physical concepts, while others examine specific scenarios, combining review and preview, including the mechanisms behind, the forecasting of, and the management of Xevents. We believe that all of the essays convey a sense of the methods, accomplishments, and challenges of contemporary Xevent science.

1.4.3 The Articles

The articles are ordered in three parts: *General Considerations* contains a general view in terms of philosophy, mathematics, and physics; *Scenarios* embraces nature, technology, and society; and *Prevention, Precaution, and Avoidance* discusses various management measures.

General Considerations

Mihai Nadin ("Anticipating Extreme Events") draws a metaphysical picture of Xevents. He aims to look behind and transcend specialist views. Pierce's semiotics serves as the starting point, in which three elements have to be considered and combined: the representation of the object, the object itself, and the interpreter (in the form of a computer, a certain method, or – more demanding – a hybrid system that couples human creativity and ability to anticipate with machine processing). Nadin states that an adequate representation of Xevents includes the prediction of its occurrence and the consequences associated with it. In other words, representing an event means that we fully capture it, in other words we are able to explain and reproduce it. If so, we can anticipate it, and this is all that society needs to meet the Xevent. What is and what is not an Xevent is equivalent to asking how it affects humanity, in quantity and quality. According to Nadin, the consequences count, nothing else. Another question raised is the problem of chance and necessity. Nadin believes in the causa finalis of Xevents, and suggests focussing on *why* a particular Xevent occurs. An Xevent creates a new state of equilibrium, with the "post" (the state after the event) being more stable than the "pre", the state before the event. He states that Xevents cannot be treated in a reductionist fashion, only in a holistic manner, meaning the inclusion of heterogeneity and interaction among systems. Thus Nadin's thinking is in a sense complementary to our assumption of laws overarching Xevents in the inanimate and animate worlds.

Sergio Albeverio and Wladimir Piterbarg ("Mathematics of extreme events") undertake an excursion into the mathematical world. They assemble extreme value statistics, dynamical systems theory, chaos and catastrophe theory, pointing out their specific contributions to the mathematics of Xevents. Quite naturally, they start with the classical statistical theory of Xevents ("extreme value theory"), summarising problems like estimation of Xevent indexes and Xevent prediction, and extending this to include dependent data, multidimensional variables and stochastic processes with continuous time. The other mathematical approaches addressed are extreme fluctuations, in particular in relation to the phase transitions studied in statistical mechanics, and singularities of maps, as studied in catastrophe theory, which may provide the ability to formulate certain natural systems in terms of variational principles. A partial unification of such apparently distinct approaches to Xevents is seen by connecting large deviations theory with saddle point methods.

As basic as the article described above, but emphasising the physics perspective, is the contribution by *Holger Kantz et al.* ("Dynamic Interpretation of Extreme Events: Predictability and Predictions"). Under the remit of Xevents, the authors discuss the notion of the predictability of a given dynamic phenomenon as compared to actual predictions. They start from the assumption of dynamic sources of Xevents (the hypothesis that past, current, and future states of a system are related by dynamic rules, namely equations of motion). Their existence implies a certain degree of predictability of the future when the current state is known. In this contribution, static and dynamic aspects of predictability are discussed and contrasted with actual predictions. A prediction requires an algorithm, a prediction scheme, whereas predictability sets the benchmark. In a model-free way, predictability quantifies how predictable a given phenomenon would be if we knew the optimal method. It might seem surprising at first that stochastic systems are also usually predictable to some extent. Predictions in such cases are probabilistic (a certain event will happen with a given probability), but this is still highly beneficial compared to no prediction at all. Due to temporal correlations, predictions for stochastic phenomena are time- (and better, state-) dependent. That is, the actual prediction could, for instance, yield a time-dependent risk of occurrence of an Xevent. In a specific example (turbulent wind gusts), the authors elaborate on a purely data-driven (and hence in some sense universal) prediction scheme, namely data-driven continuous state Markov chains, and demonstrate its performance in this case. In closing, the specifics of the prediction of Xevents compared to "ordinary" predictions are discussed.

Didier Sornette ("Endogenous versus Exogenous Origins of Crises") presents a combination of basic understanding and probing into reality. He asks himself to what extent fluctuations growing to extremes are intrinsic (or "endogenous"), and to what extent they are responses to externally caused perturbations ("exogenous"). Complex systems theory likes the idea of self-organised criticality (see above, "Dynamics"), which states that systems with threshold dynamics relax through repetitive fluctuations of all sizes. Accordingly, Xevents are seen to be endogenous. On the other hand, most natural and social systems are subject to external shocks of widely varying amplitudes. Thus it is not clear a priori if a given Xevent is due to a strong exogenous shock, to the system's internal dynamics, or to a combination of both. Dealing with this question is fundamental to our understanding of the relative importance of self-organisation versus external forces in Xevents.

Sornette discovers that the time evolution of Xevents is different for endogenous and exogenous mechanisms. He illustrates his findings through a variety of examples, including shocks in book sales and in financial markets. Book sales are an indicator of "commercial growth and success". Endogenous peaks are followed by a power law relaxation, which is slower than for exogenous peaks. The slow relaxation in sales implies that the sales dynamics are dominated by (internal) cascades rather than by the direct effects of news or advertisements. In many cases, however, both effects seem to blend. Financial markets are checked for distinguishing features during the time before a crash. Sornette develops a method for identifying shocks and for distinguishing between two different types. One is characterised by a log-periodic power law describing the price distribution of market prices, while for the other this distribution is absent. The log-periodic behaviour represents precursors characteristic of endogenous crashes (such as the speculative behaviour of stockholders). The absence of these precursors indicates an exogenous crash.

Scenarios

Klaus Lehnertz ("Epilepsy: Extreme Events in the Human Brain") considers Xevents related to disease, notably epilepsy. But why not tackle truly life-threatening events, such as heart attack or stroke, which are also much more common? First, epileptic seizures are not rare. This implies advantage and disadvantage. From the viewpoint of statistics, many events are more significant than single or isolated events (avoidance of the "curse of few observations"). However, the disadvantage is more serious. Many events mean that many persons are affected. Indeed, there is a need to help millions of people. Second, epileptic seizures are recurrent; in other words, they occur again and again but leave the patient alive (at least in the vast majority of cases). This opens up the possibility of collecting EEG data as a function of time, generating a chronology, which is usually impossible when it comes death by heart attack. EEG data can be analysed using some well-developed methodologies. Lehnertz states that the origin of seizures is endogenous (see the discussion in Sornette's paper), and that it is essentially nonlinear. Precursors are of utmost importance, as they may signal an oncoming event. Here the time difference between detection of precursors and manifestation of the event is important for anticipation (see Nadin's contribution). There is mounting evidence that a deep analysis of rich EEG data (and other relevant parameters) indeed reveals precursors ranging from minutes to hours. These can be used to design new therapies, both technical and psychological, in order to reduce the number of attacks or, in the best case, to prevent them altogether.

Jürgen Herget ("Extreme Events in the Geological Past") reviews natural events that are really big and rare: those of the geological past. Geological Xevents are archived in rocks, ice, and various organic and mineral deposits. They can be a real event (abrupt and well-defined in duration and location), such as meteorite impacts, earthquakes, floods or volcanic eruptions; or they can be the end result of a continuously changing environment, such as the formation of the Earth's surface by continental drift – implying millions of years – or the reversal of the Earth's magnetic field on a timescale of hundreds of thousands of years. Herget's own work focuses on the biggest flooding events known, caused by the repeated outbreaks of water masses from lakes that were blocked by huge ice masses. Whether or not (necessarily) limited knowledge about these events can help to assess the recurrence of outbreaks of stored water masses in the present still has to be ascertained.

Andreas Hense and Petra Friederichs ("Wind and Precipitation Extremes in the Earth's Atmosphere") describe typical spatial-temporal scales in the atmosphere that are prone to developing Xevents in the form of strong winds and heavy precipitation. Their interconnectivity is interesting: Xevents with large spatial dimensions (convective instabilities of the order of 100 km, say) can trigger small-scale extremes of vertical velocities of the order of 1 km, which in turn can cause heavy precipitation of a still smaller scale. Wind is generated by large vertical and horizontal gradients in atmospheric quantities, giving rise to instabilities. Extremes occur on all scales, ranging from 100 m to 1000 km, including tornadoes (of the order of 1 km) and tropical cyclones (of the order of 100 km). However, no scaling law seems to exist, implying that no universal mechanism (for example, in the form of self-organised criticality, with threshold dynamics included) is at work. In the case of precipitation, microprocesses of hydrology couple with macroprocesses of atmospheric circulation. It appears that scaling laws that relate rain extremes to accumulation periods in the form of a power law exist for convective precipitation, which takes place near the equator.

Eric J. Heller ("Freak Ocean Waves and Refraction of Gaussian Seas") focuses on specific Xevents in the oceans, namely isolated gigantic waves reaching heights of 20–30 m, which occur recurrently and much less rarely than one might expect. The reason they are not frequently reported lies in the lack of (surviving) witnesses. Ocean waves, ocean water streams, and wind eddies, which are the entities that cause these waves, follow well-known physical laws or empirical distributions. Hence, one deals with a physical system of moderate complexity here, so a theoretical approach has a good chance of fully capturing the physics of the phenomenon. Indeed, Heller's novel theory seems to be consistent with observations and the basic properties of giant waves, even if nonlinearities in wave formation and propagation are ignored. Apart from the sensation of goose-bumps in regard to the awesome power of ocean waves and the intellectual satisfaction of having this phenomenon explained, the reader might enjoy this contribution as a case study of how a dynamic view of Xevents can allow us to completely understand the phenomenon. As a consequence of Heller's freak index, one might be able to routinely forecast the frequency of such events in certain parts of the oceans in the future.

It would be improper to leave out the problem of the material fracture. *Matz Haaks and Karl Maier* have focussed their interest on the breaking of metals, notably steel. The authors do not develop just another model. Rather, they present a new experimental technique, suitable for reliably predicting the material's lifetime, which is synonymous with the number of loads exerted until the material breaks. This final state defines the Xevent. However, rupture is not a recurrent phenomenon, so the Xevent associated with material failure is in contrast to those considered so far: it marks the point of no return. The authors contribute not only to the problem of *prediction*, but also to the *management* of Xevents (see the essays below), since ideally, the metal would never break when following the authors' prescription to replace it in due time.

The interesting point is that fracture has something to do with the defects in the lattice. If the number of these rises, breakdown is very likely. So defects can be considered to be precursors that can be detected und quantified. The way this is done is described in the text. In essence, use is made of the well known results of positron physics, stating that if you implant positrons in solids they are attracted by the electric potential built up by the lattice defects. Once trapped, they are annihilated by electrons, resulting in γ-radiation that can be measured and interpreted in terms of progressing fatigue. How will this be handled in practice? The advantage of the proposed technique is that a few in situ measurements will suffice to answer the question of lifetime. Assume that the density of defects is measured, for example as a function of the applied load cycles. The resulting diagram will be interpreted according to the results obtained from the strain–stress and fatigue tests of the specific material, which are fully known from many independent experiments. Of course, the extrapolation employed has to be refined and verified by statistics. It would be interesting to see a verification of the authors' results in the near future; moreover, it would be desirable to develop a mathematical model that would support the extraordinary experimental work from a theoretical viewpoint.

Santo Fortunato and Dietrich Stauffer ("Computer Simulations of Opinions: Reaction on Extreme Events") are concerned with computer simulations of Xevents. As is typical of simulators, they build a model based on certain heuristics and let it run on the computer. Such models, however simple, allow insight into some of the basic mechanisms and their effects if coupled, which in general cannot be intuited. The hope is that the simulations provide some hints about what and why something happens. The authors' topic is the relation between society and Xevents. Here "relations" are interpreted as opinions: how do opinions about an Xevent change once it has happened? In Sornette's parlance (see above), the Xevent itself is the exogenous shock, while the response (in terms of opinion change) is due to internally driven network dynamics. Fortunato and Stauffer present the following scenario. Before the event, an individual (or a group of persons) trusts its own (subjective) estimation of risks and impacts of the event. After the event, the individual (or the group) changes its opinion abruptly. What has happened? Two reasons are conceivable. Reason 1: The (objective) assessment based on expert's knowledge and experience was right, while the group's (subjective) view turned out to be wrong. This includes under- and overestimation of the extremeness of the event. Reason 2: Before the event, the group didn't pay attention to early warnings, because they were often wrong. So they did not pay attention to the warning of what turned out to be a true Xevent. After the event, they conclude that it would have been better to believe these warnings. While in the first case the change of opinion is largely academic, the latter case entails a change in behaviour, which may have a large impact.

By means of a cellular automaton and related models, the authors show how the assumed change of opinion spreads among the population. The amount of people influenced is model- and parameter-dependent. The spread depends on location and time. Obviously, the further one is from the event, the less likely opinions are to change.

Philippe Blanchard and Tyll Krueger ("Networks of the extreme: a search for things exceptional") consider networks, as Fortunato and Stauffer do. While the latter used the network as a suitable background for transporting opinions, here the network itself is the object of research. Xevents come into play with the emphasis on their rarity as one of the determining factors (the other one is their event-like character, see above). Rarity is then connected to attention, and their is an inversely proportional relationship between them: "The rarer you are, the more attractive you become". Just as we all believe in this and behave accordingly, so do the authors. They show that the rare or exceptional (such as wealth, beauty, or influence) attracts others and gives rise to new connections (or "edges" in network jargon). The evolving network has interesting properties, since it does not depend upon the details of the distribution of the rare, but only on a constant shaping of a power law. This implies scale invariance, as already stated. So here is an (abstract) example in which Xevents, meaning rare events, seem to determine the evolution of structures to a large extent. It would be interesting to see if the conclusions drawn by the authors, such as the spread of epidemics and terrorist attacks on networks are just speculations or have a real basis that can be further exploited.

Prevention, Precaution, and Avoidance

Michael Lehning and Christian Wilhelm ("Risk Management: Modeling Mountain Hazards") are in charge of the Swiss Snow and Avalanche Institute in Davos. They write about risk management in snow-covered mountain areas. At times the inhabitants are surprised by avalanches and the like, representing a certain threat or risk. The problem that Lehning and Wilhelm tackle is the reduction of risk (up to avoidance of the event), under the constraint that financial costs must be low. They conclude that the best prevention is a good forecast, implying a confident description about when and where the avalanche is likely to occur. If so, low-budget measures can be taken into consideration, such as road closure in the endangered areas, which is effective since it saves lives and material and does not cost much compared to fixed protective constructions of the past. So it all boils down to the basic question: how do we improve avalanche forecasting?

Zuzana Chladná, Elena Moltchanova, and Michael Obersteiner ("Prevention of Surprise") discuss the management of Xevents, focussing mainly on hazards brought about by human society. The emphasis is on preventing hazards rather than developing protective measures (as insurance companies do). The authors propose a modified discrete version of a neoclassical macroeconomic model, describing an evolving economy in the presence of a threatening Xevent, including investment, consumption, and a probabilistic mitigation component, which can be adjusted so as to avoid the catastrophic event. Their model can be extended across several interacting societies. The results are interpreted in terms of mitigation strategies, emphasising collaboration rather than confrontation within society. This is clearly the most interesting point the authors make, and it is this aspect that should be pursued and explored in more detail in the near future.

Last but not least, *Dirk Helbing, Hendrik Ammoser, and Christian Kühnert* ("Disaster Response Management: Analyzing Networks of Events") consider the aftermath of Xevents (disasters), and emphasise the common ground between them in terms of effects. Disasters spread via networks, which in cities almost always give rise to interruptions in traffic, transportation and supplies, electrical power blackouts (with additional side-effects), and a breakdown in the information infrastructure, with a subsequent loss of quality of information and coordination. These, in turn, give rise to panic in the population, increased criminal behaviour, looting, and eventually – the most threatening – disease outbreaks. There are also long-range correlations implicit in retaining or losing political power, as was the case in Germany (after the flood in Saxony in 2002) and in Spain (the terrorist attack in 2003). Helbing et al. propose heuristics-based impact models in order to answer questions about the dynamics of impacts. These impacts manifest themselves as sub-events (such as breakdown in traffic systems) related to the occurrence of the primary event (the Xevent). The interesting question concerns the dynamics of such impacts, especially their strengths and timescales. The answer to this is a model that consists of a system of coupled first-order differential equations, based on the assumption that the rate of change of the impact (on a specific constituent of the system) with time is proportional to the impact multiplied by its spreading rate minus the mitigation rate (which is due to external management actions). The solutions can be interpreted in two ways: they give the probability of an impact event, as well as the path and the velocity with which the impact propagates through the system. So the authors rightly conclude that, provided we know the interaction scheme for impacts and correctly identify the constituents of the system, we should be in a better position to manage the event using their modelling results.

1.5 Outlook: Research Programme

What might the publication of this volume achieve? We hope that it fosters efforts aimed at understanding extreme events in different fields, including assessments of their degree of predictability, specifications of confidence intervals of prediction, estimates of risk, as well as proactive measures – anticipation and prevention. This volume should illustrate the potential, the requirement, and the success of such a proposal. Xevents arise in a variety of physical, life, and social systems in which the concepts of co-operability and self-organised criticality are universally relevant.

However, problems arising from Xevents are related and therefore must be approached through concerted actions. Moreover, a variety of methods, converging into a few models, is needed. As is the case with this book, the physical–mathematical perspective should be emphasised and expressed in the form of models and simulations. We believe that this is the most convincing and most effective approach. Other approaches from social sciences, economics, and psychology are most interesting and should be included wherever possible. Novel methods should be combined with state-of-the-art methods, while new models must be compared with models that already exist in various scientific disciplines.

Even if only a small part of this program is realised, it would not only influence the course of science and its various applications, but it would also have tremendous practical impact in that it would decrease vulnerability and increase resilience to Xevents. It is the combination of social and emotional aspects, superimposed on top of basic scientific problems, that characterise Xevents and arouse widespread interest in them.

References

1. E. Hobsdawn, *The Age of Extremes*, Abacus, London (1994)
2. E.J. Gumbel, *Statistics of Extremes*, Columbia Univ. Press, New York (1958)
3. U. Beck, *Risikogesellschaft. Auf dem Weg in eine andere Moderne*, Suhrkamp, Frankfurt a.M. (1986)
4. H. Haaken, *Synergetics, an Introduction. Nonequilibrium Phase-Transitions and Self-Organization in Physics, Chemistry and Biology*, Springer, Berlin Heidelberg New York (1977)
5. G. Nicolis, I. Prigogine, *Self-Organization in Non-Equilibrium Systems*, Wiley, New York (1977)
6. M. Eigen, P. Schuster, *The Hypercycle. A Principle of Natural Self-Organization*, Springer, Berlin (1979)
7. B.B. Mandelbrot, *The Fractal Geometry of Nature*, Freeman, New York (1983)
8. D. Ruelle, *Chance and Chaos*, Princeton University Press, Princeton (1991)
9. J.H. Holland, *Hidden Order: How Adaptation Builds Complexity*, Addison-Wesley, Reading, MA (1996)
10. M. Gell-Mann, *The Quark and the Jaguar: Adventures in the Simple and the Complex*, W.H. Freeman and Company, New York (1994)
11. P. Bak, *How Nature Works: the Science of Self-Organised Criticality*, Oxford Univ. Press, New York (1997)
12. R.N. Mantegna, H.E. Stanley, *Introduction to Econophysics: Correlations and Complexity in Finance*, Cambridge University Press, Cambridge (2000)
13. S. Wolfram, *A New Kind of Science*, Wolfram Media, Champaign, IL (2002)
14. D. Sornette, *Critical Phenomena in Natural Sciences*, Springer, Berlin Heidelberg New York (2003)
15. NCDC, *Extreme Weather and Climate Events Website*, National Climatic Data Center, Asheville, NC (2005); see http://www.ncdc.noaa.gov/oa/climate/severeweather/extremes.html
16. B. Merz, A. Heiko, *Risiken durch Naturgefahren in Deutschland: Abschlussbericht des BMBF-Verbundprojektes Deutsches Forschungsnetz Naturkatastrophen (DFNK)*, GeoForschungsZentrum Potsdam (GFZ), Berlin (2001); see http://www.gfz-potsdam.de/bib/pub/str0401/0401.htm
17. A. Bunde, J. Kropp, H.-J. Schellnhuber (eds.), *The Science of Disaster*, Springer, Berlin Heidelberg New York (2002)

Part I

General Considerations

2 Anticipating Extreme Events

Mihai Nadin

Summary. The urgency explicit in soliciting scientists to address the prediction of Xevents is understandable, but not really conducive to a foundational perspective. In the following methodological considerations, a perspective is submitted that builds upon the necessary representation of Xevents, either in mathematical or in computational terms. While only of limited functional nature, the semiotic methodology suggested is conducive to the basic questions associated with Xevent prediction: the dynamics of unfolding Xevents; the distinction between Xevents in the deterministic realm of physics and the nondeterministic realm of the living; the foundation of anticipation and the possibility of anticipatory computing; the holistic perspective. As opposed to case studies, this contribution is geared towards a model-based description that corresponds to the nonrepetitive nature of Xevents. Therefore, it advances a complementary model of science focused on singularity, providing a nondeterministic understanding of high-complexity phenomena.

2.1 The Representation of Extreme Events

Let us imagine that somehow we could fully capture an Xevent – an earthquake, a stock market crash, a terrorist attack, an epileptic seizure, a tornado, a massive oil spill, a flood, an epidemic or any other occurrence deemed worthy of the qualifier "extreme" (the kinds of measurements and other observations that qualify the result as extreme will remain unanswered for the time being). Based on what we know today – aware more than ever that everything is in flux ("Panta rhei", to quote Heraclitus [1]) – and on the scientific models that presently guide knowledge acquisition, we understand that to fully capture (represent) an event (extreme or not) involves not only explaining it, but also ultimately being able to reproduce it. This is another way of saying that if we could adequately represent an Xevent, we would be able to predict it and similar events, as well as their consequences. Implicit in this perspective is the expectation of determinism, a particular form of causality. More precisely, the representation contains the description of the cause or of the causal chain. Obviously this is no longer a case of simplistic representation of cause and effect, but one tempered by the realization that only an acknowledgement of a rich variety of causal mechanisms can explain the broad dynamics of complex phenomena. After all, the common denominator of Xevents is their complexity.

With all this in mind, let us denote the full description of an Xevent as its *representamen*, R. (The informed reader will have already noticed that this unusual word comes from Charles S. Peirce [2].) There are no limitations upon what R can be. It can be a record of quantities (numbers – the set N of natural numbers); it can be an event score (similar to a music score or to a detailed film script). It can be a completed computation – assuming that the algorithm/s behind the computation is/are tractable, that is, that they have a polynomial solution in the worst case. It can be a computation in progress, about to reach a halting stage, or reaching one in several generations (an evolutionary computation). It can be a combination of some or all of the above, plus anything else that science might come up with. Regardless of what R is and how it was obtained, if we could fully capture an event, we could also understand how the event – henceforth called the object (and denoted O) for reasons of convenience – and its representation relate to each other (in other words how a change in R, the representamen, might affect a change in O, the event reproduced or anticipated). This understanding (by a human being, a scientific community, a computer program, or a neural network procedure), called I for *interpretant process* (according to the same Peircean terminology already alluded to), is actually all that society expects from us as we dedicate out inquiry to Xevents. Indeed, we are commissioned (some explicitly, others implicitly) to conceive of methods for predicting Xevents. Based on such predictions, society hopes to avoid some of their consequences, or even to avoid the event (in the case of, say, a terrorist attack or an epileptic seizure).

The three entities introduced so far – R for representamen (the plural is *representamina*), O for object (to be defined in more detail), and I for interpretant – are derived from Peirce's semiotics. For the scientist wary of any terminology that does not result from some specialization (such as the many mathematical branches growing on the trunk of *mathema*, the various theories of physics, the biological fields of inquiry such as molecular biology or genetics, and so on), a word of caution: regard the entities introduced so far only as conceptual tools, and only in conjunction with the descriptions given so far. Actually, their relationship can be conveniently illustrated thus:

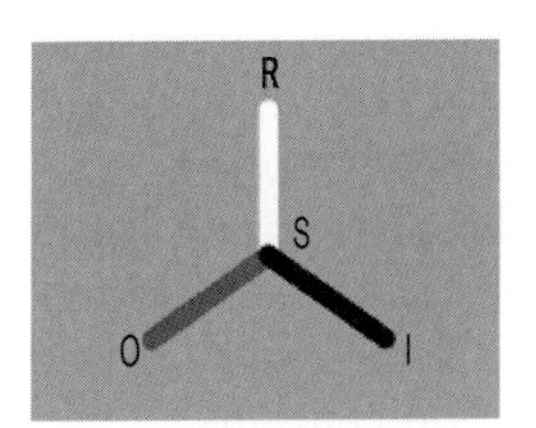

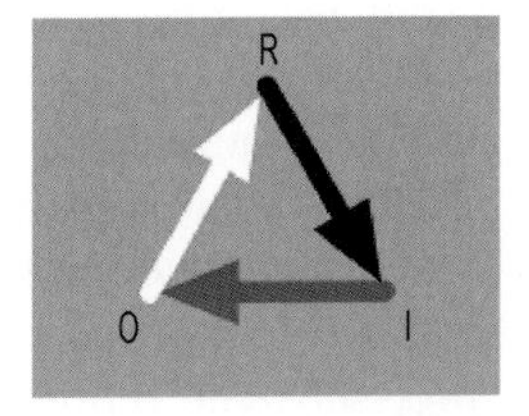

Fig. 2.1. A sign is something that stands for something to someone in some form or capacity (see C.S. Peirce, [3–5]). The two diagrams represent two views: on the *left*, the sign as a structure $S = S(O, R, I)$, and on the *right*, the sign as a process that starts with a representation (R of O) to be interpreted in a sign process

The diagrams tell a very clear story: "In signs, one sees an advantage for discovery that is greatest when they express the exact nature of a thing briefly and, as it were, picture it; then indeed, the labor of thought is wonderfully diminished" (Leibniz as cited by Schneiderman [6]).

An unusual scientist, grounded in mathematics, astronomy, chemistry, logic, and geodesic science, Peirce considered natural phenomena, as well as social events, from a meta perspective. Indeed, semiotics is a metadiscipline, transcending all those partial representations that are the focus of the object sciences. This confers upon semiotics an epistemological status different from that of particular sciences. That is, it is a "science of sciences" as Charles Morris [7] called it. Its generalizations in semiotic theory are not conducive to technological innovation as such, but rather guide the effort, such as in the design of user-computer interfaces, or the conception of languages, such as those used for programming or those based on the DNA code. However, semiotic generalizations are extremely effective at helping specialized research to maintain a reference outside the specialization pursued. They help scientists realize the relation between what is represented – in our case, Xevents (the representation), in whatever scientific theory and by whichever means, including mathematical formulae and computer programs – and the interpretation process associated with it. When many disciplinary and societal views are produced, which is certainly the case for this book and for which my contribution is conceived, we realize the need for a comprehensive transdisciplinary framework that can guide the individuals involved towards realizing the meaning of all of the specialized views and methods advanced. An effective framework for further research in Xevents ought to facilitate integration of knowledge, as well as the conception of new ways of disseminating knowledge, leading to decision-making and action.

The formalisms associated with semiotics are varied. They are of logical origin. In the late 1970s, I worked on a mathematical formulation of semiotic operations [8]. Animated by his interest in Peirce's sign classes, Robert Marty [9–11] pursued a similar goal. Joseph Goguen [12] finally worked towards the explicit goal of an algebraic semiotics, facilitating applied work. Neither of us considered that the study of Xevents might benefit from semiotics, formalized or not. But in the final analysis, what brought up the semiotic perspective in these introductory lines was the broad motivation of our effort: how to make semiotics useful beyond the contemplative dimension of every theory. One avenue, as it now turns out, is the path towards the foundation of anticipation, the anticipation of Xevents, in particular.

2.2 From Signs to Anticipation

Let me quote from the Introduction to the Report of the workshop entitled *Extreme Events: Developing a Research Agenda for the 21st Century [13]:*

> It is no overstatement to suggest that humanity's future will be shaped by its capacity to *anticipate* [italics are mine], prepare for, respond to, and, when possible, even prevent Xevents.

The notions of anticipation and of the sign are coextensive. Representations come into existence as the living – from the simplest level (monocell) to the most complex known to us (the human being) – act. Vittorio Gallese [14] brings proof that acting and perceiving cannot be effectively distinguished. He starts with an obvious example: the difficult task of reducing one's heartbeat is made easier once a representation – an electrocardiogram in real time – is made available to the subject. Indeed, biofeedback provides an efficient way of controlling a given variable (heart rhythm in the example mentioned, see [14]).

Representations, which are the subject of semiotics, are relational instruments. Every human action – and for that matter, every action in what is called the living – is goal driven. Gallese reports on single-neuron recordings in the premotor cortex of behaving monkeys. What drives the neurons is the goal of the action. He states:

> To observe objects is therefore equivalent to automatically evoking the most suitable motor program required to interact with them. Looking at objects means to unconsciously 'simulate' a potential action. In other words, the object representation is transiently integrated with the action-simulation [14].

Quite some time before Gallese's experiments, my own elaborations [15] on what drives the human being – the actions through which they self-constitute; their pragmatics (we are what we do, no more, no less) – reached a point that is the fundamental thesis of this article.

Thesis 1. *Xevents should be qualified in relation to how they affect human life and work.*

Let me explain: Xevents, regardless of their specific nature, are not simply acknowledged by virtue of their syntax (the formal characteristics as we read them on various recordings of seismic activity, brain activity, wind direction and intensity, and so on). Xevents are not reducible to the semantics defining them as such; the label applied in the form of a category of hurricane, or assigned seismic intensity on a standardized scale, or a seizure type, for example. The defining quantifier regards how they affect human activity: the pragmatics of existence.

It is the dynamic relation between the event and those experiencing it (directly or through some form of mediation) that counts. Moreover, the plurality of relations, corresponding to the various ways in which we interact with the world in which we constitute our identity, is what interests us. That we can quantify the effects of Xevents (in the number of lost lives, in the costs

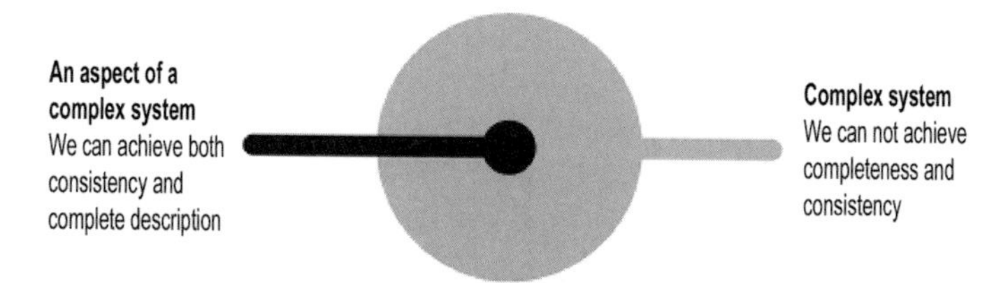

Fig. 2.2. Consistency and completeness are complementary. In order to circumvent the intrinsic characteristics of complex systems, we can focus on partitioned aspects. The challenge is to perform an adequate partitioning

of preparation, recovery, and damaged infrastructure, in ecological impact, and so on) does not mean that the numbers represent the impact of the event. Human life and activity are subject not only to quantity descriptions, but also to deep quality consequences.

What guides the exposition so far is the realization that while everyone wants to anticipate, or at least somehow, even in a limited way, to predict Xevents, we must remind ourselves of Gödel's warning [16], that we can at best expect partial results: a complex system cannot be described in both a complete and consistent manner at the same time. As a theorem in formal logic, it has often been misinterpreted. The reason we bring it up here is the methodological need to find out the extent to which it predicts the necessary failure of all attempts to anticipate Xevents, or whether it only suggests that we need to consider ways to segment or partition the various aspects of Xevents and concentrate on partial representations (see Fig. 2.2).

2.3 Descartes Rehabilitated

Seen from this perspective, Descartes' reductionism and determinism – the foundation of humankind's enormous scientific and technological progress in the last 400 years – makes more sense than his critics would like to credit him with. The question of whether Descartes knew well ahead of Gödel that complex systems are impossible to handle in their entirety, or whether he only asked himself (obviously in the jargon of the time) how to reduce complexity without compromising the entire effort of knowing will never be unequivocally answered. What we do know is that reductionism and determinism operate in a major section of perceived reality: everything there is (reality) is reduced to that subset of reality that constitutes the subject of physics. And everything that functions, including the living – minus the human being, for religious reasons that had more to do with Descartes' caution than with scientific reasoning – is seen as equivalent to a machine. That Descartes' understanding of the physical world and our current understanding of physics are quite different needs no elaboration. Science advanced our understanding of determinism and causality in ways that at times appear to be counter-intuitive. Think of the quantum mechanical description of the microcosm. Think of the dynamic system models in which self-organization, among other dynamic characteristics, plays an important role in maintaining the system's coherence. Hence, it would be unwise not to distinguish between

1. Xevents in the realm of the physical world, for which the science inspired by the Cartesian model is, if not entirely adequate, the best we have.
2. Xevents in the living, for which the Cartesian perspective is only partially relevant.

2.4 Time, Clocks, Rhythms

If the representation of an Xevent as a representamen R were possible, it would necessarily involve a time dimension. After Descartes, time was associated with the simplest machine of his age – the pendulum clock – and reduced to an interval. If not Descartes, then at least some of his contemporaries already knew that to associate gravity with rhythm is convenient, but not unproblematic. At the poles (north or south), time in this embodiment is quite different from the time in Paris or in Dallas, Texas. And on a satellite, depending upon its orbit, it is a different time again. This problem was addressed by adopting oscillations (mechanical, as in clocks and watches, or atomic) and resonance as a "time machine", and then declaring a standard – that of the cesium atom – which was easy to maintain and to reference. But with Einstein and relativity theory, we came to realize that the "atomic clock" is only a good reference as long as it is not subjected to a trip on a fast-moving carrier. Some physical phenomena take place along a timeline for which the day-and-night cycle in the western hemisphere, or the pendulum's gravity-driven rhythm, is either too fine-grained (think about cycles of millions of years), or too coarse (fast processes at nanoseconds and scales below this). Even more dramatic is our realization that many different clocks operate at the same time within the living, and many synchronization mechanisms are apparent. If they are affected, the system can undergo extreme changes. It turns out that the linear representation of time, through an irreversible vector, is a useful procedure so long as the time it describes is relatively uniform and scale-independent. But time is neither uniform nor independent of the frame of reference.

Once we ask what it would take for the representamen R of an Xevent to become a complete, effective description of the event, we implicitly ask what it would take to anticipate it. Indeed, a complete description can only be fully predictive if it makes a time difference mechanism possible:

$$\text{Event } (E) \text{ as a function of time } t_E$$
$$\text{Prediction } (P) \text{ as a function of time } t_p$$

Evidently, the two times t_E and t_P are not identical: $t_E \neq t_P$. Moreover, t_P must be faster than t_E in order to allow for the possibility of prediction or anticipation.

An anticipatory system is a system whose current state depends not only on a previous state, but also on future states [17, 18]. In contrast to a predictive mechanism that infers probabilistically from the past, an anticipatory procedure integrates past experiences, but weighs them against possible future realizations. One of the better-known operative definitions of an anticipatory system is: "An anticipatory system is a system that contains a model of itself unfolding in faster than real time" [19]. What this description says is that simulations are the low end of anticipation. What it does not say is that although we can execute different operations (for instance, computations) in parallel within physical systems (machines, in particular), and even perform some operations (computations) faster than others, without a mechanism for interpreting the meaning of the difference between "real-time" operations (computation) and "faster than real-time" operations (computations), we still do not have an effective anticipatory mechanism. Indeed, only an understanding of the difference in outcome between so-called real-time and faster-than-real-time operations can afford anticipation. Two conditions must be fulfilled:

1. The effective model should be complete.
2. An effective mechanism for discrimination between the process and its model must be implemented.

Some would argue that the model does not have to be complete (or that it cannot be complete). If this were true, we might as well make the conception of the incomplete but still useful model the task of predicting, as though we knew which part of the dynamics of the system is more relevant than what is left out. Others argue that all it takes is some intelligence in order to understand the meaning of the difference. From all we know so far in dealing with anticipation and the human being, intelligence is marginal, if it plays any role at all. Let us discuss some classical examples.

Anticipation of moving stimuli (see Berry et al. [20]) is recorded in the form of spike trains of many ganglion cells in the retina. The facial action coding system (see Ekman and Friesen [21]) is a record of "character" that we spontaneously "read" as we perceive faces in some unusual situations (the trusting hand extended when there is need). Proactive understanding of surprising events is the result of associative cognitive activity (see Fletcher et al. [22]. More recently, Ishida and Sawada [23] confirmed that the hand motion precedes the target motion. (Remember when you last caught a falling object before you "saw" it?) Intelligence is not traceable in the process or in the quantitative observations. As a matter of fact, high performance anticipation, such as that seen in skiing, tennis, hockey and soccer, is not associated with a high IQ or with any other feature of intelligence. What is identifiable is learning (and implicitly the dimension of training anticipatory attributes) although of a precise type. It is not explanatory learning; it is rather procedural, internalized rather than externalized.

In view of this, a representamen R of an Xevent understood as being its full operational description makes sense only in association with an interpretant process I. This itself can be conceived as a machine that is context-sensitive and able to learn. As the representamen unfolds in a neverending prediction sequence, the interpretant not only relates it to the Xevent it captured, but also to other events as they take place in the world. In order to achieve this dynamic behavior, it has to be conceived as a distributed computation; actually, as a grid process that takes as input the knowledge acquired so far (representamen) as well as the new knowledge resulting from the representation of Xevents taking place in real time. And even in this possible implementation, the interpretant process will not be more than a surrogate to a living interpretant process.

2.5 The Hybrid Solution

Thesis 2. *Since the living is not reducible to a machine, our best chance of understanding our own knowledge regarding Xevents, and thus provide for effective anticipation, are hybrid systems that integrate the human being.*

I am aware that this thesis runs counter to the dominant expectation of fully automated anticipation, or at least prediction. Although we deplore the enormous cost of the consequences of Xevents – often including death and bodily impairment, disease and suffering – we are, so it seems, not willing to take on board that the most expensive machinery imaginable today will not fully replace the interactions of minds (see Nadin [24] as a component of the interpretant process. Our obsession is still with the interaction (see Fig. 2.3) of the human – machine interface in particular.

This focus is not unjustified to the extent that we entertain the illusion that machines will eventually carry out any and every form of human activity. After all, Minsky [25] was not alone in stating that

> In from three to eight years, we will have a machine with the general intelligence of an average human being. I mean a machine that will be able to read Shakespeare, grease a car, play office politics, tell a joke,

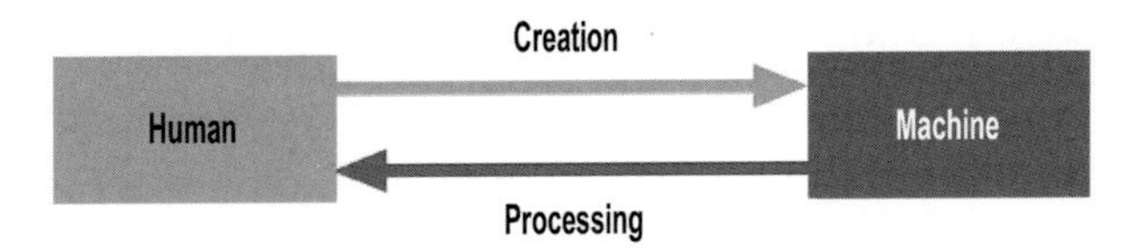

Fig. 2.3. Human–Machine interaction. The process is intensely asymmetric/asynchronic

have a fight. At that point, the machine will begin to educate itself with fantastic speed. In a few months, it will be at genius level, and a few months after that, its power will be incalculable.

Despite this reductionist-mechanistic viewpoint, we are now discovering that, given the continuous diversification of human activity, the machine is bound to be at least one step behind human discovery: it does not articulate questions. Expressed in other words, we are discovering new ways through which we can increase the efficiency of our efforts (physical, mental, emotional). Therefore, the logical alternative is not to transfer human functions and capabilities to machines, but to provide for an alternative model: the integration of the human being and the machine. What results is a very complex entity, ultimately characterized by its degree of integration. Instead of limiting ourselves to the Human–Machine interaction, we should concentrate on the very complex entity that results from integration (see Fig. 2.4).

Fig. 2.4. The "Human–Machine" living machine

Let us contemplate simple examples of implementation:

1. The "mind" driving the machine (see the experiments, so far performed with monkeys [26]), which avoids the "bottleneck" of current user interfaces, which are notoriously asynchronic. We know that a lot comes out of machines, but very little – mainly interrupt commands – pass from the user to the machine.
2. The coupling of the nondeterministic "state-of-the-human" informational space (containing many parameters, with heterogenous data types such as temperature, color, pressure, rhythm, and so on) with the deterministic machine state, such as in hybrid control mechanisms. The data bus in the machine part is connected to the "living bus"; rich learning and forgetting affect interactions between the human and the machine.

These examples reflect the "state of the art" currently reached. If we could further integrate the living (not only human) and a machine endowed with pseudo-living properties (such as evolutionary programs), we would be better positioned to achieve a semiotic machine in the proper sense of the expression, and thus we might expect anticipatory characteristics augmented by computation.

Let us revisit the introductory hypothesis: the possibility of achieving a full record of an Xevent. We denoted the Xevent as object O without considering its condition. In reality, an Xevent appears to us – as we are part of it, experiencing it – as an immediate object: the meteorologist is inundated with data from trackers, radar readings, and sensor information. (Similar readings are made by a physician examining a patient who might have an epileptic seizure; or by seismologists as they consider issuing a warning of a catastrophic earthquake that would require massive emergency measures.) The immediate object O_i, which can be characterized through rich data, is only suggestive, but not fully indicative, of the dynamic object. After all, the tornado might not take place, despite all the readings; the seizure might not occur, or might take a mild form, indistinguishable from normal brain activity; or the seismic wave might be ambiguous.

Associated with the immediate object O_{i} is the immediate, although at times less than precise, understanding of what the description (representamen) conveys (the immediate interpretant I_{i}). In what we all we call prediction (including forecasts), most of the time this understanding is based on previous experience, that is, on probabilities. For example, in the past, a radar echo and a triple point on a surface chart suggested tornados. The dynamic interpretant I_{d}, not unlike a neural network propagation, corresponds to inferences from what is apparent to what might happen – to the space of possibilities. This consists of all that can happen (events associated with meteorology data such as weak shear, moisture, stationary front in vicinity). Integrating the probabilistic and the possibilistic dimensions of Xevent forecast is the final interpretant I_{f}: "If you haven't thought about it before it develops, you probably won't recognize it when it does". This comes from a professional in weather forecasting, Charles A. Doswell, III, as reported by Quoetone and Huckabee [27]. The transcripts of the various conversations among traffic controllers, airline representatives, and the military personnel in charge of guarding USA air space during the events surrounding the terrorist attacks of what has come to be known as 9/11 clearly reveal that nobody thought about the possibility of an operation at the scale of and with the means conceived of by the terrorists. The diagram given below captures the intricate relation (corresponding to a triadic-trichotomic sign relation) of the entities under consideration:

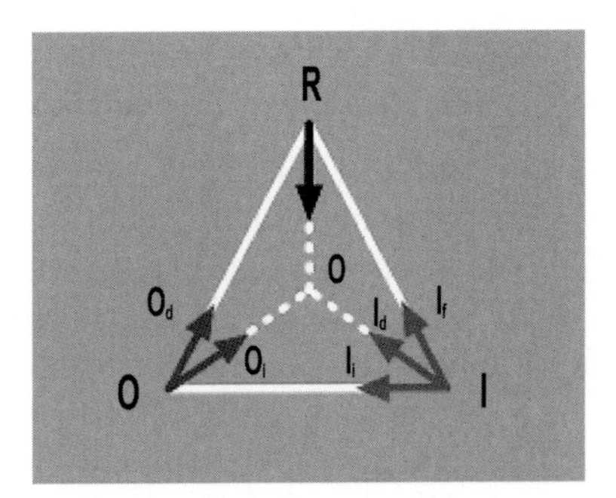

Fig. 2.5. The triadic-trichotomic sign relation

It is no accident that the infatuation with the sign originates in medicine; more precisely, in the practical endeavor known as diagnostics. But the "symptoms" of Xevents (also known as foretelling signals, whether in seismology, meteorology, medicine, terrorist activity, market analysis, and so on) point to a very complex sign process. The triadic-trichotomic representation of the sign suggests the need to distinguish between the appearance (immediate object O_{i}) and the evolving object of our attention (dynamic object O_{d}) It also makes us aware that the process of interpretation starts with the perception of appearance (I_{i}) and continues with the formulation of a theory (I_{f}), which in turn can be further interpreted (the state of knowledge regarding an Xevent at some moment in time). Quite often, we examine a representamen R as a symptom (for instance a seismogram, or some representative data pertinent to physical events) and infer from symptoms to possibilities, that is, a quantified record of what can be expected.

Xevents are notorious for casting doubt on forecast verification statistics. Due to the nonlinearities characteristic of Xevents, random factors lead to an ever-increasing difference between the statistically driven prediction and the observed event. Combining probabilistic and possibilistic descriptions allows new modeling perspectives. The fact that probability and possibility are not independent of each other (nothing can be probable unless it is possible, and not every possibility can be associated with a probability before an event) makes the task even more difficult. Since a vast body of literature on probability is available, I will make only a brief reference to possibility distributions.

Actually, we know that there is no "generally accepted formula for the mean of uncertainty or ignorance induced by a possibility distribution" [28]. The best that, to my knowledge, has been proposed so far is an E-possibilistic entropy measure. If $\Lambda = \alpha_1, \ldots \alpha_k$ is a set of outcomes (for example the effects of an earthquake, or of a storm), and $\Pi = \pi_1, \ldots \pi_k$ is a possibility distribution (with $\pi_i = 1, i = 1, \ldots k$), the measure of uncertainty (or ignorance) is the optimal value of the nonlinear equation:

$$\max H(p) = -\sum_{i=1}^{k} pi \log pi \tag{2.1}$$

subject to the limitations:

$$\sum_{i=1}^{k} pi = 1 \,, \quad \sum_{i=1}^{k} \pi_i \quad pi \geq \varepsilon \,, \quad \text{and} \quad pi \geq 0 \,.$$

Indeed, we are always informed, at least partially, about what has already happened; but we are ignorant in respect to what might happen (the possible event). The measure of our ignorance is always dependent on how well defined the possibility space is. The consequences for which α_i stands are hypothetical, and are sometimes (such as in financial crashes) affected by the perception of those involved (the investors), but are usually not obviously dependent upon their activities. A house constructed in the vicinity of a fault-line does not augment the intensity of the earthquake (should one take place) but it does affect the impact (human, social, economic, and so on). The possibility distribution is therefore a representation of the various correlations expressed in the Xevent. Possibilistic entropy does not depend only on the possibility distribution. Together with probability considerations, these intricate relations are implicit in the R expression and are indicative of Xevents both in the physical and in the living.

We ought to note that these entities (R, O, I) are not abstractions, but a logical guide to constructing an effective system of anticipation. Accordingly, we need to proceed by giving life to this diagram, such as by specifying the relation between the data and the possibility distribution. We also need to define all of its components, and furthermore, to proceed with a semiotic calculus that will generate an anticipatory self-mapping system. To give just one example, let us define an Xevent as an expression of dynamics. ("Expression" is used here in analogy to gene expression.) If we accumulate data (such as meteorological, geological, brain activity, financial market transactions; each associated with a possible Xevent) our goal would be to extract R patterns of expression (patterns of dynamics, or patterns of change) inherent in the data from the representamen. Mathematical techniques for identifying underlying patterns in complex data (in complex representamina) have already been developed for object recognition by computer-supported vision systems, for phoneme identification in speech processing, for bandwidth compression in electrocardiography and sleep research. These are clustering techniques (hierarchic, Bayesian, possibilistic, and so on). Among these techniques, so-called self-organizing maps [38] can be defined to correspond to a semiotic self-mapping. Such maps use visualization techniques to reduce the data space with the help of self-organizing neural networks. In effect, similarities in the data are evidenced by grouping similar data items. This involves a high number of iterations. In the final analysis, an SOM is associated with a grid. The rectangular grid used is somewhat analogous to an entomologist's drawer (adjacent compartments hold similar insects), although I actually prefer the analogy to a philatelic collection (adjacent pages in the album hold similar stamps or series). The SOM of a possible Xevent is a representation of all the "insects" or "stamps" not yet collected, or the "stamps" not yet printed. For an iteration i, the position of a variable V_e is denoted $f_i(V_E)$. The formula

$$f_{i+1}(V_\mathrm{E}) = f_i(V_\mathrm{E}) + \tau(d(V_\mathrm{e}, V_\mathrm{EP})I)(P - f_i(V_\mathrm{E})) \tag{2.2}$$

describes the next position of the Xevent variable considered. Notice that the position corresponding to possible data point P (which is a node V_{EP} in the grid) and the variable V_e are subject to a distance evaluation $d(V_e, V_{EP})$. Learning is involved in the process (τ is the learning rate); the learning rate decreases as the distance between V_e and V_{EP} decreases. Indeed, if the difference tends to zero between the possible value and the observed value, there is nothing left to learn. There is randomness, accounted for in order to replace the living component. In such a procedure, semiotic considerations are no longer meta-statements, but become operational.

Human beings operate naturally in the semiotic realm. They generate cognitive maps as they act in relation to the Xevent variable. These mental maps guide our actions. Even the influence of predictions and forecasts affect these self-generated maps. We do not process chairs or electrons or thunder in our minds, but rather their representamina. Accordingly, a semiotic machine combines the perception of signs with the production of signs (see Fig. 2.6):

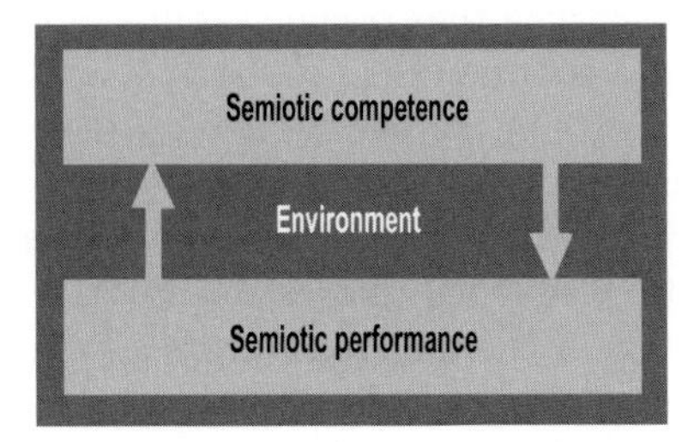

Fig. 2.6. The unity of action and perception

What the diagram suggests is that the human being's self-constitution (how we become what we are through what we do) implies the unity of action – driving our perception of the world – and reflection. Therefore, to do something, such as to deal with Xevents – reflect upon them, cope with their impact, predict them – actually means to anticipate the consequences of our actions. In effect, this translates neither into an anticipation method nor into specific means, but rather into the realization that anticipation is an evolution-immanent characteristic. Should we ever be able to build an evolutionary machine (to create a living entity), it will have to display anticipatory characteristics. For all it's worth, the realization that anticipation is an evolution-immanent characteristic means that anticipation of Xevents is possible, but not guaranteed. Evolution itself is not a contract with nature for individual survival or survival of the species. The Xevent that led to the extinction of the dinosaurs is only one example among many others.

If instead of considering the R of a natural Xevent (an earthquake and the like) we look at the plans on whose basis the A-bomb was built, or on which chemical and biological weapons are produced, or the new "smart" weapons (producing targeted Xevents!), we still remain in the semiotic realm. The R_{A} (for atomic bomb), or the R_{C} (for some chemical weapon), or the R_{B} (for biological weapons), or the R_{S} (for smart weapons) are an effective description of a potential Xevent, which we can fully predict within an acceptable margin of error. Bombing the desert (which used to be called nuclear testing) is quite different from bombing a populated area. We can also, within other margins of error, predict what might happen with respect to R_{C} or R_{B}, and even R_{S}.

The Xevent (O) – the unity between O_i and O_d, or the appearance and the dynamic unfolding of the event – contained in the description (R) becomes subject to a process of interpretation. This extends from scientific analysis and planning, to engineering and testing, as well as to media reports, fiction, and movies, not to mention the production of interpretations, true and false, of secret services intent on confusing the potential users of such devices. Indeed, Xevents, whether natural or artificial, become part of the political experience, and thus their prediction also impacts politics. Xevents lead to a whole bureaucracy (emergency funds set up to meet needs), and to new laws (including ones intended to prevent market crashes or terrorist attacks).

It has often been remarked that social systems (and for that matter, systems pertinent to living communities, human or not) display anticipation. The less constrained a system is, the higher its resiliency. Meaning comes into existence with hindsight: "What happened to the subway that came to a screeching halt? What does it mean that an airplane hit a skyscraper? What does it mean that someone has a seizure?" A logistic map can inform us about a direction of change. Market processes exemplify the process. Feedback and feedforward work together; production, supply, demand, and all other factors are underlying factors in market dynamics. A crash – an Xevent – is not dependent upon the anticipatory actions of informed or uninformed individuals, but rather upon aggregate behavior. Cellular automata able to operate on two different timescales (real time vs. faster than real time) could, in principle, capture the recursive nature of those who make up the market.

But in real life (whatever that means), we can only act in the present (as events are triggered). As such, the interpretant process for which a cellular automaton stands appears as a funnel to us:

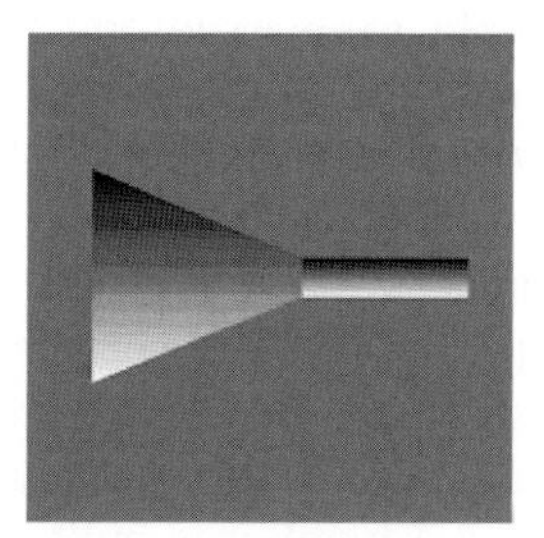

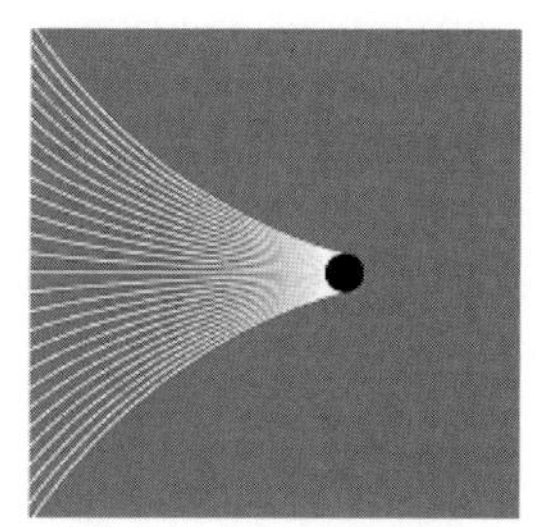

Fig. 2.7. An Xevent as a realization in a possibilistic space along a time axis

The immediate object (the Xevent) unfolds in the huge space of possibilities that one can conceive of and pursue systematically. To be successfully anticipatory means to progressively reduce this space until the convergence of the open cone-shaped object.

2.6 Can a Computer Simulate Anticipation?

In effect, to anticipate is to move along the time vector from the event (tornado, flood, seizure ...) – the neck of the funnel – as it unfolds, to its initial conditions. In the language of dynamic systems, this means to move from the strange attractors embodying the Xevent to the conditions feeding the dynamics of the system. Nonlinear processes affect the "edges" of the dynamic distribution (for example: how wide a swing a stock market can take, what the most extreme temperatures are, what the atmospheric pressure values are, what the seismic parameters are, the level at which a system's stability is affected). But there is no indication whatsoever that these processes display any regularity.

Scientists such as Sornette, Helbing, and Lehnertz – to name three among those published in this volume – are dedicated to this approach. For instance, Sornette is well respected for considering self-organized criticality and out-of-equilibrium conditions. He advanced the hypothesis that Xevents are due to the system's endogenous self-organization. In contrast to prevailing views, he covers a very large area of public interest (from geological aspects to the future of humankind on earth). The abstraction in Helbing's model of collective behavior goes back to self-driven many-particle systems. Malfunctions in the form of abnormal synchronization of a large number of neurons catch the attention of those (such as Lehnertz) looking for the prediction potential of apparati (such as multichannel EEC recorders), if an appropriate determination of the abnormality, detected through statistical evaluation, is performed.

Each time they, and others who follow a similar physics-driven path to discovery, come upon patterns in the data subject to their examinations, they pursue the thought of identifying regularities that can ultimately justify prediction. Some are on record – a very courageous scientific attitude – with predictions (regarding, say, the economy, financial markets) that the public can evaluate. Others have commercialized their observations (for example on crowd behavior). It would be out of character and out of the question for me to cast doubt on models mentioned here for their elegance and innovation. But the reader already knows where I stand epistemologically. And given this stand, I can only suggest that, for the particular aspects on which my colleagues focus, acceptable predictions are possible. What is not possible is a good discrimination procedure, one that allows us to compare, ahead of the

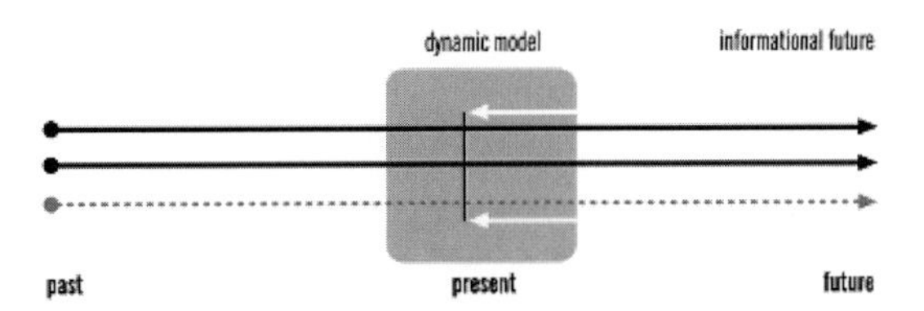

Fig. 2.8. A model unfolding in faster than real time appears to the observer as an informational future

future, between good and bad, appropriate and inappropriate predictions. For the past, which they assume to be repeated in some form or shape, the prediction is usually good (or at least acceptable). But once complexities increase, even within the physical, the Xevent starts to look like a "living" monster. We cannot afford to ignore this fact.

This journey from the Xevent states to the states leading to it is, for all practical purposes, a reverse computation (regardless of whether the computer is implemented in silicon, DNA, quantum states, and so on). In Richard Feynman's words [29], this is equivalent to asking, "Can a computer simulate physics exactly?" Reversibility is in fact the characteristic of a computation in which each step can be executed and unexecuted. Making and unmaking an omelette is one way of suggesting what we are referring to here. Increase this to the scale of an earthquake and imagine the weird computation of the earthquake as output, and its reverse. But even at the scale of an epileptic seizure or financial crash, the film played in reverse is not easy to conceptualize – and it is not at all clear whether it is feasible.

That physical laws are generally reversible automatically allows for a reversible computer (with all of the costs associated with the erasure of information). But what is not clear is whether an earthquake, a heart attack, a tornado, or a seizure is the result of a deterministic process, or at least one of deeper levels of order. If the computation of an earthquake involved the condition of the process leading to the earthquake – that is, if one could define an "earthquake machine" – we would probably profit from the reversibility of the computation. It is very exciting to compute in the medium we examine, provided that we examine events of a regular nature (no matter how deep the regularity is hidden). But, not unlike the infinite interpretant process characteristic of semiotics (each interpretation becomes a new sign, ad infinitum), Xevents seem either unique (irreducible to anything else) or only an instance of a longer development that goes beyond what we call tractable.

2.7 A New Equilibrium

Thesis 3. *Xevents are actually the preliminary phase leading to a necessary new state of balance leading to the next Xevent* (see Fig. 2.9).

What I am saying here is that the epileptic seizure is, in its own way, a process that preserves life, since it leads to the post-seizure condition that replaces the endangering state prior to it. Or, that the earthquake – a tremendously energetic peak – ends up in the post-quake condition of relatively energetic balance, and of infinitely less destructive potential earthquake. Otherwise, the potential future event would grow and grow until the earthquake's resources are exhausted. With this in mind, I suggest here that we are dealing with what physics has stubbornly rejected for the sake of homogeneity and determinism: the causa finalis as the necessary path of dynamic unfolding.

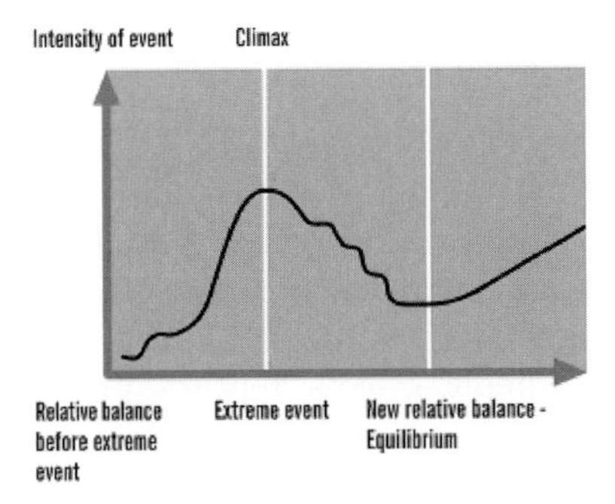

Fig. 2.9. The relevance of the post-extreme event time

Aristotle distinguished among four categories of causation: material cause, formal cause, efficient cause, and final cause. In contemporary jargon, they correspond to different kinds of information. If we apply these categories to a house, it is evident that materials – cement, brick, wood, nails – pertain to the material cause. Builders (think about the many types of workers involved in excavating, mixing and pouring concrete, bricklaying, and so on) make the efficient cause clear through their work, while the plans they go by (blueprints and various regulations) represent the formal cause. The final cause is clear and simple: someone needs or wants to live in such an edifice.

Now take a work of art. The materials, the work of an artist, and the various sketches are well defined. But who needed or wanted the work? (Who needs or wants an Xevent?) In some cases, there is one person who acted as commissioner. In the majority of cases, the action (make the artwork) is driven by the artist dedicated to expressing himself or herself, to ascertaining a view or perspective, to unveiling an aspect of reality unknown to others or perceived in a non-artistic way. The final cause is the work itself, as it justifies itself within a culture and within a social context. The pragmatics of art is the answer to the "Why?" of a work of art, not to the "How?" (which pertains to its efficient cause). As science eliminated the legitimacy of anything even slightly related to causa finalis, it raised the infatuation with "How?" to the detriment of the question essential to any artistic experience: "Why?" Indeed, the "Why?" of an Xevent should interest us at least as much as the "How?" if we want to get closer to the prediction, and to the anticipation of Xevents.

It is less suspicious to affirm such an idea today now that bifurcations and attractors were introduced into scientific jargon (see Feigenbaum et al. [30]). The equilibrium following Xevents makes us aware of the variety of ways in which the physical substratum of everything is preserved through infinite processes. However – and this goes back to the major distinction I have advanced so far – Xevents in the physical universe compared to Xevents in the living are subject to predictive actions only to the extent that a pre- and a post-phase are identifiable. In this sense, time appears as a component of life, not just as one of its descriptions.

This prompts the next thesis of this article.

Thesis 4. *The action of anticipation cannot be distinguished from the perception of the anticipated.*

This applies in particular to Xevents as episodes in the self-constitution of the living, of the human being in particular. Testimony from folklore and anthropological evidence make us aware of the variety of anticipatory behaviors of the living (animals, insects, reptiles, plants, bacteria) in relation to Xevents. This evidence has always been subjected to scientific scrutiny: can it be that in some cultures important information regarding Xevents (earthquakes, floods, seizures, epidemics) has been derived through the interpretation of animal behavior and characteristics by the people sharing the environment with them? And if so, can we derive anything useful for predictions of Xevents from this information? After all, the living is endowed with anticipation, and accordingly, the anticipation of Xevents in the natural realm cannot be excluded. Moreover, there is sufficient anecdotal evidence to suggest that epileptic seizures in humans are signaled ahead of time by dogs. Similar anecdotal reports are often mentioned, even in scientific publications.

As non-natural factors – such as anthropogenic forces related to urban development, land conversion, water diversion, pollutio – increasingly affect the environment, animals, birds and plants exhibit new patterns of behavior. Even these changes are indicative of the tight connection between all of the components of the ecosphere. Ecological consequences of Xevents are rapidly becoming the focus of many scientists who realize the need for a holistic approach (such as the British Ecological Society [31]). We are losing important sources of information as we create artificial circumstances for nonlinearities that, instead of eliminating the risks associated with Xevents, actually increase their impact, and sometimes their probability and possibility. Numerous dams that only marginally adequately function under extreme weather conditions have made us aware of the Xevent potential their failure can entail. Buildings of all types, devices we place on mountains or under water, satellites circling the earth – these have all amplified the possibility space of Xevents. The possibility of hitting skyscrapers with airplanes did not exist before we started to fly using "mechanical birds", and built them high into the sky. In this context, interestingly enough, we are forcing nature (and ourselves) towards machine behavior. Farms become food factories; workers are expected to act like machines; institutions become machines with specialized functions (doctors are human body mechanics, hospitals are spare parts factories, the state is a machine for maintaining the coherence of the social system, the police are machines for maintaining order). The expectation is regularity, and all the measures undertaken worldwide following 9/11 are meant to maximize the predictability of the irregular (including the Xevents subject to the scrutiny of Homeland Security).

This expectation is fed by a scientific model of prediction and reproducibility corresponding to the world of physics. Indeed, dropping a stone

Fig. 2.10. Given the same conditions, a stone will fall the same way

from the same position, under the same circumstances (humidity, wind) will always result in the same measurements (of speed, position at any moment in time, impact upon landing). It is a predictable experiment; it is reproducible. Even if we change the topology of the landing surface, the outcome does not change.

Let a cat fall and derive the pertinent knowledge from the experiment. This is no longer a reproducible event. The outcome varies a great deal, not the least from one hour to another, or if the landing topology changes. The stone will never get tired, annoyed, or excited by the exercise.

Applied differential geometry allows for the approximate description of an object flipping itself right side up, even though its angular momentum is zero. In order to accomplish this, it changes shape (no stone changes shape in the air). In terms of gauge theory, the shape-space of a principal SO(3)-bundle, and the statement "angular momentum equals zero" defines a connection on this bundle [32]. The particular movement of paws and tail conserves the zero angular momentum. The final upright state has the same value. This is the "geometric phase effect", or monodrony. Heisenberg's [33] mathematics suggests that, although such descriptions are particularly accurate, we are, in observing the falling of a cat, not isolated viewers, but coproducers of the event. The coherence of the process, not unlike the coherence of the apparently incoherent class of events we call extreme, is the major characteristic.

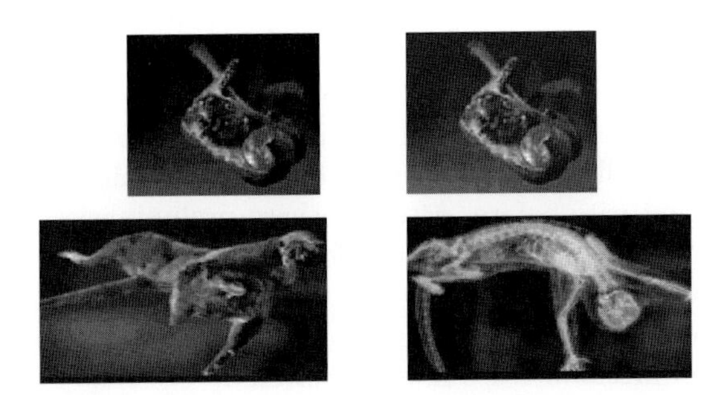

Fig. 2.11. The cat never falls the same way

This is where the need to consider the living as different from the inanimate physical becomes more obvious. In order to address this, I will make reference to the work of an established physicist, Walter Elsasser, who worked in quantum mechanics (with Niels Bohr) and was very familiar with Heisenberg. He dedicated the second part of his academic career to a scientific foundation of biology. These considerations are appropriate in this context if the fundamental distinction between Xevents that are physical in nature and Xevents that are peculiar to the living are to be pursued effectively. They can guide us further if we realize that there is more than a one-way interaction between Xevents, as they emerge, and our perception. We are not just spectators at a performance (sometimes scary), but also, in many ways, coproducers.

2.8 A Holistic View

A physicist of distinguished reputation, Walter Elsasser [34] became very interested in the living from an epistemological perspective. As in Rosen's case – Rosen being the mathematician most dedicated to the attempt to understand what life is – it would be an illusion at best to think that we could satisfactorily summarize Elsasser's attempt to reconcile physics with what he correctly perceived as a necessary theory of organisms. Rosen and Elsasser had a focus on complexity in common. But in contrast to Rosen, Elsasser was willing to pay his dues to the scientific matrix within which he found his own way: "The successful modern advance of reductionism rests on certain presuppositions which at this time are no longer questioned by any serious scientist". Moreover, and here I quote again, "There is no evidence whatever that the laws of quantum mechanics are ever wrong or stand in need of modification when applied to living organisms". All this sounds quite dogmatic and, for those versed in science theory, almost trivial given the fact that theories are ultimately coherent cognitive constructs, not continents waiting to be discovered. Physics, in its succeeding expressions, is no exception. For the reader not willing to delve into the depths of the argument, the position mentioned is not really inspiring. Opportunistically, and as Rosen did too, he refutes vitalism, "the idea that the laws of nature [that is, physics] need to be modified in organisms as compared to inanimate nature". Serious scientists in all fields and of all orientations have discarded vitalism, just as alchemy was discarded centuries before. After all these preliminaries, Elsasser finally articulated a clear point of departure for his own scientific journey, which justifies continued interest in his work: "Close reasoning indicates the existence of an alternative to reductionism. This is so despite the fact that the laws of quantum mechanics are never violated".

From this point on, we have quite an exciting journey ahead of us. Indeed, biology is a "non-Cartesian science". The "master concept" in describing the holistic properties of the living is complexity [34]; more precisely, what he describes as unfathomable complexity. This concept dominates the entire endeavor; therefore an extended quotation is probably justified. Unfathomable complexity "implies that there is no series of actual experiments, and not even a set of suitably realistic thought-experiments such that it would be possible to demonstrate the way which all the properties of an organism ... can be reduced to consequences of molecular structure and dynamics ... ". Furthermore, he defines properties those that remain unaccounted for by physics and chemistry as morphological. Four principles and a "basic assumption" stand at the foundation of his biology. The assumption refers to the holistic view adopted – the living cannot be understood and described other than as a whole: "the organism is a source (or sometimes a sink) of causal chains which cannot be traced beyond a terminal point"; that is, they are ultimately expressed in the unfathomable complexity of the organism.

According to this viewpoint, Xevents in the living cannot be meaningfully addressed on the basis of reductionism (not even at the level of detail of single neuron functions or genetic expression), but only globally, in a holistic manner.

The first principle that Elsasser further articulates is known as *ordered heterogeneity*. It states that, as opposed to the homogenous nature of physical and chemical entities (all electrons are the same), the living consists of structurally different cells. There is order at the cellular level, and heterogeneity at the molecular level. Heterogeneity corresponds to individuality, a term that has no meaning in the physical world. The principle of creative selection focuses on the richness of living forms. For homogenous systems, the variation of structure (if there is such a variation) averages out. For heterogenous systems (the living), an immense multitude of possible states is open to realization (selection). The property of selection is attributed to matter alone – a more refined mathematics of dynamic systems, which to date has not been formulated, would probably define some specific self-organizing action here. The selection as such is based on the third principle, of holistic memory. The new morphological pattern actually selected resembles earlier patterns, but is not the realization of stored information. Elsasser is quite convincing in arguing for a "memory without storage – the touchstone of the theoretical scheme proposed" [34]. The argument is based on the distinction between two processes: homogenous replication (the assembly of identical DNA molecules) and heterogenous reproduction (self-generation of similar though distinct forms). Always different, the living practice creativity as a modus vivendi. Replication is a "dynamic process" [34] resulting in what we perceive as regularities in the realm of the living. Replication and reproduction need to be conceived together. What makes this possible is the fourth principle, of *operative symbolism*. The discrete, genetic message is represented by a symbol that stands for the integrated reproductive process. Elsasser himself realized that this operative symbolism is merely a *tag* for all processes through which the living experiences its own dynamics. He looked for a triggering element, a *releaser*, as he called it, that could start a restructuring process. From a piece of genetic code, the releaser will trigger the generation of the complete message necessary for the reconstruction of a new organism. We can imagine this releaser – the operative symbol – as able to start a "program" that will result in a new biological form, as an alternative to storing and transmitting the form itself. The biological information is stored as data (in the homogenous replication) and as an immense number of alternate states from which one will eventually be realized (in the heterogeneous reproduction). This latter assumption implies that biological phenomena are "in part" autonomous.

Again, when studying Xevents in the living – to which not only epileptic seizures and strokes belong, but also cancer and heart attacks – these observations are a good guide for prevention and anticipation. That earthquakes and hurricanes are always different, as are strokes and financial market crashes, speaks for the adoption of the over-arching notion of heterogenous replication.

Instead of searching for laws, Elsasser highlights regularities. Where reductionists would expect that "the gametes contain *all* the information required to build a new adult", a non-reductionist biology would rely on holistic memory and his *Rule* of repetition: "Holistic information transfer involves ... the reproduction of states or processes that have existed previously in the individual or species as the case may be" [34]. Of special interest to him is the re-evaluation of the meaning of the Second Law of Thermodynamics (and the associated Shannon law of information loss). Elsasser argued that since paleontology produced data proving the stability of the species (over many millions of years of existence), and since the Second Law of Thermodynamics points in the opposite direction, only a different integration of both these perspectives can allow us to understand the nature of the living. Therefore, two types of order were introduced, in a way such that they never contradict each other. This is what he called biological duality: "living things can be described by a different theory as compared to inanimate ones". As a consequence, if one attempted to verify holistic properties, a different kind of experiment from the one conventionally used in physics would be required. It is worth mentioning here that Wilhelm Windelband [35] made the distinction between nomothetic and idiographic sciences. The latter focused on singularity: "der Gegensatz des Immergleichen und des Einmaligen", (the contradiction of the invariable/unchanging and the unique).

It is at this juncture that Rosen's thinking and Elsasser's meet – I doubt that they had a chance to study each other's work on the living and life in depth. Rosen was "entirely dedicated to the idea that modeling is the essence of science" [36]; Elsasser realized that no experiment, in the sense of experiments in physics, could capture the holistic nature of the living. Moreover, both asked the fundamental question: what does it take to make an organism? If the representamen R for an organism (such as the stem cell) were available, we would be able to anticipate Xevents in the living in relation to the end of life (return to physicality). But Xevents can also be viewed from the perspective of the same question: what does it take to make an earthquake?

Or an act of terrorism? Elsasser, not unlike Rosen, concluded: "The synthesis of life in vitro encounters insuperable difficulties" [34]. It is quite possible that such a strong statement corresponds to the realization that anticipation, as the final characteristic of the living, might be very difficult to describe (the analytic step) but probably impossible to reproduce (the synthesis). So, we might even be able to say what it takes to create a certain Xevent, but that does not mean that we could literally make it. Even induced seizures are not exactly like the ones experienced by individuals who go through real seizures. Low-scale earthquakes (caused by experiments and tests that researchers conduct) are by their nature on a different scale and quality than the ones that people experience on the Islands of Japan, in California, in China, or in Turkey. It is therefore of particular interest to take a closer look at the various factors involved in what, from a holistic perspective, appears to us as anticipatory.

We learn from this that there is no anticipation in the realm of physics. Accordingly, if we are dedicated to addressing Xevents – whether in the physical world or in the realm of the living (birth and death are themselves Xevents) – we need to realize that answers to what preoccupies us will result from understanding how the living anticipates. (We know why, since this results from the dynamics of evolution.)

The final thesis of this article is of less significance to prediction of Xevents and more to our anticipatory condition in the universe.

Thesis 5. *The project of extreme scientific ambition, of creating life from the physical, can succeed only to the extent that a physical substratum can be endowed with anticipatory characteristics.*

This conclusion is not a conjecture; it is strongly related to a better understanding of Xevents. It ascertains that in addressing questions pertinent to Xevents, we are bound to address (differently to Rosen) the notion of what life is. After all, nothing is extreme and nothing is an event unless it pertains to life.

References

1. Heraclitus, as mentioned in Plato: *Cratylus*, 402A
2. Peirce, C.S.: "Indeed, representation necessarily involves a genuine triad. For it involves a sign, or representamen, of some kind, inward or outward, mediating between an object and an interpreting thought...". The Logic of Mathematics (1896). In: Hartshorne, C., Weiss, P., eds., *Collected Papers of Charles Sanders Peirce*, Harvard University Press, Cambridge, MA (1931), pp 1–480
3. Peirce, C.S.: "A sign is something which stands for another thing to a mind." In: *Of Logic as a Study of Signs*, MS 380 (1873). See also: The Writings of Charles S. Peirce, Volume 3, 1872-1878, pp. 82–83. Peirce Edition Project, Christiane S.W. Kloesel et al, Eds. Bloomington IN: Indiana University Press, 1998

4. Peirce, C.S.: "Anything which determines something else (its interpretant) to refer to an object to which itself refers (its object) in the same way, the interpretant becoming in turn a sign, and so on ad infinitum" (1902). In: Hartshorne, C., Weiss, P., eds., *Collected Papers of Charles Sanders Peirce*, Harvard University Press, Cambridge, MA (1931), pp 2–303
5. Peirce, C.S.: "A sign is intended to correspond to a real thing." In: *Foundations of Mathematics*, MS 9 (1903) 76 Definitions of the Sign by C.S. Peirce, researched and collected by Robert Marty. cf. http://members.door.net/arisbe/menu/library/rsources/76defs/76defs.htm
6. Schneiderman, B.: *Designing the User Interface*, Addison Wesley, Boston, MA (1998), p. 185
7. Morris, C.: *Foundations of the Theory of Signs*, In: Neurath, O., ed., *International Encyclopedia of Unified Science*, vol. 1, no. 2, University of Chicago Press, Chicago, IL (1938)
8. Nadin, M.: *Zeichen und Wert*, Gunter Narr Verlag, Tübingen (1981)
9. Marty, R.: *Catégories et foncteurs en sémiotique*, Semiotica 6, Agis Verlag, Baden-Baden (1977), pp 5–15
10. Marty, R.: *Une formalisation de la sémiotique de C.S. Peirce á l'aide de la théorie de categories*, Ars Semeiotica, vol. II, no. 3, John Benjamins BV, Amsterdam (1977), pp 275–294
11. Marty, R.: *L'algebre des signes*, A John Benjamins, Amsterdam (1990)
12. Goguen, J.: *Algebraic Semiotics Homepage* (see http://www.cs.ucsd.edu/users/goguen/projs/semio.html), last accessed August 2005
13. NSF/CSPO/NCAR, *Extreme Events: Developing a Research Agenda for the 21st Century*, conference sponsored by the National Science Foundation (NSF), organized by the Center for Science, Policy, and Outcomes (CSPO) at Columbia University and the Environmental and Societal Impacts Group of the National Center for Atmospheric Research (NCAR), Boulder, CO, 7–9 June (2000)
14. Gallese, V.: *The Inner Sense of Action. Agency and Motor Representation*, In: Journal of Consciousness Studies 7, vol. 10 (2000), pp 23–40
15. Nadin, M.: *The Civilization of Illiteracy*, Dresden University Press, Dresden (1997)
16. Gödel, K.: *Über formal unentscheidbare sätze der Principia mathematica und verwandter systeme.* In: Monatshefte für Mathematik und Physik, 38 (1931), pp 173–198; translated in van Heijenoort, *From Frege to Gödel*, Harvard University Press, Cambridge (1971). The exact wording of his first theorem is "In any consistent formalization of mathematics that is sufficiently strong to define the concept of natural numbers, one can construct a statement that can be neither proved nor disproved within the system." His second theorem states: "No consistent system can be used to prove its own consistency."
17. Rosen, R.: *Anticipatory Systems*, Pergamon Press, New York (1985)
18. Nadin, M.: *Anticipation: The End Is Where We Start From*, Lars Müller Verlag, Basel (2003)
19. Rosen, R.: *Life Itself: A Comprehensive Inquiry into the Nature, Origin, and Fabrication of Life*, Columbia University Press, New York (1991)
20. Berry, II, M.J., Brivanlou, I.H., Jordan, T.A., Meister, M.: *Anticipation of Moving Stimuli by the Retina*, In: Nature, vol. 398 (1999), pp 334–338
21. Ekman, P., Friesen, W.V.: *Facial Action Coding System: A Technique for the Measurement of Facial Movement*, Consulting Psychologists Press, Palo Alto (1978)

22. Fletcher, P.C., Anderson, J.M., Shanks, D.R., Honey, R., Carpenter, T.A., Donovan, T., Papdakis, N., Bullmore, E.T.: *Responses of Human Frontal Cortex to Surprising Events are Predicted by Formal Associative Learning Theory*, In: Nature Neurosciences, vol. 4, no. 10 (2001), pp 1043–1048
23. Ishida, F., Sawada, Y.F: *Human Hand Moves Proactively to the External Stimulus: An Evolutional Strategy for Minimizing Transient Error*, In: Physical Review Letters, vol. 93, no. 16 (2004)
24. Nadin, M.: *Mind-Anticipation and Chaos*, Belser Verlag, Stuttgart (1991)
25. Minsky, M.: *The Virtual Duck and the Endangered Nightingale*, In: Digital Media, June 5 (1995), pp 68–74
26. Nicolis, M., Dragan Dimitrov, A.L., Carmena, J.M., Christ, R., Lehew, G., Kralik, J.D., Wise, S.P.: *Chronic, Multisite, Multielectrode Recordings in Macaque Monkeys*, In: Proceedings of the National Academy of Sciences of the Unites States of America (PNAS Online), Sept. 5 (2003)
27. Quoetone, E., Huckabee, K.L.: *Anatomy of an Effective Warning: Event Anticipation, Data Integration, Feature Recognition*, Preprints, 14th Conf. On Weather Analysis and Forecasting, Amer. Meteor. Soc., Dallas, TX (1995), pp 420–425
28. Guiasu, S.: *Comment on a Paper on Possibilistic Entropies*, In: International Journal of Uncertainty, Fuzziness and Knowledge-Based Systems, vol. 10, no. 6 (2002), pp 655–657
29. Feynman, R.: *Potentialities and Limitations of Computing Machines*, Lecture series, California Institute of Technology, CA (1983–1986)
30. Feigenbaum, M.: *Universal Behavior in Non-linear Systems*, In: Los Alamos Science, vol. 1 (1980), pp 4–27
31. BES, *The Scientific Rationale of the British Ecological Society (BES) Symposium*, March–April (2005)
32. Montgomery, R.: *Nonholonomic Control and Gauge Theory*, In: Canny, J., Li, Z., eds., *Nonholonomic Motion Planning*, Kluwer Academic, Dordrecht (1993), pp 343–378
33. Heisenberg, W.: *Uncertainty Principle*, first published in Zeitschrift für Physik, vol. 43 (1927), pp 172–198
34. Elsasser, W.M.: *Reflections on a Theory of Organisms*, Johns Hopkins University Press, Baltimore, MD (1998). Originally published as *Reflections on a Theory of Organisms. Hoilism in Biology*, Orbis, Frelighsburg, Quebec (1987)
35. Windelband, W.: *Geschichte und Naturwissenschaft* (1894) (cf. Strassburger Rektoratsrede. http://www.fh-augsburg.de/~harsch/germanica/Chronologie/19Jh/Windelband/win_rede.html)
36. Rosen, R.: *Essays on Life Itself. Complexity in Ecological Systems*, Columbia University Press, New York (2000), pp 324–324
37. Kohonen, T.: *Self-Organized Formation of Topologically Correct Feature Maps*, In: Biological Cybernetics, vol. 43 (1982), pp 43–69
38. Kohonen, T.: *Self-Organizing Maps*, Springer, Berlin Heidelberg New York (1997)

3 Mathematical Methods and Concepts for the Analysis of Extreme Events

Sergio Albeverio and Vladimir Piterbarg

Summary. Mathematical tools for the analysis of Xevents, maxima of processes and rare events are presented. Methods and concepts of classical statistical extreme value theory are described, as well as those of large deviation theory. Techniques from other areas such as statistical mechanics, the theory of dynamical systems and the theory of singularities are also briefly discussed.

3.1 Introduction

In everyday parlance, Xevents are associated with characteristics such as seldomness, extremality (in the sense of being "larger" than "usual events"), as well as having important, often catastrophic consequences. Phenomena with these characteristics arise in mathematics in several areas. One of these derives from classical statistics, namely the theory of extreme values. These are, in their simplest form, described by asymptotic distributions of suitably normalized maxima of sequences of random variables. We discuss this theory, which is well developed and has many areas of applications, in Sect. 3.2 below.

In Sect. 3.3 we move on to briefly discuss an extension of this theory to the case of continuous time, in particular continuous time Markov chains and processes, as well as to the case of random fields. We also discuss the relationship between the discrete time and the continuous time cases and address the question of predicting Xevents.

In Sects. 3.4 and 3.5 we briefly discuss some interpretations of the intuitive concept of Xevents in terms of statistical mechanics, the related theory of self-organized criticality and the theory of dynamical systems. Our presentation here has strong links with other contributions in this book, in particular with the paper of H. Kantz and coworkers, although we place more emphasis on mathematical aspects rather than concrete modelling.

In Sect. 3.6 we briefly mention other areas of mathematics where singularities that cause "extreme behaviour" arise; these include the theory of singularity of mapping, catastrophe theory, and turbulence theory.

This article should serve as an introduction to a broad area of mathematical research, and it should also (hopefully) motivate the reader to learn more about this research field. We have concentrated on presenting general aspects rather than concrete case studies, which are covered by other contributions in this book, and so they are complementary to our presentation.

To avoid referencing a huge number of works, we have mainly cited books and survey articles rather than original articles. In this respect, when citing reference X, we implicitly refer to X and references therein.

3.2 Statistical Extreme Value Theory

Any statistical theory is based on a probabilistic model that provides, at the very least, a plausible description of the data. In this section we will concentrate on the probabilistic modelling of Xevents, referring to statistical issues for each of the models considered. This will involve the following: 1) a brief description of the classical theory and its sources; 2) an excursion through models, methods and tools, balancing the interests of readers and authors; 3) mentioning important recent developments and applications; 4) providing a discussion of important open problems in the statistical analysis of Xevents.

3.2.1 Origins: Classical Univariate Case

Extreme value theory is an established area in probability theory and mathematical statistics. It originated from the asymptotic study of maxima and minima (extremes) of finite time series provided by random variables, assumed to be independent and identically distributed (i.i.d.). The story of the origin of this theory is itself interesting, interwoven as it is with the practice of statistics and with strong personalities. Let us mention in particular that one of the main founders of the theory and the author of the first and most influential book on extreme value statistics, E.T. Gumbel, was a scientist with a clear social and political agenda against the Nazi regime and for the pacifist cause [1]. E. Gumbel was also a pioneer in the application of extreme value statistical theory, particularly in the fields of climatology and hydrology.

The main tenet of extreme value theory is that if $(X_n)_{n \in \mathbb{N}}$ is a sequence of i.i.d. real-valued random variables, and if the maximum $Y_n \equiv \max(X_1, \ldots, X_n)$, suitably "standardized", has a limit distribution as $n \to \infty$, then this distribution belongs to one of three standard types (characterized by the names Fréchet, Weibull and Gumbel). Extreme value distributions were initially applied to failure distributions of tensiles [2] and materials [32] and distributions of floods, droughts, extreme winds, storms and other meteorological phenomena. The book of Gumbel, [1], is still one of the best sources of applications (see also [3], Tirozzi et al., Time series and neural networks. An application to reconstruction and Xevent analysis of sea time series, in preparation).

The problem of evaluating parameters from extreme value distributions is dealt with in [1] (see also [3] for the connection to order statistics; in particular, there are results based on studying the $m \leq n$ largest observations of a series of n observations, for example in the limit as $m \to \infty$ with $\frac{m}{n} \to 0$)

After a long evolution, including contributions from Dodd in 1923, von Mises in 1923 and 1936, Fréchet in 1927, Fisher and Tippett in 1928, de Finetti in 1932, Gumbel in 1935 and 1958, Gnedenko in 1943, and de Haan in 1970, the *maximum limit theorem* took the following form:

Theorem 1. *The class of limit distributions for the law P^n of $a_n Y_n + b_n$, where $a_n > 0$, b_n are suitable chosen constants, contains only laws with densities*

a) $P_{Fr}(x) = \begin{cases} 0\,, & x \leq 0\,, \\ e^{-x^{-\alpha}}\,, \quad x > 0\,, & \alpha > 0\,; \end{cases}$ *(Fréchet)*

b) $P_{Gu}(x) = e^{-e^{-x}}\,, \quad x \in \mathbb{R}\,;$ *(Gumbel)*

c) $P_{Wei}(x) = \begin{cases} e^{-(-x)^{-\alpha}}\,, \quad x \leq 0\,, & \alpha > 0\,, \\ 1\,, & x > 0\,; \end{cases}$ *(Weibull)*

In 1936 Von Mises suggested combining these three extreme value distributions, which we shall call standard extreme value distributions, into a generalized extreme value distribution. Up to a linear transformation, one can see that the above three types of extreme value distributions laws can be rearranged into the common form:

$$P_\gamma(x) = \exp\left(-\left(1 + \gamma\frac{x-m}{\sigma}x\right)_+^{-1/\gamma}\right)\,, \qquad x \in \mathbb{R}$$

where $a_+ = \max(0, a)$, so that $\gamma > 0$ corresponds to the Fréchet distribution, $\gamma < 0$ corresponds to the Weibull distribution, and since $(1+0\cdot x)^{-1/0} = e^{-x}$, the value $\gamma = 0$ corresponds to the Gumbel distribution. The parameter γ is called the *extremum value index,* and it provides important information about the tail of the underlying distribution P_γ.

A necessary and sufficient condition for a general distribution P to belong to the Fréchet domain of attraction ($\gamma > 0$) is the requirement that $1 - P(x)$ is varying regularly at infinity. Further, P belongs to the Weibull domain of attraction ($\gamma > 0$) if and only if P has a finite right endpoint, x^*, and $P(x^* - 1/x)$ varies regularly at infinity. For the case of the Gumbel distribution, the necessary and sufficient conditions are more complicated, see [9], [28]. It should be clear that there are many distributions which do not belong to the three mentioned domains of attraction. For example, the maximum distributions for geometric and Poisson distributions cannot be well approximated by the standard extreme value distributions. Nevertheless, most applied distributions, such as Pareto-like distributions (Cauchy), normal, beta, and arcsin fall in these domains of attraction [24].

Rates of convergence have also been discussed [4], as well as convergence in the sense of moments [3].

At the beginning of the 1970s, A.A. Balkema, L. de Haan and J. Pickands III (and later R. Smith, T. Davidson and M.R. Leadbetter) considered an alternative approach to extreme value statistical analysis. In this approach, one prescribes a high critical level u, and then studies each occasion the level is exceeded by the time series $(X_n), n \in \mathbb{N}$.

When an underlying distribution P belongs to the domain of max-attraction of a standard extreme value distribution, the distribution of an exceed value, that is, the conditional distribution of the height of $X_k - u$ given $X_k \geq u$, can be approximated up to a linear transformation by a generalized Pareto distribution

$$H(x) = \begin{cases} 1 - (1 + \gamma x)_+^{-1/\gamma} \, , & \gamma \neq 0 \, , \\ 1 - e^{-x} \, , & \gamma = 0 \, , \end{cases}$$

(This is the work of A.A. Blakema and L. de Haan in 1974 and J. Pickands in 1975.) Such an approach allows us to consider not only a single absolute maximum (extreme) but to look at all "disadvantage" events, and even to consider them together with their respective heights. It is natural to suppose that, provided the threshold level u is high, the common distribution of the exceed points, as rare events, can be approximated by a distribution of a Poisson marked process, [13]. For mathematical background concerning this approach, see [8] and [9]. The general theory of random measures [7] closely relates these two approaches. For infinitely divisible point processes $\mu(\cdot)$, such as a Poisson process, the probability set $\{P(\mu(A) = 0), \ A \in \mathcal{A}$ fully determines the distribution of the point process provided $\mathcal{A}$ is a sufficiently rich collection of sets, such as all closed intervals and their unions on the real line.

Another topic of investigation in extreme value analysis is the asymmetric distribution of linearity (see [10]).

For other references on extreme value theory and its applications see also [5, 6, 15–18, 21, 23, 31, 44, 45, 48, 49, 64].

3.2.2 Dependent Data

In reality the data are very often statistically dependent. This requires a more elaborate theoretical framework. The first question to answer is to what extent the above-described approximations of extreme value distributions are valid for statistically dependent data. One solves this by introducing appropriate mixing conditions. The first result was obtained by Volkonski and Rozanov in 1961, who proved a Poisson limit theorem for high excursions of a random process with strong (Rosenblatt) mixing conditions. As introduced by M.R. Leadbetter in 1973, high-level mixing conditions are much weaker and more suited to the study of high level excursions, because they only involve events generated by the exceedances; see [8] for details and references.

It may be (see e.g. [8]) that for a stationary time series $X_t,\ t \in N$ with a marginal distribution function $P(x)$ one has

$$P(Y_n < x) \sim P^{\theta n}(x) ,$$

(not $P(Y_n < x) \sim P^n(x)$) as n becomes large, where $\theta \in (0, 1]$, θ is called the extremal index. The value $\theta = 1$ corresponds to the "almost independent" case. If $\theta < 1$, then, under some additional conditions, the large extremes gather in clusters in the sense that the limit of the point process of height excursions is not a simple Poisson process, but rather a marked Poisson process, where marks indicate sizes of clusters. This key parameter for extending extreme value theory for i.i.d. random variables to the case of stationary time series was originated by Loyens in 1965 and O'Brien in 1974. In 1983, M.R. Leadbetter developed the concept of the extremal index in detail. It characterized the dependence of the degree of clustering of extremes on the data. It can be shown that the limit averaging value of a cluster is equal (under some conditions) to $1/\theta$. This result is a step towards a prediction theory for extreme values.

The high-level exceedances approach in extreme value analysis of dependent data presents new challenges in the analysis of "moderately high" extremes, where the Poisson character of their distribution changes towards a normal distribution. Let $\{X_k,\ k \geq 0\}$ be a strongly mixing stationary time series, ψ_n be a set of non-negative functions and u_n be a sequence of high levels, $u_n \to \infty$ as $n \to \infty$. In work published by H. Rootzin, M.R. Leadbetter and L. de Haan in 1998, limit theorems for distributions of the array sums $\sum_k \psi_n(X_k - u_n)$ are given, for high u_n, when the limit is a compound Poisson distribution, and (very relevant to the prediction of extremes) when u_n are moderately high, so that the limit is a normal distribution. Work by V. Piterbarg and I.I. Rychlik in 1999 proves the central limit theorem for wave functionals of Gaussian processes.

The following two examples show how Gaussian models may be (and may not be) used to model processes with predictable extremes.

Example 1. Extremes of Gaussian stationary sequences as extremes of i.i.d. Gaussian variables.

A Gaussian stationary sequence with a correlation function r_k such that $r_k \log k \to 0$ as $k \to \infty$ (Berman's condition) obeys the Leadbetter mixing condition. This follows from an extension of Berman's inequality to the general algebra of events generated by excursions. Such a sequence is not necessarily due to strong mixing, because its spectral density may be exactly zero at intervals, which contradicts the necessary conditions for strong mixing (see [14]). Thus Leadbetter's condition is weaker than the strong mixing condition. Furthermore, distributions of high extremes of Gaussian sequences (even for slowly decreasing correlations) behave like extremes of i.i.d. Gaussian random variables (result from S. Berman in 1964). Notice that when r_k

slowly tends to zero, $r_k \log k \to a > 0$ as $k \to \infty$ for $T \to \infty$, so the limit distribution of the maximum is no longer an extreme value distribution but a mix of Gumbel distributions (Y. Mittal in 1974, see [12] for details and references). As a complete mathematical solution to the problem of finding necessary and sufficient conditions in the Poisson limit theorem for extremes of Gaussian stationary sequences, we point to a result from V. Piterbarg, [12]: *The number of extremes of a Gaussian stationary sequence on an interval tends in distribution to a Poisson random variable if and only if the mean value and the variance of the number tend to the same positive number, as the interval increases beyond any limit.*

Example 2. A storage process with fractional Brownian motion as input.

Let $B_H(t)$, $t \geq 0$, be fractional Brownian motion with a Hurst parameter $H \in (0, 1]$, that is, a Gaussian a.s. continuous zero mean process with stationary increments, starting at zero, $B_H(0) = 0$, and $EB_H(t)^2 = t^{2H}$. A storage process is defined, following work by I. Norros in 1975, by:

$$S(t) = \sup_{s \geq t}(B_H(s) - B_H(t) - c(s - t)) , \quad c > 0 . \tag{3.1}$$

This process is stationary and continuous. It was introduced to describe overloads in teletraffic systems, and can also be used to model any storage or traffic systems. Consider the storage process in discrete time, $S_k = S(hk)$, $h > 0$. If $H > 1/2$, then for any n,

$$P\left(\sup_{0 \leq k \leq n} S_k > u \right) \sim P(S_0 > u) . \tag{3.2}$$

Here it is not necessary for n to be fixed; the equivalence is still valid when $n \to \infty$ but $n = O(e^{\varepsilon u^2})$ for a small $\varepsilon > 0$ (as shown in work by V. Piterbarg in 2001). This equivalence indicates a high dependence of extremes. When S_k is high, it changes very slowly. It seems that the extremal index for S_k equals 0. Since $S(t)$ is constructed using the maximum of a Gaussian field with a “good” correlation function, it is very probable that S_k obeys the Leadbetter mixing condition. In 2004 Albin and Samorodnitsky proved similar properties for storage processes with some self-similar and infinitely divisible processes as input.

For problems of multivariable extreme value theory, the reader should refer to [3,9,11,50,53]. We do not consider the issues of statistically estimating parameters and statistical inferences in relation to extreme value theory here. For these, see [10, 11, 15, 21, 23, 24, 44, 45, 48, 54, 69, 70].

3.3 Extremes in Continuous Time: Stochastic Processes, Random Fields

Many models applied to the reliability of mechanical structures, environmental engineering, pollution control engineering, insurance mathematics (to quantify probabilities of large claims), and financial risk management describe the development of observations continuously in time. Statistical extremum value theory in continuous time can often be reformulated in terms of discrete time. For example, let $X(t)$ be a stationary continuous time stochastic process, $t \in \mathbb{R}$, and suppose we are interested in the limit behaviour of the probability

$$P\left(a(T)\left(\max_{t\in[0,T]} X(t) - b(T)\right) < x\right),$$

for appropriate normalizing functions $a(T)$ and $b(T)$.

Set $X_k = \max_{t\in[kh,(k+1)h]} X(t)$, $h > 0$. It is natural to expect that

$$P\left(a(T)\left(\max_{t\in[0,T]} X(t) - b(T)\right) < x\right) \sim P\left(a(n)\left(\max_{k=1,\dots,n} X_k - b(n)\right) < x\right),$$

as $n \to \infty$, where $n = [T/h]$, (the integer part). Furthermore, it is natural to expect that high level mixing conditions for $X(t)$ could be extended to corresponding mixing conditions for X_k. Thus, one expects that the limit of the above probability is an extreme value distribution described by Theorem 1. However, one very non-trivial problem often encountered is to evaluate the asymptotic behaviour of the tail probability

$$P_u := P(X_k > u) = P\left(\max_{t\in[0,h]} X(t) > u\right) \tag{3.3}$$

when $u \to \infty$. One needs the exact asymptotic behaviour to determinate which extreme value domain of max-attraction the distribution of $X(t)$ belongs to and to evaluate the normalizations $a(T)$ and $b(T)$. We give now a brief review of existing methods for evaluating the asymptotic behaviour of the probability (3.3).

3.3.1 Probabilities of Large Deviations: Exact Behaviour

The problem of evaluating and estimating the probability (3.3) is, without any doubt, one of the central problems in the theory of stochastic processes. First of all we refer the readers to a central component of the theory, the Stroock–Varadhan large deviation theory, and the (related) Freidlin–Wentzell minimum action principle, [20], [25], [26]. A direct consequence of this theory is the ability to evaluate the asymptotic behaviour of the logarithm of P_u for processes besides diffusion processes. By extending the Laplace asymptotic method to Banach spaces, this method can also be applied to evaluate the exact asymptotic behaviour of P_u, which is desirable for extreme values statistical analysis. See, [30] for references, and also work by S. Kusuoka and Y. Tamura in 1991. In 2004, E. Weiman provided a good example of the direct numerical application of the Wentzel–Freidlin theory to the probabilistic analysis of rare events.

Method of Differential Equations. The probability P_u can be studied for large u for some diffusion processes by considering the corresponding boundary problem and/or using well-developed techniques from diffusion process theory. A review, new developments and a bibliography are given in the doctoral thesis of A. Kunz. Many recent works on extreme value analysis of diffusion processes exploit well developed Gaussian techniques of asymptotic analysis (see works by M. Borkovec and C. Klüppelberg in 1998, and B. Buchmann and C. Klüppelberg).

Methods of evaluating the exact asymptotic behaviour of P_u are well developed not only for diffusion processes but also for processes with "good" finite-dimensional distribution properties, such as Gaussian, other infinitely divisible distributions, or processes with smooth trajectories and computable finite dimensional distributions.

Rice (Moments) Method. This method is based on the Kac–Rice formula for the average value of the number of up-crossings of a level by a random process; for history and references see [27]. We denote by $N_u(0,h)$ the number of up-crossings of the level u by the process X that occur in the interval $[0,h]$. It can be proved in many cases that

$$P_u \sim EN_u(0,h) \,, \quad \text{as} \quad u \to \infty \,. \tag{3.4}$$

The physical sense of this approximation formula is that up-crossings of a high level occur very rarely, so no more that one high level up-crossing is observed for a fixed interval $[0,h]$. Using the Kac–Rice formula,

$$EN_u(0,h) = \int_0^h \int_0^\infty y p_t(u,y) dy dt \,,$$

where p_t is the joint density distribution of $(X(t), X'(t))$. Furthermore, the formula

$$P_u \approx EN_u(0,h) + P(X(0) > u) \tag{3.5}$$

is often pretty precise and gives the second term of an asymptotic expansion for P_u. One can also give some physical arguments for this approximation. It can happen (as a rule with smaller probability) that $X(0) > u$ but $N_u(0,h) = 0$. This, in turn, provides correction terms for the limit theorems for maxima. This method has been elaborated in detail for Gaussian processes. The Kac–Rice formula has been generalized to random fields (work by Yu. Belyaev), in other words random functions with several arguments. Moreover, the relations (3.4),(3.5) have multidimensional analogues in the Gaussian case, [12]. Further development of the Rice method for Gaussian and close to Gaussian processes and fields is associated with J.M. Azaïs, V. Piterbarg, I. Rychlik (including software), and M. Wchebor.

Pickands' Double Sum Method. The observation in the previous paragraphs, that up-crossings of high level occur rarely for processes that are very random, is crucial. In addition, each excursion (which is a part of the trajectory above the level) is typically very short. These observations, applied to processes with non-smooth trajectories, lead to a powerful method of evaluating the asymptotic behaviour of $P_{u.}$ J. Pickands was the first to really apply this to nondifferentiable processes. Let $X(t)$ be a Gaussian stationary zero-mean process with correlation function r_t, and assume that

$$r_t = 1 - |t|^\alpha + o(|t|^\alpha) \,, \quad t \to 0 \,. \tag{3.6}$$

Then, for any $\lambda > 0$, and $h = \lambda u^{-2/\alpha}$,

$$P_u \sim H_\alpha(\lambda) P(X(0) > u) \,, \quad u \to \infty \,,$$

where $H_\alpha(\lambda) = E\exp(\max_{[0,\lambda]}(\sqrt{2}B_{\alpha/2}(t) - t^\alpha))$. This "local" result generated many far-reaching consequences, not just for Gaussian processes. Dividing the interval $[0, h]$ into small intervals of length $\lambda u^{-2/\alpha}$, and proving that two intervals with excursions above u occur with a negligible probability (say, using the Bonferroni inequality), one obtains $P_u \sim H_\alpha h P(X(0) > u)$, where $H_\alpha = \lim_{\lambda\to\infty} H_\alpha(\lambda)/\lambda$ with $H_\alpha \in (0, \infty)$. This, known as Pickands' localization principle, has been substantially developed and extended to wide classes of random processes. For example, S. Berman extended such results, to other classes, such as processes with independent increments, in particular stable processes (see [22]). Further development of Berman's method has been mainly performed by J.M.P. Albin. The book given in [12], Chap. 2, presents an extension of the Double Sum method to a wide class of Gaussian processes and fields.

Comparison Method. This method has only been developed for the Gaussian case and has been widely discussed in relation to Gaussian random fields. The mathematical background of the method is a study of the geometrical properties of the excursion set $\{\mathbf{t} : X(\mathbf{t}) \geq u\}$. Let $X_0(\mathbf{t})$ and $X(\mathbf{t})$, $\mathbf{t} \in \mathbb{R}^d$, be two Gaussian zero mean homogeneous fields with smooth trajectories, unit variance, and similar behaviour of their correlation functions $r(\mathbf{t})$ and $r_0(\mathbf{t})$ at zero,

$$\left.\frac{\partial^2 r(\mathbf{t})}{\partial t_i \partial t_j}\right|_{\mathbf{t}=\mathbf{0}} = \left.\frac{\partial^2 r_0(\mathbf{t})}{\partial t_i \partial t_j}\right|_{\mathbf{t}=\mathbf{0}} \,, \quad i, j = 1, \ldots, d$$

(any two homogeneous smooth Gaussian fields can be standardized to satisfy this equality). Then, under some additional nondegeneracy and smoothness conditions, there exists an $a > 1$ such that for any finite convex cell complex $M \subset \mathbb{R}^d$ [12],

$$\left| P\left(\max_{\mathbf{t}\in M} X(\mathbf{t}) > u\right) - P\left(\max_{\mathbf{t}\in M} X_0(\mathbf{t}) > u\right)\right| = O\left(e^{-au^2/2}\right) \,, \quad u \to \infty \,,$$

`for some` $a \geq 0$ (3.7)

Thus, upon computing the asymptotic behaviour of P_u for a "simple" Gaussian field, one also obtains it for other fields with close correlation near zero, up to an exponentially smaller order. Notice that $\log P_u \sim -u^2/2$. Taking a field with a finite Karunen–Fourier expansion, the level sets of which can be pretty exactly described, and using Hadwiger expansions for the Minkowski product of sets, one gets an asymptotic expansion for the probability P_u on degrees of u in terms of Hermite polynomials $H_\nu(u)$ and Minkowski functionals $W_\nu(M)$ of M,

$$P\left(\max_{\mathbf{t}\in M} X(\mathbf{t}) > u\right) = \phi(u)\sum_{\nu=1}^{d-1} \frac{\binom{d}{\nu}}{\omega_\nu(2\pi d)^{(d-\nu)/2}} H_{d-1-\nu}(u) W_\nu(M)$$
$$+P(X(\mathbf{0}) > u) + O\left(e^{-au^2/2}\right), \quad u \to \infty, \tag{3.8}$$

where ϕ is the standard Gaussian density and ω_ν is the corresponding volume of the $\nu-$ dimensional unit ball, with $a > 0$ being some constant. R. Adler considered the Euler characteristics of the excursion sets. Like the number of up-crossings in the one-dimensional case, the expected Euler characteristics can be represented by integrals of finite dimensional distributions of the Gaussian field, so that one can make inferences about the number of excursions connected to parts of the excursion set. Let $\varphi_u(M)$ be the Euler characteristics of the excursion set $\{X(\mathbf{t}) \geq u\} \cap M$, then, from the above comparison arguments it follows that

$$\left| P\left(\max_{\mathbf{t}\in M} X(\mathbf{t} > u\right) - E\varphi_u(M)\right| = O\left(e^{-au^2/2}\right), \quad u \to \infty. \tag{3.9}$$

This relation has been proved several times in different cases and by different methods; for a purely geometrical and elegant proof, see work by A. Takenura and S. Kuriki in 2002.

3.3.2 Maxima and Excursions of Gaussian and Related Processes and Fields

As has already been mentioned, if we know the asymptotic behaviour of $P(\max_{[0,h]} X(t) > u)$ as $u \to \infty$, and have the mixing conditions for the stationary process $X(t)$, we can get a limit theorem for the maximum of the process $X(t)$. We can also use Pickands' limit theorem for the maximum of a Gaussian stationary process and its generalizations to Gaussian fields, as well as to Gaussian processes and fields with a covariance that has a slower decrease at infinity. The comparison method provides the ability to obtain corrections to this theorem for Gaussian smooth processes and fields. Here we provide a formulation of a corresponding result from [12], given by H.E. Krogstad from NTNU (a private communication, see also http://www.ifremer.fr/web-com/stw2004/rw/fullpapers/krogstad.pdf). Let $X(\mathbf{t})$ be a Gaussian homogeneous field. We standardize it, $\widehat{X}(\mathbf{t}) = \sigma^{-1}X\ (A^{-1}\mathbf{t})$, so that the variance of $\widehat{X}(\mathbf{t})$ equals 1 and the covariance matrix of its gradient, ∇X, equals $d^{-1}I$, where I is the unit matrix. Then, under some conditions of nondegeneracy, one can use the approximation

$$P\left(\max_{\mathbf{t}\in T} \widehat{X}(\mathbf{t}) \le u\right) \approx \exp\left(-(2\pi)^{(d-1)/2} e^{-u^2/2} H_{d-1}(u)|T|\right), \qquad (3.10)$$

where $|T|$ is the "non-dimensional size" of T, defined by $|T| = (2\pi)^{-d}d^{-d/2}$ $\mathrm{Vol}(T)$. The "limit theorem form" of this approximation, stated in [12], applies when $T \uparrow \mathbb{R}^d$ in a regular way (see details in [12], Definition 14.1) and $u \to \infty$ in such a way that the expression under the exponent tends to a constant and the covariance function $r(\mathbf{t})$ of X decreases at infinity faster than the degree of $|\mathbf{t}|$. In such a situation, the approximation error is $O(|T|^{-a})$ for some $a > 0$. As discussed in the manuscript cited above, this formula gives good approximations for Gaussian fields with ocean-like spectra. This formula, likes any other limit theorem for a maximum, can be transformed into a Poisson limit theorem for the number of high crossings, extremes, high crests, and so on. Using very simple estimations and Kallenberg theorems about convergence to infinitely divisible point processes, the Poisson limit theorem follows almost immediately from limit theorems like Pickand's limit theorem (or formula (3.10)). This is why formula (3.10) can be called a Poisson approximation. If the underlying process is not differentiable, so that one cannot define up-crossings to count excursions, it is often convenient to consider ε-upcrossings; in other words points t such that $X(t) = u$ and $X(s) < u$ for all $s \in [t-\varepsilon, t)$. From Pickand's theorem it immediately follows that the point process of ε-upcrossings converges in distributions to a Poisson point process.

The Gaussian technique described above for studying the probability of high extremes can also be applied to non-Gaussian processes. One of most investigated examples of this is the χ^2 random process, which is the sum of the squares of independent identically distributed Gaussian processes. Using the duality $|\mathbf{x}|^2 = \max_{|\mathbf{u}|^2=1}\langle \mathbf{u}, \mathbf{x}\rangle$, where $\langle .,.\rangle$ is a scalar product and $|\mathbf{x}|^2 = \langle \mathbf{x}, \mathbf{x}\rangle$, the problem of evaluating the distribution of the maximum of a χ^2-process, which is non-Gaussian, can be reduced to the problem of evaluating the maximum distribution of a Gaussian field on a cylinder. This work was performed for stationary and cycle stationary processes by D. Konstantinidis, S. Stamatovich and V. Piterbarg in 2004. The above duality works for any norm in Euclidean space, so that one may consider a wide range of non-Gaussian stationary processes and apply Gaussian asymptotic extreme value analysis to them.

One may directly consider a point process consisting of crossings of a large surface, say a cylinder of large radius, by a random vector process. One can prove a Poisson limit theorem for this point process. It is also important to consider distributions of excursions of trajectories above this large region. This direction is investigated in connection with many applications, such as the structure and distribution of sea wave crests, fatigue of materials, and other problems. Even the crossing rate (the mean value of number of crossings) of a stationary vector process is a valuable tool when studying high crest distributions and maxima of sea level elevations. It has been shown, in work by U. Machado and I. Rychlik published in 2003, that the sea elevation at a fixed point can be modelled as the sum of a Gaussian process and a quadratic random correction term. It has also been shown that the correction term process can be written as a quadratic form of a vector-valued Gaussian process with an arbitrary mean. See also work done by K. Breitung in 1988, Rychlik and Leadbetter in 2000, and other developments by K. Breitung, R. Illsley, O. Hagberg, A. Rusakov, and I. Rychlik.

3.3.3 Relationship Between Continuous and Discrete Time: Prediction of Extremes

The relationship between continuous time and discrete time modelling of irregular and random data is very important in extreme value analysis. First of all, it is connected with numerical simulations of trajectories of random processes when high extremes are taken into account. On the other hand, extreme value analysis works well when one tries to estimate errors in discrete time modelling; see work done by J.M.P. Albin in 2004, and J. Hüsler, V. Piterbarg and O. Seleznjev in 2003.

This problem has also been extensively discussed in the financial literature (see the work of D. Duffie, and P. Potter from 1992, and those of D. Brigo and F. Mercurio in 2000 and Y. Fang in 2000). The advanced development of both computers and information technology (particularly of internet technologies) has led to the availability of high-frequency financial data, that is, various financial time series recorded with very high resolution. This novel feature suggests more accurate and efficient continuous time models and has thus provided data analysts with the opportunity to apply powerful mathematical techniques developed in continuous time theory.

However, modelling of "nearly continuous" data poses new challenges, since the extent to which both purely continuous and purely discrete methods are applicable in the high-frequency domain is not apparent and needs to be explored. In particular, such a problem typical applies when modelling large extremes, where the quality of the discrete time estimators for values at risk may depend heavily on the recording frequency chosen (see [10]).

The importance of the above problem goes well beyond the area of financial mathematics and extends to many applications where Xevents play a crucial role. In particular, an important (often vital) problem is that of efficient forecasting of extremes. In many real-life situations, the times separating successive observations may be irregular or even random. Correlations between high- and low-frequency data are of great practical and theoretical interest, and these should be investigated in order to optimize the observational discipline and the measurement process. Hence, an efficient analytical comparison of both types of results should be possible in principle, although a complete solution to this problem has not yet been obtained.

These problems are tackled primarily using the Gaussian framework, since there are well-developed and powerful analytic tools that can be used to study the distributions of extremes in both continuous and discrete time settings for Gaussian and related processes. A complete solution for Gaussian stationary process is provided in work by V. Piterbarg (2004). Let (3.6) and Berman's condition be fulfilled for a Gaussian stationary process $X(t)$. We denote the maximum of $X(t)$ over $[0, t]$ taken over a uniform discrete grid with the step $\delta > 0$, by $M_\delta(t$, and the maximum over the interval $[0, t]$ by $M(t)$. If $\delta(\log T)^{-1/\alpha} \to \infty$ with T the limit distribution of the vector $(M_\delta(T), M(T))$, in suitable normalization, tends to a conditional distribution of two independent Gumbel random variables given first is at the most the second. It means that maxima in discrete and continuous time are asymptotically independent, so that one can say nothing about the location of the absolute maximum, upon observing the process on the grid. If $\delta(\log T)^{-1/\alpha} \to 0$, the limit distribution concerns two identical Gumbel random variables; that is, the maxima are asymptotically completely dependent. In the boundary case, $\delta(\log T)^{-1/\alpha} \to a > 0$, the limit distribution is not trivial but it can be expressed in terms of functionals of fractional Brownian motion, so that one can locate the point of absolute maximum based on the discrete time observations using this distribution. The Gaussian technique, used in the proof of this result (see [12,22]), can also be applied to more general Gaussian processes and fields. Using these techniques for the storage process with fractional Brownian motion as input, one can get a rather different result. Now let the vector $(M_\delta(T), M(T))$ represent the maxima of the storage process over continuous time and on a grid with step size δ. Then, if $\delta = o((\log T)^{(2H-1)/(2H(1-H))})$, where H is the Hurst parameter of the fractional Brownian motion, the maxima are asymptotically completely dependent. For $H > 1/2$, notice that the value of δ may tend to infinity. That is, although one may observe the process $S(t)$ only very rarely, it is nevertheless possible to obtain full information about the location and the value of the maximum! In other words, the latter model is a model with predictable extremes.

Gaussian processes, although very convenient analytically, are not quite realistic, as many real-life processes display a long-range dependence ("long memory") leading to tails heavier than those of Gaussian distributions. In an attempt to keep the analytical efficiency of the Gaussian class but to find models that incorporate these kinds of effects, one is led to consider a fairly broad class of conditionally Gaussian random processes, whose distribution is Gaussian given their mean and correlation function. One well-known model of conditionally Gaussian processes is the sub-Gaussian process, see [29]. In 1993, Adler, Samorodnitsky, and Gadrich evaluated the expected number of up-crossings of a fixed level for a sub-Gaussian process and studied the asymptotic behaviour of this quantity when the level is high. The main result of their paper, the asymptotic behaviour of the rate of up-crossings of a high level, can be easily obtained using conditioned Gaussian processes and Gaussian tools. Such results provide a good starting point for the asymptotic analysis of a wide class of conditionally Gaussian processes. Following this direction, one may hope to obtain models with predictable extremes that can be elaborated with well developed asymptotic techniques for Gaussian processes [12]. Processes in random environments could provide a good basis for models of stochastic processes with predictable extremes. Predictable random media (such as stochastic volatility) provide a platform to predict extremes and other rare events. For example, a Gaussian process with a random predictable variation could be a desirable model for predicting extremes. It is well known that extremes are more likely to occur in a Gaussian process in the vicinities of the absolute maxima of its variance. Predicting the positions of these maxima of the random variance therefore leads to a prediction of the hazard presented by extremes.

Other models that could be used to develop a extreme prediction theory can also be considered. It turns out that shock noise models yield an extreme value analysis; see for example the papers published in 2003 by A.V. Lebedev where extremum value analysis was performed on stationary shock noise processes with heavy tailed (Pareto-type) marginal distributions.

3.3.4 Other Problems

Physical Extremes. Each excursion of a "typical" stochastic process above a high level is, as a rule, very short and very low. This is indeed the case, at least for Gaussian processes. It is wrong, however, for the storage process. Let us define a positive number v. We wish to estimate the probability that an excursion above a high level occurs, which supported by a set with a volume at least v. This is an actual problem in tomography – separating pictures generated by a noise from real images. This problem is difficult to address, even for well elaborated Gaussian processes. The simplest formulation may be as follows. Let $X(t)$ be a Gaussian process, and look for the asymptotic behaviour of the probability $P_{v,u} := P(\exists [a,b] \subset [0,T] : \ b-a \geq v, \ \min_{[a,b]} X(t) \geq u)$, as $u \to \infty$, which is the probability of a "powerful" or physical extreme. It seems that

$$\log P_{v,u} \sim \log \sup_{c(t)\geq 0: \int_a^{a+v} c(t)dt=1} \sup_{a\in[0,T-v]} P\left(\int_a^{a+v} c(t)X(t)dt \geq vu \right) ,$$

Using variational analysis it should be possible to get the exact asymptotic behaviour.

Clusters of Extremes. An excursion in discrete time is simply the event $\{X_k > u\}$. In continuous time it looks like a connected part of a trajectory situated above a level u. Again, in "typical" cases one can observe, with dominant probability, at most one excursion above a high level. What about the probability of two or more excursions? This problem, even for Gaussian processes, is only beginning to be tackled (see [72]).

3.4 Extremes and Statistical Mechanics

Classical statistical mechanics describes the behaviour of large classical systems of particles (or, more generally, degrees of freedom, such as classical ferromagnets in a lattice).

Phase transitions, which are typically observed in the systems studied by classical statistical mechanics, can easily be related to the somewhat vague intuitive concept of "extreme events". For example, at the critical points seen with phase transitions, large fluctuations of observables can be shown to exist. These imply large cooperative phenomena in the systems and this may, under certain circumstances, imply consequences that are "catastrophic" in the intuitive sense of the word. For example, volcano explosions have been related to such phenomena.

However, generally, phase transitions only occur for particular values of parameters, and in this sense they are highly nongeneric. According to ideas of self-organized criticality (SOC) [51, 52, 62, 63], phase transitions may also be large systems that are far from equilibrium. This can lead to "catastrophic events", which are due not to external influences (like those that would tune the parameters towards a critical point, where phase transitions occur) but rather inherent to the system itself.

This hypothesis of SOC seems to apply particularly well to granular systems, where it has been studied mainly using physics tools and numerical computations, although some mathematical results have been published [55, 61]. Extreme catastrophic events that the theory has been applied to include earth slides, the formation of earthquakes, the development of avalanches, or catastrophic price fluctuations in financial markets; see [33, 34, 63, 65].

The phenomenon of large fluctuations in financial markets has also been investigated via power-law distributions, which have been studied empirically (see [47, 52, 59, 65, 66]).

Dynamical models involving a large number of traders have also been introduced [59].

A model describing large stock market crashes involving critical points of statistical mechanical systems has also been developed [65]. In this model and related ones, the main feature is the presence of log-periodic behaviour generated endogenously from the system, see [46, 60]. This view on financial crashes is quite different from the one pursued on the basis of extreme value theory, see [65] versus [10].

Other approaches to the statistical mechanics of systems far from equilibrium have been developed using the theory of dissipative structures, see [67], and using synergetics, see [73]. The latter theory also has connections with catastrophe theory, which we shall briefly discuss in Sect. 6. In general, although the systems described by these approaches can develop "large fluctuations" that connect them to the theories of phase transitions, critical phenomena and catastrophic events, these connections are yet to be fully understood in a mathematical sense.

3.5 Extremes and Dynamical Systems

It is now well known that simple systems, described by classical mechanics, can exhibit very complex behaviour, due to nonlinearities in their equations of motion. This behaviour is particularly apparent in chaotic motion, making predictions of the behaviour of such deterministic systems only really possible using tools from probability theory, much the same as for systems with an "external" random input to the dynamics.

As an example, we recall the detailed study of the iterated logistic equation, and the dependence of its parameter: as the parameter increases, the asymptotic behaviour is first dictated by the increasing number of point attractors, and finally, after a critical value, it becomes similar to the one observed for coin tossing systems.

In continuous time, examples of such chaotic behaviour can be found for nonlinear systems with at least three degrees of freedom; Lorentz equations (a truncation of the Navier–Stokes equations of hydrodynamics) being perhaps the most well known case here (see, [56, 57]).

The asymptotics of such chaotic systems are best studied using tools from the probabilistic theory of dynamical systems, where concepts like invariant measures, ergodicity, mixing properties, entropy and fractality play a basic role.

Such concepts are also at the root of classical statistical mechanics, which however traditionally handles systems with a large number of degrees of freedom.

Because of this, one can try to apply concepts like transience, intermittency, recurrence and concepts of large deviation theory (such as those that occur in the framework of Markov chain theory) to describe the "extreme" asymptotic behaviour of nonlinear dynamical systems of the above type. This is yet to be performed systematically, but see [56, 57, 67] for some examples and references.

Another way of looking at extremes in dynamical systems is to study singularities of the solutions of such systems (including values of the relevant variables describing the systems that diverge to infinity) This is the case in certain systems described by hyperbolic partial differential equations; see for example [Alb-Sh]. But singularities can also arise in systems governed by equations of parabolic type, like those describing the propagation of water waves; for example, see [36] for a discussion of singularities in classical Navier–Stokes or Euler-equations, and (Dobrochotov et al., in preparation) for recent work on deriving singularities related to tsunami waves.

In some deterministic models of population dynamics and ecology, singularities can arise in relation to species extinctions, which are, of course, catastrophic events. For recent work on modelling ecological or population dynamical systems, and a corresponding discussion of catastrophic events, see [58, 68].

3.6 Mapping Singularities and Catastrophe Theory: How Can They Be Related to Xevents?

When modelling many phenomena in natural and socioeconomical sciences, solutions of dynamical systems are often represented by probabilistic or oscillatory integrals performed over large dimensions or even infinitely sized dimensions. Examples are:

- (a) Classical statistical mechanics: here the integrals are of the form $\int_{\Gamma} e^{-\beta H(\gamma)} f(\gamma) d\gamma$, where Γ is the configuration space of particles (for example $\mathbb{R}^{3N}$ for N particles moving in $\mathbb{R}^3$, $\left(\mathbb{Z}^d\right)^S$ for S degrees of freedom on a lattice $\mathbb{Z}^d$ ...), where:
  f is an observable of interest,
  H is an energy functional, describing the interaction of the particles,
  $\beta > 0$ is the inverse temperature,
  $d\gamma$ is the "flat measure" of Γ.
- (b) Several problems in financial mathematics and macroeconomics can be tackled in this form; see for example [37, 74]. The quantum statistical mechanics expectation in Gibbs (temperatures) states can be obtained in the above form, using probabilistic methods (Albeverio et al., book in preparation), Γ being there a space of paths (e.g. $C([0, \beta]; \mathbb{R}^{3N})$ for N quantum particles in $\mathbb{R}^3$).
- (c) Quantum mechanical systems. For example, here the solutions of the Schrödinger equations of N particles can be obtained in the form $\int_{\Gamma} e^{-\beta H(\gamma)} f(\gamma) d\gamma$, with $-\beta H$ replaced by $\frac{i}{\hbar} S$, where S is the classical action of the underlying classical system evolving in the time interval $[0, t]$. Here $\Gamma = C([0, t]; \mathbb{R}^{3N})$.

Similar oscillatory integrals also arise in electromagnetic theory, particularly in optics. Other examples can be given, ranging from SPDE theory to economics and biology. For further discussion and references, we refer the reader to books about path integrals, such as [38, 39], while [40] gives a more physical approach. It is useful to find the asymptotics of the above integrals for certain parameters (such as large or small β, large N, small or large $t \ldots$), since they can be used in various contexts. Powerful methods have been developed for this, which have their roots in nineteenth century Laplace methods for deriving the asymptotics of finite dimensional integrals of the form $\int_{\Gamma} e^{-\beta H(\gamma)} f(\gamma) d\gamma$, or a stationary phase method for integrals like those arising in example (c) above (and saddle point methods for where H is complex).

Extensions of these classical methods to the case where Γ is infinitely dimensional, which is the most relevant for the above examples, have been developed in the last 30 years. These include Wentzell–Freidlin theory [25], large deviation theory [20,26], Donsker-Varadhan asymptotics [39], the method for a stationary phase in infinite dimensions [38], and the Laplace method [30,43]. The asymptotics are given in terms of developments around points in the critical manifold $\Gamma_c = (\gamma \in \Gamma | d\psi(\gamma)) = 0\}$, where ψ stands for H in example (a) or S in (c).

For the examples mentioned above, Γ_c is interesting in itself, since it depends on the problem at hand. For example, in the study of the $\beta \to \infty$ limit, Γ_c is a space of classical minimal configurations, in example (c) it consists of classical orbits, and in the case of example (b) it consists of classical periodic orbits.

The contributions from each power of the relevant parameter in the expansion (for example β^{-1} for $\beta \to \infty$) depend on the geometric structure of Γ_c.

For example (c), the structure of Γ_c is related to the degeneracy of the classical action functional. Morse theory and more generally singularity theory (of which catastrophe theory can be seen as a particular case) describe the possible forms of Γ_c. This gives an extremely interesting connection between catastrophe theory and integrals describing statistical mechanical, optical or quantum mechanical systems. Catastrophe theory in itself was originally developed as a pure mathematical theory for investigating Γ_c for certain ψ. The founder, R. Thom, saw certain particular Γ_c as interesting in themselves, expressing certain typical catastrophes (such as caustics in optics). The range of the method was quickly extended to other contexts and areas (particularly by Zeeman), and it received a lot of attention (and also criticism); see for example [41]. In the context of the above integrals, catastrophe theory is related to dynamical systems, and it can indeed aid discussions of extremes arising in natural and societal systems.

The contributions obtained from the above expansions are classified using catastrophe theory terms, and should be linked to theories like the statistical mechanics of systems with a large number of components (see Sects. 3.4 or 3.5).

Mathematically, it is also possible to relate these developments to large deviation theory and extreme value theory, although we are still far from a unified presentation; see [30, 42, 43, 74] for some work done in this direction.

References

1. Gumbel E.J., Statistics of extremes. Columbia University Press, New York (1958)
2. Lieblein J., Two early papers on the relation between extreme valued and tensile strength. Biometrika, 41, 559–560 (1954)
3. David H.A., Order statistics. Wiley, New York (1981)
4. Galambos J., The asymptotic theory of extreme order statistics, 2nd edn. Krieger, Malabar, FL (1987)
5. Pfeifer D., Einführung in die Extremwertstatistik. Teubner, Stuttgart (1989)
6. Gottschalk L., Olivry J.C., Reed D., Rosbjerg D., eds, Hydrological extremes: understanding, predicting, mitigating. IAHS, Wallingford, UK, Publication 255 (1999)
7. Kallenberg O., Random measures. Akademie-Verlag, Berlin (1983)
8. Leadbetter M.R., Lindgren G., Rootzén H., Extremes and related properties of random sequences and process. Springer, Berlin Heidelberg New York (1983)
9. Resnick S.I., Extreme values, regular variation, and point processes. Springer, Berlin Heidelberg New York (1987)
10. Embrechts P., Klüppelberg C., Mikosch T., Modelling extremal events for insurance and finance. Springer, Berlin Heidelberg New York (1997)
11. Falk M., Hüsler J., Reiss R.D., Laws of small numbers: extremes and rare events. Birkhäuser, Basel (1994)
12. Piterbarg V.I., Asymptotic methods in the theory of Gaussian processes and fields. Translation Monographs AMS Series, 148 (1996) (processed edition; original in Russian, 1988, Moscow UP)
13. Aldous D., Probability approximations via the Poisson clumping heuristic. Springer, Berlin Heidelberg New York (1989)
14. Ibragimov I.A., Rosanov Yu.A., Gaussian random processes. Springer, Berlin Heidelberg New York (1978)
15. Coles S., An introduction to statistical modeling of extreme values. Springer, Berlin Heidelberg New York (2001)
16. Beirlant J., Goegebeur Y., Segers J., Teugels J., Statistics of extremes: theory and applications. Wiley, Chichester, UK (2004)
17. Kotz S., Nadarajah S., Extreme value distributions: theory and applications. Imperial College Press, London (2000)
18. Castillo E., Extreme value theory in engineering (Statistical Modeling and Decision Science Series). Academic, New York (1988)
19. Benz W., Gumbel E.J., Die Karriere eines deutschen Pazifisten. In: Walberer U. (Hrsg), 10. Mai 1933: Bücherverbrennung in Deutschland und die Folgen. Fischer Taschenbuch, Frankfurt (1983)
20. Deuschel J.D., Stroock D.W., Large deviations. Academic, New York (1989)
21. Harter H.L., A bibliography of extreme value theory. lnt Stat Rev, 46, 279–306 (1978)
22. Berman S.M., Sojourns and extremes of stochastic processes. Wadsworth, Belmont, CA (1992)
23. Reiss R.-D., Thomas M., Statistical analysis of extreme values: with applications to insurance, finance, hydrology, and other fields, extended 2nd edn. Birkhäuser, Basel (2001)

24. Einmahl H.J., de Haan L., Empirical processes and statistics of extreme values, 1 and 2. AIO Course, available at www.few.eur.nl/few/people/ldehaan/aio/aio1.ps, www.few.eur.nl/few/people/ldehaan/aio/aio2.ps
25. Freidlin M., Wentzell A., Random perturbations of dynamical systems. Springer, Berlin Heidelberg New York (1984)
26. Dembo A., Zietouni O., Large deviation techniques and applications, Springer, Berlin Heidelberg New York (1998)
27. Cramer H., Leadbetter M.R., Stationary and related stochastic processes. Wiley, Chichester, UK (1967)
28. de Haan L., A spectral representation for max-stable processes. Ann Probab 12, 1194–1204 (1984)
29. Samorodnitsky G., Taqqu M., Stable non-Gaussian random processes. Chapman & Hall, New York (1994)
30. Piterbarg V.I., Fatalov V.R., The Laplace method for probability measures in Banach spaces. Uspekhi Mat Nauk, 50, 9, 57–150 (1995) (English translation: Russian Math Surveys 50, 1151–1239 (1995))
31. Kinnison R.R., *Applied extreme value statistics.* Battelle, Columbus, OH (1985)
32. Weibull W., A statistical theory of the strength of materials. Ing Akad Handl, 151 (1939)
33. Scafetta N., West B.J., Multiscaling corporative analysis of time series and a discussion on earthquake conversations. Phys Rev Lett 92, 138501–138504 (2004)
34. Hergarten S., Self-organized criticality in earth systems. Springer, Berlin Heidelberg New York (2002)
35. Albeverio S., Shelkovich V., Delta shock waves in multidimensional non-conservative systems of zero pressure gas dynamics. Bonn Preprint 176 (2004)
36. Albeverio S., Ferrario B., Uniqueness of solutions of the stochastic Navier-Stokes equation with invariant measure given by the enstrophy. Ann Prob S2, 1632–1649 (2004)
37. Gomez M.E., Dynamic probabilistic models and social structure. Kluwer, Dordrecht (1992)
38. Albeverio S., Hoegh-Krohn R., Mazzucchi S., Mathematical theory of Feynman path integrals – An introduction, 2nd edn. Springer, Berlin Heidelberg New York (2005)
39. Simon B., Functional integration and quantum physics. Academic, New York (1979)
40. Kleinert H., Path integrals in quantum mechanics, statistics, polymer physics and finance. World Scientific, Singapore (2004)
41. Arnold V.I., Catastrophe theory. Springer, Berlin (1984)
42. Albeverio S., Liang S., Asymptotic expansions for the Laplace approximation of sums of Banach space-valued random variables. Ann Prob, 33, 1, 300–336 (2005)
43. Albeverio S., Röckner H., Steblovskaya V., Asymptotic expansions for Ornstein-Uhlenbeck semigroups perturbed by potential over Banach spaces. Stoch Repts, 69, 41–100 (2000)
44. Tiago de Oliveira J., Statistical experiments and applications. D. Ruidel, Docheckz (1984)

45. Beinlart J., Tengeb J.L., Vynckier P., Practical analysis of extreme values. Leuven University Press, Belgium (1996)
46. Chang I, Stauffer D., Pandey R.B., Asymmetries, correlations and fat tails in percolation market model. eprint arXiv cond-mat/0108345V1 (2001)
47. Matthews R., Far out forecasting. New Scientist, 37–40 (1996)
48. Smith R., Extreme value theory. In: Handbook of applicable mathematics, Supplement. Wiley, Chichester, UK, 437–439 (1990)
49. De Haan L., On regular variation and its application to the weak convergence of sample extremes. Math Center Facts, Amsterdam (1971)
50. Coles S.G., Tawn J.A., Modelling extreme multivariate events. J R Stat Soc B, 53, 377–392 (1991)
51. Badii R., Politi A., Complexity: Hierarchical structures and scaling in physics. Cambridge Univ Press, Cambridge (1997)
52. Sornette D., Critical phenomena in natural sciences. Chaos, fractal, self-organization and disorder: concepts and tools. Springer, Berlin Heidelberg New York (2000)
53. Barnett V., The ordering of multivariate data. J R Stat Soc A, 139, 318–345 (1976)
54. Christopeit N., Estimating parameters of an extreme value distribution by the method of moments. J Nat Plann Infer 41, 173–186 (1994)
55. Cessac B., Blanchard Ph., Krüger T., Meunier J.L., Self-organized criticality, thermodynamic formalism. J Stat Phys, 11, 1283–1326 (2004)
56. Ruelle D., Chance and chaos. Princeton University Press, New York (1991)
57. Berger A., Chaos and chance. Walter de Gruter, Berlin (2001)
58. Jain S., Krishna S., Graph theory and the evolution of antocatalytic networks. In: Handbook of graphs and networks. Wiley-VCH, Weinheim, 355–395 (2003)
59. Gabaix X., Gopikrishnan P., Pleron V., Stanley H.E., A theory of power-law distribution in financial market situation. Nature, 423, 267–276 (2003)
60. Arneido A., Muzy J.F., Sornette D., Direct causal cascade in the stock market. Eur Phy J B 2, 277–282 (1998)
61. Blanchard Ph., Hongler M.O., How many blocks can children pile up – some analytical results. J Phys Soc Jap, 71, 9–11 (2002)
62. Bak P., How nature works. Springer, Berlin Heidelberg New York (1998)
63. Jensen H.J., Self-organized criticality: emerging complex behaviour in physical and biological systems. Cambridge Univ Press, Cambridge (1998)
64. Hahn M.G., Mason D.M., Wiener D.C., eds, Sums, trimmed sums and extremes. Birkhäuser, Boston, MA (1991)
65. Sornette D., Why stock markets crash. Critical events in complex financial systems. Princeton Univ Press, New York (2003)
66. Mandelbrot B.B., Fractals and scaling in finance: discontinuity, concentration, risk. Springer, Berlin Heidelberg New York (1997)
67. Balakrishnan V., Nicolis C., Nicolis G., Extreme value distributions in chaotic dynamics. J Stat Phys, 80, 307–336 (1995)
68. Lamper D., Howison S.D., Johnson N.F., Predictability of large future changes in a competitive evolving population. Phys Rev Lett, 88, 017902-1-4 (2002)
69. Einmahl J., de Haan L., Piterbarg V.I., Non-parametric estimation of the spectral measure of an extreme value distribution. Ann Stat, 29, 1401–1423 (2001)

70. Gomes M.I., Martins M.J., Neves M., Generalized jacknife semi-parametric estimation of the tail index. Pas Math, 59, 393–406 (2002)
71. Kunz A., Personal webpage. See http://www-m4.ma.tum.de/pers/kunz/
72. Ladneva A., Piterbarg V., On double extremes of Gaussian processes. EURANDOM, Technische Universiteit Eindhoven, The Netherlands; document is available at http://www.eurandom.tue.nl (see eurandom reports/2000-027)
73. Güttinger W., Eikemeier H., eds, Structural stability in physics. Springer, Berlin Heidelberg New York (1980)
74. Albeverio S., Lütkebohmert E., in preparation

4 Dynamical Interpretation of Extreme Events: Predictability and Predictions

Holger Kantz, Eduardo G. Altmann, Sarah Hallerberg, Detlef Holstein and Anja Riegert

Summary. Due to their great impact on human life, Xevents require prediction. We discuss scenarios and recent results on predictions and the predictability of Xevents, focusing on nonlinear stochastic processes since they are assumed to provide the basis for extremes. These predictions are usually of a probabilistic nature, so the benefit of this type of uncertain prediction is an additional issue. As a specific example, we report on the prediction of turbulent wind gusts in surface wind.

4.1 Introduction

In a contribution to a book entitled "Extreme Events" there is no need to motivate our interest in this issue – ample motivation is supplied by the Introduction and by those other contributions, which address specific phenomena such as weather, fracture, epilepsy, or gigantic ocean waves. In all types of phenomena where the event magnitude can assume any value inside some interval, one has to decide beyond which magnitude we call an event "extreme". In our everyday understanding, we do this implicitly by thinking of "extreme impact events" when we say "extreme events". This means that we consider an event to have overcome the critical magnitude to be extreme if it causes damage or harm to us. This implies the rareness of Xevents. Events that happen frequently call for either counter-measures or for adaptation. In terms of diseases this means that either our immune system can cope with them, or we need vaccination, or life on earth is really threatened (as discussed later in the context of avian flu). If trees were not able to stand "normal" autumn storms (adaptation), there would be no forests anywhere in northern Europe. A meteorite hitting the earth could extinguish all life, so this kind of Xevent must be rare enough to allow evolution to form human life.

Hence, the rareness of extreme impact events is a direct consequence of their impact and it is implied by their definition. A similar argument may be used to require some form of irregularity in the occurrence of Xevents. Even if the event magnitude cannot be reduced by a mere prediction, anticipation of the event might reduce its impact and hence its extremity. A simple but nonetheless characteristic example is the ocean tide. Dams for coastal protection have a certain height which protects against normal tides, including spring tides, where the alignment of sun and moon amplify the effect. However, weather conditions (in particular, storms) can create exceptionally high water levels. Since humans try to be efficient, a balance is usually obtained where the cost of protecting against an Xevent does not exceed the cost of the damage of those events that break through the protection. Since the latter costs are proportional to the frequency of occurrence of these "above-threshold events", these events must be rare (see above). Nonetheless, exactly this reasoning tells us that above-threshold events (or extreme impact events) are expected to occur for every phenomenon, since perfect protection is usually much too ambitious. And this brings us to our point: if there are events which overcome our normal precautions/protections, then the best approach we can take (and this approach is also urgently needed) is to attempt to predict them.

In this article, we will first try to identify the objects of our study, namely Xevents, in a proper way, which should then form the basis of a general scientific approach. We do not want to focus on particular phenomena but instead consider Xevents to be large deviations from the average behaviour in temporally evolving systems. Before this, we will give a physical and mathematical framework to the class of systems that will be addressed by our reasoning. For this class of systems we will then discuss predictability from an information theory point of view and predictions from an algorithmic point of view. The prediction method will then be illustrated with a kind of case study. We will finish the chapter with by comparing the prediction of Xevents to other prediction tasks.

We are dealing here with a notion that comes from daily life, and so it has many different facets. Not surprisingly, a precise definition will be either restrictive or worthless (because it is too general). We will use a restrictive notion, being well aware of the fact that it will exclude many phenomena where one could well argue that these are Xevents as well. Moreover, it also turns out that the definition of Xevents and the mathematical/physical concept for the system that generates such events are closely linked.

In physics, the concept of a *state* of a system has proven itself to be very powerful. Hence we restrict here the discussion to systems where this makes sense. In order to introduce temporal evolution, (dynamics), one needs equations of motion. Such equations might be purely deterministic (as in Newtonian equations of motion), they might be stochastic (as in Langevin equations), and they might even act not on the states themselves but on the probabilities of finding the system in a given state at a given time (as in Fokker–Planck equations or Master equations). Even if "real systems" live in continuous time and continuous space, then in order to simplify (in particular when modelling) one might discretise time, space, or even the possible states (in the latter case one arrives at deterministic or stochastic cellular automata or lattice gas models). Hence, the assumption in the following will be that there exists an often unknown abstract representation of our system where the current state is well defined and where the rules of how all possible future states emerge from the current state are fixed. Even in cases where this is the case (say, in hydrodynamic turbulence, where the state is the velocity field in some volume and the dynamics are given by the Navier–Stokes equations), it is often impossible to identify the current state of the system by observation (one cannot measure the velocity of a fluid at the same time at every point in some container).

The required irregularity in the occurrence of Xevents implies that the underlying system responsible for the particular events possesses some complex dynamics. Complex means that despite the potential simplicity of the equations of motion their solutions are highly irregular and seemingly unpredictable. In an idealised physical model, the system might be purely deterministic but chaotic or it might be driven by noise. Of course, uncorrelated stochastic processes can also generate Xevents. One example of this class is the lottery, where the sudden gain of a million € is clearly an Xevent in the life of the gambler. However, since the randomly selected set of numbers that determines the winner of the lottery is not extreme in the slightest when viewed as a set of numbers, we exclude this class from our discussion. We hence require that the property of being extreme is reflected by the fact that we can define an observable which assumes an extreme value when the Xevent takes place.

Additional distinctions are useful. In some cases, the extremity of an event is directly linked to the extremity of a dynamical state of the underlying system. In other cases, the extremity of the event is caused by the way the system is observed; in this case one could define another observable and the dynamical states that show up as extremes would change (and with them the temporal sequence of when the system generates an Xevent). To illustrate this, let us understand weather as a state of the atmosphere. Both temperature and precipitation are observables that may be used to study its dynamics. Evidently, extreme precipitation is not strongly correlated to extreme temperatures, so it is unclear whether we can say that the atmosphere is in an extreme state when it is exceptionally hot. Many examples could be a combination of both types. As said before, humans often recognise Xevents as bringing some potential damage. Hence, there are clear physical effects that suggest that the system is in an extreme state (such as the release of energy in an earthquake), but there are also human-made thresholds and constraints (such as the number of victims or the financial damage due to the quake). A thirty-day drought would not be considered to be an Xevent at an airport, but a farmer working on the agricultural fields next to the airport would lament the loss of his harvest. The issue of whether an Xevent is also an extreme state of an underlying system or just the extreme value of a given observable has implications for its predictability, and we will discuss this later.

Another issue is whether or not an Xevent appears endogenously, generated by the system dynamics, or exogenously, being induced by some external perturbation. We will assume a slightly different point of view to Sornette [1], since ultimately we will not distinguish between deterministic or stochastic systems. Hence, small perturbations that act like noise are then considered as inherent to the system, and only externally controlled changes of system parameters or macroscopic perturbations that are much stronger than noise will be considered to be exogenous. The stock market is an example of externally triggered extremes, although it does contain some complications: a severe external event (political, military, social) can have strong impacts on the stock market, but this impact should not be taken for granted. Only together with an appropriate state of the market (intrinsic instability) can a strong external perturbation cause a crash. However, it is evidently impossible to predict such events just from studying the system's dynamics, since the system itself does not contain information about the next strong perturbation from outside. Therefore, we will restrict ourselves to endogenous events.

Finally, a classification can be made in terms of recurrence. In many systems, Xevents occur recurrently: after an Xevent, the system continues with unmodified dynamics and thus it has the potential to generate other Xevents. Typical non-recurrent Xevents terminate the lifetime of a system (such as material fracture). Again, there are some intermediate possibilities: a system might be altered by an Xevent in such a way that after the event the system has different dynamics, but nonetheless can suffer from new Xevents. Examples are the evolution and extinction of species (in biology), but also of political entities (related to revolutions and wars).

To summarise these introductory remarks, we will focus on Xevents that are generated by systems with complex (deterministic or stochastic) dynamics. They should be generated by the system dynamics itself (being in some sense endogenous), and they should appear recurrently. In any case, the events are assumed to occur rarely and with some kind of irregularity in time. Such events should be forecasted. Some lines of thought might be easily transferred to Xevents in different settings, but this is beyond the scope of this article.

4.2 Prediction versus Predictability

What one usually expects from a prediction is that it gives the value of some observable at some specific time in the future. The prediction error is then the difference between the predicted value and the observed value. If one repeatedly performs predictions for a given temporally evolving system, then the quality of the predictions can be quantified by the average prediction error, which is usually taken as the root mean square of the individual errors.

When we wish to predict Xevents, the task is usually somewhat different: the precise value is less important than whether the value is above or below a given threshold. Also, timing then plays a role. Consider events that, on average, occur once a year: if we predict such an Xevent for tomorrow, but it happens one day later, how can we quantify this rather small error? There are currently no generally accepted concepts for quantifying the precision of these kinds of predictions.

However, in many cases such individual predictions are not a realistic goal. Nevertheless, when Xevents are considered, much weaker predictions are already very useful, such as the *probability* of an observable overcoming a given threshold during a certain time interval. In order to include this more general kind of prediction, we consider in this paper a prediction to be the *forecast of a probability distribution* of the future state of the system or of the future value of a specific observable. If desired and useful, the predicted probability distribution can be converted into a specific value for the observable of interest by computing its mean. In this case the prediction is optimal in the sense of the maximum likelihood principle. It is also clear that if we are able to resolve the deterministic nature of a deterministic system, then the predicted probability of the future states is very narrow (ideally a δ-peak). To make predictions, irrespective of its individual or probabilistic nature, one needs some prediction algorithm which converts our knowledge of the system into a statement about its future, including possibly knowledge of the current state of the system (in a deterministic setting, this could correspond to the equations of motion plus the initial conditions). The precision of our predictions depends not only upon choosing the optimal prediction algorithm, but also on just how much predictability is present in the system.

Predictability is the potential of the system to allow us to perform a prediction, in principle. Predictability can be quantified by how precisely we could predict the future at a given time interval in the future if we knew everything that could be known about a system, and if we knew its current state with a given finite precision (if we assume unrealistically that we know the present state precisely, then every deterministic system is perfectly predictable, which is not a useful statement). Under additional assumptions, this knowledge can be converted into a statement about how far into the future we could make a prediction with a given precision.

As an example, the best prediction for the outcome of the next lottery draw is simply to say that every possible number is equally probable. We will say that in this case there is no dynamical predictability, since the next drawing is uncorrelated to previous lottery draws, and studying the record of past draws will not improve our prediction. On the other hand, without astronomical skills and precise observational facilities, the next total eclipse of the sun would appear to be unpredictable, whereas its predictability is perfect.

In this section, we will first discuss predictability in general and why it should be present in many complex systems, and then we will introduce the algorithms that might enable us to extract the relevant knowledge from a system needed to perform a prediction.

4.2.1 Predictability

In the following we want to quantify the existence of structures that in principle enable us to know something about the future of a system (for a similar and very inspiring discussion see [2]). If such structure exists, we speak about predictability. This predictability has two components: the *static part* comes from the probability distribution of all possible future events, and the *dynamical part* comes from temporal correlations. The static part will enable us to make predictions that are always the same, so they are static. The dynamic part leads to predictions that are time-dependent, and hence dynamic.

Both aspects require stationarity, which means that we can only make statements about the future learning from the past if we require that the future will obey exactly the same dynamical laws, with the same parameters, as the past did. In realistic settings, stationarity is usually violated. Since there are currently no good theoretical concepts for handling nonstationarity, violation of stationarity is ignored in practice, but one has to bear in mind that nonstationarity might lead to a future evolution of a given system that is very different from the evolution that can be predicted from the past.

The following discussion on predictability applies with only minor modifications to two settings. In the ideal setting, we speak about a state space of a system, and we can obtain stationary probability distributions or invariant measures in the state space and on sequences of state vectors. In the time series setting, we speak about the corresponding distributions of the values of our observable, and about probabilities of finding certain sequences of values in successive observations.

Static Part of Predictability

The *static part of predictability* is related to the probability distribution of either the state vector or the observable. In terms of predictability, it makes a big difference whether we have one regular dodecahedral die where eleven faces are labelled by the numbers from two to twelve (the twelfth face saying "repeat"), or whether we have two standard dice that are thrown simultaneously. In the first case, each number has an equally probable outcome, whereas in the second case the number seven has a six-fold greater probability of occurring than two or twelve. Hence, predicting the value of seven at every trial will on average lead to a much higher hit rate for the two dice than for the dodecahedron. The precision of such predictions is quantified by the root mean squared error $\bar{e} = \sqrt{\langle (x - \hat{x})^2 \rangle}$, where x is an outcome,

$\hat{x}$ is our prediction for this outcome and $\langle . \rangle$ denotes the average over very many trials. In the example of the dice, the optimal prediction minimising $\bar{e}$ is $\hat{x} = 7$ (static prediction), the mean value of the distribution of the possible outcomes. The same will hold for the dodecahedron. However, the average error $\bar{e}$ is different: It is $\sqrt{10}$ for the dodecahedron and $\sqrt{5.83}$ for the two dice. The reason lies in the fact that large deviations from the prediction $\hat{x} = 7$, such as two and twelve, are rare in the case of the two dice, but occur more frequently in the case of the dodecahedron. This mean error $\bar{e}$ could be given as an "error bar" together with the prediction, as a kind of mean uncertainty about the true outcome.

In information theory, the uncertainty of the outcome, when the distribution is known, is quantified by the Shannon entropy, $H_S = -\sum_{i=1}^{N} p_i \ln p_i$, where p_i denotes the probability that the event labelled with i will occur. For a finite number N of possible outcomes, the uniform distribution $p_i = 1/N$ generates the largest entropy, $H_S = \ln N$, so it leaves us with the largest uncertainty about the outcome. Every distribution with nontrivial structure (where the different outcomes are not equally probable) leads to smaller values, as we saw in the example above. The properties of the Shannon entropy and why it quantifies predictability can be found in, for example, [3]. The Shannon entropy can be generalised in a straightforward way to a continuum of states, provided a probability density function (pdf) exists.

The probabilities that the different outcomes will occur can also be converted into static temporal information: an event that has a probability p of occurring within a given time interval will on average take place every $1/p$ time intervals. Of particular interest is the largest possible event: no events larger than this will ever occur, and there is no need to concern ourselves with them. It is evident that such an upper limit is unknown for many natural phenomena, and this illustrates that this information, if available, is a real prediction. In summary, knowing the probability distribution of all possible outcomes enables us to predict not only what will happen on average, but also the frequency of occurrence of specific events.

In order to make use of this source of predictability, we usually have to estimate the probability distribution from data (if model equations exist, one can try to determine them from these equations). In view of Xevents, there is a severe complication. We argued before that Xevents are usually rare. Hence, if we are particularly interested in the prediction of Xevents, we are referring to the tails of a probability distribution, which are badly resolved by a given finite sample of observations. Obtaining an accurate estimate is therefore a highly nontrivial task and the subject of many recent scientific activities, partly summarised as "extreme value statistics" [4, 5]. Sometimes a suitable model allows one to predict the functional form of the distribution, so that fitting its parameters in the bulk supplies full knowledge about its tails [6]. A complication beyond the finite sample effects is given by temporal correlations: we wish to find correlations in data since they are the second, even more relevant, source of predictability. However, they can introduce some bias when estimating the tails of a probability distribution from a finite sample, leading to either over- or underestimates of the relative frequencies of Xevents [7].

Dynamical Part of Predictability

Xevents are not usually generated as if someone were throwing the dice, but instead the system is governed by equations of motion that propagate the actual state of the system forward in time. Correlation and other temporal structures appear in the time series as fingerprints of the hidden dynamics. In order to fully compute the predictability of the system, the dynamical constraints upon the values the system may achieve have to be taken into account. Enhanced predictability may be present even if the whole process is stochastic, because the probability distribution of future values may depend upon the past values. In contrast to the static part of predictability, where prediction can be made without knowing the current state of the system, the aspect we want to quantify here links successively observed states. It is called the *dynamic part of predictability* because the actual predictions will depend upon the current state of the system and hence on time, and they make use of what is known about the dynamics of the system. Dynamical predictability will quantify how uniquely the dynamical laws governing our system can relate a future state to a present state. However, such knowledge can only be exploited if we also have some knowledge about the current state. Hence, the precision of our prediction will depend upon the precision with which we know the current state. This interpretation leads to a quantitative concept: we will investigate how the accuracy of the prediction depends on the accuracy of the knowledge of the current state.

If a system's dynamics are perfectly periodic, predictability is maximal since there is no uncertainty about the future. In order to distinguish once again between predictability and predictions, let us emphasise that even though the predictability of a periodic process is perfect, an actual prediction is only possible after we have observed the periodicity of the process (so we need a historical record which covers at least two periods of the process in order to support the hypothesis of periodicity), and in order to be able to predict the continuation we need to know the current position along the periodic cycle. A simple everyday example of this is the phases of the moon: seeing a half-moon in the sky, and knowing that it has a period of 28 days, we can only predict whether it will be full moon or new moon in a week's time if we are able to interprete whether the moon is crescent or waning. The generalisation of the Shannon entropy to the characterisation of temporal structure eventually leads to the concept of Kolmogorov–Sinai (KS) entropy h_{KS} (see again [3]). Essentially, the KS entropy is a Shannon entropy based on the conditional probability of observing an event i after having observed a sequence of events $k, l, m, \ldots$ at one, two, three, $\ldots$ time steps in the past.

In the case of a periodic phenomenon, the KS entropy is zero, meaning perfect predictability. An important rephrasing of this situation is that even if we are slightly uncertain about the actual state of a system, the error in our prediction will not grow faster than proportionally to the time into the future for which we make the prediction.

In a still deterministic but chaotic system, the far future is unpredictable: every time step ahead magnifies a given error in our current observation by a certain factor, so that the uncertainty about the future (caused by the uncertainty about the present) grows exponentially with time. In this case the KS entropy has a finite nonzero value. Its interpretation is that we have to add an amount of information quantified by the KS entropy in order to predict the future with the same accuracy as we have observed the past (assuming perfect knowledge of the equations of motion). Rephrased, this means that when we know the current state with a given uncertainty, then the smallest possible error in our prediction (on average) will be larger than this uncertainty by a factor $\exp(t \cdot h_{KS})$.

For all nondeterministic, random processes, the KS entropy is infinite and does not allow us to distinguish them. Evidently, depending on the strength of the temporal correlations, random processes also differ in their predictability. However, we have to be more specific and to discuss the level of accuracy to which we are working. Let us assume that we know the current state with an uncertainty ϵ (for instance, given by our measurement errors). Let us assume that we also know the (stochastic or deterministic) time evolution perfectly, so that, given the state x, we can obtain the exact probability that another state x' will be obtained one time step ahead. Now, since by assumption we only know the current state with an error ϵ, we could create an ensemble of initial states that are distributed around the given value with a standard deviation ϵ, and then consider the probability distribution of all of the future values that might succeed each initial value from this ensemble. This distribution might be much broader than ϵ. Hence, in order to reduce the uncertainty of the future to the same level ϵ to which we assume to know the current state, we need extra information. This lack of information is quantified by a KS-like entropy $h(\epsilon)$ (called ϵ-entropy; see [8]) for the scale ϵ. Of course, this extra information is unavailable. So we can invert the argument and say again that the precision of our prediction is less, by a factor of $\exp(t \cdot h(\epsilon))$, than the precision of the current state. What differs here from the (chaotic) deterministic case is simply that $h(\epsilon)$ depends on the accuracy ϵ, and it diverges for $\epsilon \to 0$ if stochasticity is involved, whereas it converges to h_{KS} in the perfectly deterministic case. So in order to use this concept, no knowledge of the true nature of the system is needed. As soon as some randomness enters the dynamics, then for some range of ϵ it is still profitable to increase the accuracy of the measurements of the current state, but below a system specific value, extra accuracy cannot be converted into extra prediction accuracy.

Let us illustrate what we have said using two very simple examples. The first one is a deterministic chaotic system, $x_{n+1} = 1 - 2x_n^2$, with $x_0 \in [-1:1]$. Assume that we know the last observation x_N with an uncertainty (measurement error, computer precision, finite representation of real numbers) of ϵ. Then the optimal way to use the knowledge of an infinite past sequence of observations cannot be better than applying the map itself to the last observation, hence we compute $\hat{x}_{N+1} = 1 - 2x_N^2$. The uncertainty of the present state x_N translates into the uncertainty of our prediction, which for this particular system is twice as large on average [3]. This error amplification is exactly the source of chaos. The KS entropy of this process is $h_{KS} = \ln 2$, and our estimate about the uncertainty of the prediction from above is $\epsilon' = \epsilon e^{h_{KS}} = 2\epsilon$. If ϵ were 2^{-10}, then after ten steps the uncertainty would grow to unity and we would have reached the maximum prediction horizon.

The second example concerns the same system with additive dynamical noise, $x_{n+1} = 1 - 2x_n^2 + \xi_n$. For technical reasons (the resulting value x_{n+1} must always be inside the interval $[-1 : 1]$ in order to stop it escaping to infinity) the noise terms ξ_n are not independent of x_n, but this dependence is weak and hence not relevant for our considerations. As before, the optimal prediction $\hat{x}$ for the observation $N+1$ (in the maximum likelihood sense and assuming $\langle \xi_n \rangle = 0$) would be to say $\hat{x}_{N+1} = 1 - 2x_N^2$, taking the standard deviation of the unknown noise input ξ_N as the inevitable average prediction error, in other words as the theoretical lower limit of the uncertainty about the true future. If our knowledge of x_N is imperfect, its error increases. Hence, the more precisely we know the current state x_N, the closer we come to this theoretical limit. However, the improvement of the forecast gets less the better we know x_N. In other words, from a coarse-grained viewpoint we see (almost) the deterministic law $x_{N+1} = 1 - 2x_N^2$. If, on the other hand, we have high-precision measurements, then the noise in the time evolution prevents us from making an adequately accurate prediction (in contrast to the previous example, where the measurement error was doubled independently of its magnitude). This is visualised in Fig. 4.1: We show the probability distribution of future values x_{N+1} for a sample of given x_N which, in the assumed precision of the measurement, are identical. Whereas reducing the measurement error from 1/4 to 1/8 reduces the width of the distribution (and hence the uncertainty of the prediction) by almost 1/2, a further reduction down to $\epsilon = 1/32$ does not create a similar improvement. So one can argue that the dynamical predictability, quantified as the precision of knowledge of the actual state divided by the error in the prediction, goes to zero when the precision improves.

The concept of entropies hence quantifies predictability on average. As we emphasised, taking into account the dynamical part of predictability will lead to time-dependent predictions (we would usually expect a prediction to be time-dependent). What is masked by these average considerations is the fact that predictability might vary as a function of the actual state of the

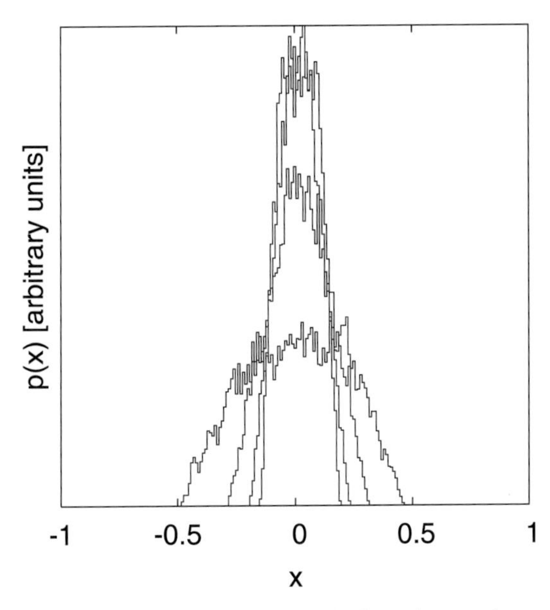

Fig. 4.1. Ensemble forecasts. Assuming that the last observation x_N is known with uncertainty ϵ, and applying the map $x_{N+1} = 1-2x_N+\xi_N$ with randomly chosen ξ_N, yields the shown distributions for x_{N+1}. The widths of these distributions are the uncertainties of the prediction. Whereas it clearly decreases when reducing ϵ from 1/4 (broadest distribution) to 1/8, a further increase in the precision to 1/16 and 1/32 leads to saturation. The latter is a consequence of the unpredictable stochastic component in the equation of motion ($x = -0.7$, the ensemble was created by randomly selecting points in $[x-\epsilon, x+\epsilon]$)

system. It is an everyday experience that there are weather conditions (such as a summer high in central Europe) where the prediction "tomorrow the weather will be as fine as today" has a very high hit rate, and there are other seasons in the year where the weather forecast is notoriously unreliable. This can also be seen in the simple example $x_{n+1} = 1 - 2x_n$: if $x_n \approx 0$, there is no error amplification; but instead, any inaccuracy in the knowledge of the current state is translated into reduced uncertainty about the future. An attempt to resolve this state-dependent predictability has been made in [9]. The fact that individual states might lead to prediction errors that are either much smaller or much larger than the mean prediction error to be expected from entropy analysis has clear implications for the prediction of Xevents. It includes the two possibilities that extremes might be much less or much more predictable than the "average" states.

To summarise this section, we state that predictability can be quantified by KS-like entropies. Their values tell us how much greater the uncertainty (mean error) of the optimal prediction is than the uncertainty about the current state of the system. The generalisation of the entropy concept to stochastic processes through the ϵ-entropies yields the benefit that we know that more precise knowledge of the current state will pay off when we have only vague knowledge (large measurement errors), but that there will usually be some limit $\epsilon_{\min}$ below which it does not help any further. Moreover, we tried to emphasise that predictability is, of course, also present in stochastic processes. There is a distinction between the static part of predictability, given by the knowledge of the distribution of future values of the observable, and which does not require any knowledge about the current state, and the dynamical part of predictability, which exploits the possible sequence of future states based on the present state and on the dynamical rules.

The predictability of a process quantifies how much uncertainty we have to accept on average when we exploit all existing information about a process *in the optimal way without violating causality.* Unfortunately, it does not tell us *how* to extract this information from past observations and how to make use of it. Nevertheless, if we compute these slightly non-intuitive quantities $h(\epsilon)$ from data, we will have a clear benchmark for every prediction algorithm. In what follows, we discuss prediction schemes that are appropriate for time series originated from deterministic and from stochastic processes. They are in some sense optimal, since they exploit exactly the type of information that enters the definition of predictability.

4.2.2 Prediction Schemes for Deterministic and Stochastic Time Series

In most real world systems, the underlying equations of motion are only known to some approximation (if at all; there is, for example, no evident and generally accepted model for the stock market), and even if they were known perfectly, in many cases the current state of the system cannot be determined in an appropriate way. Hence, we will assume a time series approach here; we will discuss a prediction scheme which starts from a long record of observations, $\{s_1, \ldots, s_N\}$, where the sampling rate is appropriate, the time span covered by the observations is long compared to the internal timescale of the phenomenon, and where the number of observations N is also large. In Sect. 4.3 we will discuss predictions of turbulent gusts based on surface wind time series recordings, where the terms "large, long, appropriate" will be illustrated.

Deterministic Time Series

Expecting a purely deterministic origin in a real world phenomenon is unrealistic. Nonetheless, for didactical reasons, we assume exactly this here. If the observed quantity represents a deterministic dynamical system with a low dimensional phase space (so the system has only a few independent variables which are governed by deterministic equations of motion), then a mathematical theorem from Takens [10, 11] tells us the following: a general observable applied to the (unknown) phase space variables of the system can be used to reconstruct an auxiliary space in which we again find a kind of deterministic equation of motion. We simply have to combine successive measurements to delay vectors, $\boldsymbol{s}_n := (s_n, s_{n-1}, \ldots, s_{n-m+1})$, where the dimensionality m is a parameter that has to be larger than twice the number of active degrees of freedom for the theorem to hold. If so, then every delay vector can be mapped uniquely and smoothly onto an unobserved state vector of the system, so that in particular the future of a delay vector is uniquely determined by the future of the corresponding state vector. Hence, using this embedding procedure, the deterministic properties of the underlying unknown dynamical system are transported into our auxiliary space, the delay embedding space.

Now, since we know that the future $\boldsymbol{s}_{N+1}$ of the most recent delay vector $\boldsymbol{s}_N$ is deterministically and uniquely encoded by $\boldsymbol{s}_N$, we "only" have to extract the information about this future from what we know about the system (from the large set of past delay vectors). Here, the additional relevant assumption to be made is the smooth dependence of the future on the present state – we have to assume that similar states will have a similar future. Then one can employ what is called the "Lorenz method of analogues" [12], or the "locally constant approximation" [13], in other words one searches for all similar states in the past (for which $|\boldsymbol{s}_k - \boldsymbol{s}_N| < \epsilon$ holds), and the prediction $\hat{s}_{N+1}$ is the average over the futures of the neighbours, $\hat{s}_{N+1} = 1/K \sum_{\{k\}} s_{k+1}$, where K is the number of indices in the set $\{k\}$ fulfilling the similarity condition. The smaller the tolerance level ϵ the closer each individual s_{k+1} is to its mean value, but of course ϵ must be chosen to be large enough such that in the given data set there are still some points that fulfil the similarity condition. More sophisticated ways to determine the deterministic mapping $\boldsymbol{s}_N \rightarrow \hat{s}_{N+1}$ exist and have been used successfully [9].

In such a deterministic setting, the prediction error, $e = s_{N+1} - \hat{s}_{N+1}$, is related to modelling errors and to measurement noise on the data. The latter reduces our knowledge about the present state $\boldsymbol{s}_N$. If we knew these effects, we could predict the future value with a given uncertainty in its magnitude. It is evident that the complete failure of a single prediction with such a scheme could only happen with very low probability, since it can only be explained by exceptional measurement errors.

Stochastic Time Series

Typical phenomena of our world, such as weather, climate, the economy, and daily life, are much too complex for a simple deterministic description to exist. More precisely, even if there is no doubt about the deterministic evolution of, say, the atmosphere, the current state (whose knowledge would be needed for a deterministic prediction) contains too many variables in order to be measurable with sufficient accuracy. Hence, our knowledge does not usually suffice for a deterministic model. Instead, very often a stochastic approach is more suited. Ignoring the unobservable details of a system, we accept a lack of knowledge. Depending on the unobserved details, the observable part may evolve in different ways. However, if we assume a given probability distribution for the unobserved details, then the different evolutions of the observables also appear with specific probabilities. Hence, the lack of knowledge about the system prevents us from deterministic predictions, but allows us to assign probabilities to the different possible future states, at least in principle. It is then the task of a time series analysis approach to extract this information from past data, and we will outline this scheme in next section.

Before describing the Markov chain model for prediction, we want to arguing once more for the benefits of a probabilistic prediction. First of all, the probability that a certain event will occur will depend on the current state of the system, so we will explore the dynamical part of predictability. If this probability varies considerably from state to state, then knowledge of this probability can be very helpful. We all make use of such information every day, since under serious considerations the weather forecast is a probabilistic forecast, even if it sounds like it is deterministic. Secondly, we have to transform the value of the probability of a certain event occurring into a reaction. This is also something that we are used to: we are all trained to consider the need for an umbrella when leaving the house when it is *not* raining. Although probabilistic predictions are not what we are really after, they contain much more information than no prediction at all. Of course, in contrast to the deterministic prediction, the system might evolve into a state which is extremely improbable according to our prediction without causing any inconsistency. The formal way to determine the optimal reaction to a predicted probability is to minimise a cost function. We can, for example, try to quantify the cost if I carry my umbrella but it does not rain, and the cost of not having an umbrella and so getting wet or having to wait in a shelter. Depending on the costs for each case, the result of the minimisation would define a specific probability p_c. If the predicted probability is below p_c, not taking the umbrella will be advantageous, and beyond p_c one should take the umbrella. Hence, the optimal reaction to a probabilistic prediction requires us to think about the costs, and to solve an optimisation problem. These two issues are widely discussed in the literature and are therefore not the topic of this contribution. We will restrict it to how to predict the probability of a certain event occurring.

4.2.3 Predictions Based on Markov Chain Models

The data-driven Markov chain is a nonparametric model that exploits the information used to compute entropies. In this sense we expect it to be optimal, irrespectively of whether the system is stochastic or deterministic. In fact, in an ideal, low dimensional deterministic setting, it reduces to the Lorenz method of analogues explained above. Since this model is very data intensive, parametric models may be superior in practice.

In Sect. 4.2.2 we outlined how the time delay embedding method enables one to extract a deterministic relationship between past and future states from time series data. Determinism means that a time window m exists such that $s_{N+1} = f(s_N, s_{N-1}, \ldots, s_{N-m})$, and the task is to find the function f from the past data. This is, in principle, very easy, since we have a huge number of "learn pairs" $(s_{k+1}; s_k, s_{k-1}, \ldots, s_{k-m})$ which, by assumption, represent this function f. There are many different ways to extract f from these, such as simple interpolation schemes, locally linear approximations, kernel estimators, neural networks as global models, or a representation based on the linear superposition of nonlinear basis functions. These methods have in fact proven their usefulness in a wide range of applications.

If a process is intrinsically stochastic, s_{N+1} is not a unique function of $s_N, \ldots s_{N-m+1}$. Instead, one can hope that a value m exists such that the *probability* that the next measurement s_{N+1} assumes the value s' is fully determined by $(s_N, \ldots, s_{N-m+1})$. If this is true, then the stochastic process is a continuous state Markov chain of order m, and we have complete knowledge of it if we know the conditional probabilities $p(s'|\boldsymbol{s})$, where $\boldsymbol{s}$ is a Takens-like m-dimensional reconstructed state vector of the system. As discussed in [14], such a finite value m will not generally exist, but a finite value m usually yields a very good approximation to the non-Markovian process, so this offers an efficient prediction method in practice.

The conditional probabilities can be estimated from a long record of observations, under two assumptions. The first is the standard assumption in time series analysis, which is stationarity: all probabilities should be constant in time. This corresponds to the requirement that physical parameters influencing the dynamics of a system must be time-independent, but it also requires that the process has had enough time to relax into a stationary state. Despite the fact that stationarity cannot be proven to hold for a finite data set, and even worse, that almost all data sets are nonstationary to some extent, let us assume that stationarity holds.

As a second assumption, we require that $p(s'|\boldsymbol{s})$ is a smooth function of $\boldsymbol{s}$, so the conditional probability varies smoothly as the condition $\boldsymbol{s}$ varies. Then we can generalise the Lorenz method of analogues. In every previous situation in the recorded time series data where $\boldsymbol{s}_k \simeq \boldsymbol{s}_N$, the future value s_{k+1} is drawn according to the same probability distribution $p(s'|\boldsymbol{s}_N)$ as the unknown future value s_{N+1} will be drawn. Hence, collecting a larger sample of such values provides a sample estimate of this probability distribution.

The probability distribution of the values to be measured in the next measurement can now be exploited in different ways. If we insist on making a prediction in terms of a single value, then the mean value of this distribution is optimal in the sense that it minimises the root mean squared prediction error $\sqrt{\langle(\hat{s}_{N+1} - s_{N+1})^2\rangle}$, where the average is to be taken over many prediction trials N, and $\hat{s}$ denotes the predicted value, whereas s, as before, denotes the actual measurement. More appropriate for a stochastic process would be the extraction of probabilities from the given probability distribution, for example the probability that the next observation will lie inside a certain interval $[s_{\min}, s_{\max}]$, or that it will overcome a given threshold. Hence, determining a probability distribution for the future allows us to adapt the prediction to our needs. Let us repeat here that the assumed stochastic nature of the process allows outcomes for the next actual measurement that are in the tails of the predicted probability and are hence unlikely to happen, without causing any contradiction.

An evident and severe drawback of this method in terms of Xevents lies, however, in the fact that by construction it cannot predict anything that has not been observed in the past. In the best case this method can detect that the current situation has never been observed in the past and therefore there are not any sufficiently similar states $\boldsymbol{s}_k$ to allow any prediction to be made. In this sense, this method is weaker than the statistics (static part of predictability), where one can always try to extrapolate from a given histogram to larger values of the observable.

We should stress that a Markov chain model can be modified in many ways. We were using the finest possible resolution ϵ and an extended, m-step memory. One can easily introduce some coarse graining, hence reducing the number of possible system states to a small finite number. Then the transition probabilities can be expressed as a transition matrix. This was applied for wind speed predictions in [15], and it clearly resembles the earthquake predictions classified as "time-dependent hazards" [16]. A technique that is very similar to Markov chain prediction was used when modelling in [17].

4.3 An Example: Turbulent Wind Gusts

Surface wind is turbulent even at moderate average wind speeds. This implies that the wind speed measured at a given point above the surface of the earth fluctuates irregularly, and exhibits strong increases from time to time. A typical wind speed time series is shown in Fig. 4.2. Humans make use of the power contained in wind through sailboats, gliders, and through wind turbines. In such cases, the sudden increase in wind speed can cause problems. Hence, as an example, we study turbulent gusts as Xevents. Since a quantitative definition of "turbulent gust" is missing, we consider here the difference between the average wind speed in two successive time windows of 2 s. If this increment is positive and of g m/s, we talk of a gust of strength g. In Fig. 4.3 we show empirical distributions of the frequencies of increments over different days. Evidently, the distributions differ strongly from day to day: there are days which contain many strong gusts and other days without strong gusts. What is more relevant is that whatever the day, these distributions can be well approximated by exponentials, $p(g) \approx const \cdot e^{-|g|/g_0}$, where g_0 depends on the particular day (ignoring the slight asymmetry between positive and negative increments). Hence, Xevents (large g) are much

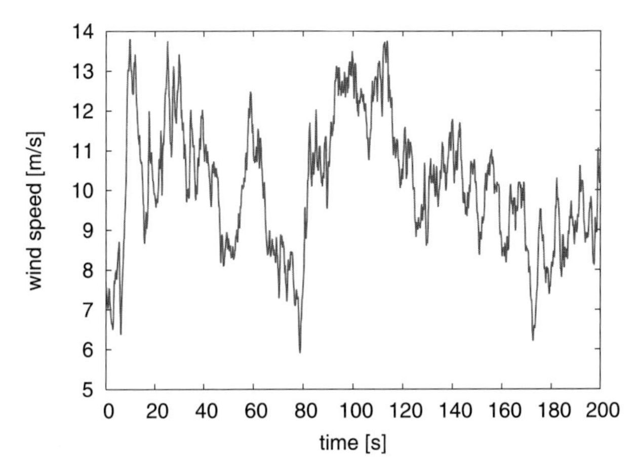

Fig. 4.2. Typical time series of near-to-surface wind speed measurements

more likely here than in data that are Gaussian-distributed and have the same variance. Considering the static part of predictability, the hypothesis of the functional form of the probability distribution $p(g)$ together with an estimate of the parameter g_0 would be fully sufficient. If the hypothesis of an exponential distribution is correct, using the estimate of g_0, one can even predict the probability of occurrence of events which are too rare to be contained in the finite sample.

In this example, we want to make use of the dynamical part of predictability. We hence look for temporal correlations and perform an analysis of entropies along the lines of Sect. 4.2.1. We expect some positive results simply because wind speed data are persistent: the best prediction for the next observation is the current observation. In Fig. 4.4 we show the results from an entropy analysis of wind speed data using the correlation integral [9, 18]. A one-step prediction (in our data, one time step corresponds to 1/8 s) based on the past ten values could, if done in the optimal way, reduce the uncertainty of the future to 1/10 of the width of the distribution of the data, and even ten steps into the future allows, in principle, for an improvement to 1/3, if the velocities are measured with a precision of 0.1 m/s. Let us recall that this result is obtained following the considerations of Sect. 4.2.1, so not a single prediction has been made, and at this point of the analysis it is also not known whether we could find an algorithm that really generates correspondingly accurate predictions for us. Unfortunately, the dynamical part of the predictability of velocity *increments*, which are of interest in terms of gust prediction, is less. But still, the conditioned probability distribution using a resolution of about $\epsilon \approx 1$–$2\ \mathrm{m/s^2}$ and a memory of about $m = 5$–10 reduces the uncertainty of the future velocity increment by half. Here, however, predictability decays very fast if we consider predictions further into the future than one time step (1/8 s).

Some interesting questions are: what is the typical profile of a wind gust, and can it serve as a precursor? We define a gust as a situation where two successive 2 s-mean values of the wind speed differ by more than g m/s. When we superimpose time series segments that fulfil this condition, we also see clear structure before the actual jump in wind speed (the continuous lines in Fig. 4.5). It is tempting to interpret this (the downward dip) as precursor. However, very similar precursors can be found in a simple AR(1) model with the same correlation time. As one can easily verify, even for uncorrelated random numbers with zero mean, this procedure will yield a signature that is different from the mean: through the selection criterion we find that inside the first 2 s window such a curve assumes the value $-g/2$, and in the second 2 s window it assumes the value $+g/2$, being zero elsewhere. Hence, precursor-like structures are a natural consequence of the fact that defining an event and selecting data sequences if they obey the definition of the event means to form a more or less specific subsample, whose value is evidently nontrivial due to the conditioning. The proper shape of the precursor depends a great deal on the way how the data are selected – how the event is defined. If there are also temporal correlations (as for the AR(1) data and the wind speed data) then the conditioning also enforces nontrivial structure outside the time window where it is imposed. Therefore, the existence of precursor-like structure on its own does not imply the predictability of the specific

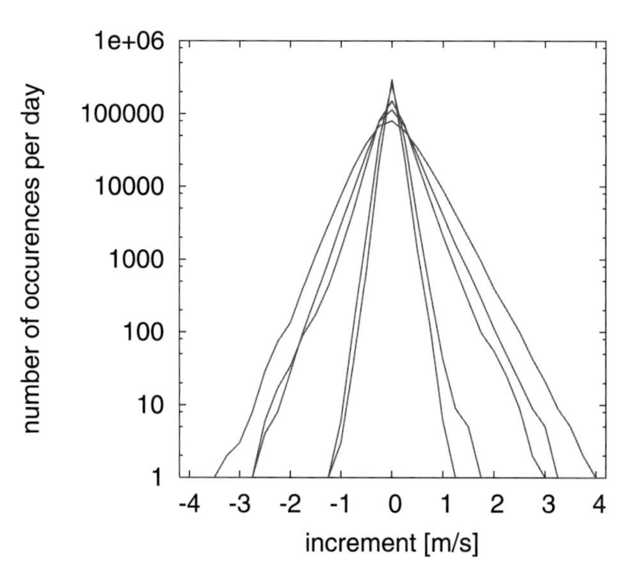

Fig. 4.3. The frequency of occurrence of wind speed increments of different sizes during a particular day (five different days are shown, 8 Hz data). The distributions are clearly non-Gaussian, their widths differ strongly from day to day, and they are slightly asymmetric

event; for example, by definition no relation between successive states exists in uncorrelated random data.

We can also explain this latter fact in a different way: finding precursor-like structure means that once we have observed the event we know that *on average* the signal underwent a typical structure before. However, what we need is the opposite: if the signal undergoes a specific structure, then the event of interest should follow with a high probability. To make a much stronger statement like this, more structure is required in the data, the existence of which can be probed by additional statistical tests. The simplest is to collect data segments that are similar to the supposed precursor over a time interval right before the gust, and to study their average future. For brevity, we call these data segments followers. What can we expect to find? In a solely deterministic setting we can expect that the mean of the followers again exhibits the gust structure. In uncorrelated random data, the followers will not show any structure outside the time interval where we require similarity to the precursor. But even if we do not see significant structure in the mean value, there might be useful hidden information: in terms of the conditional

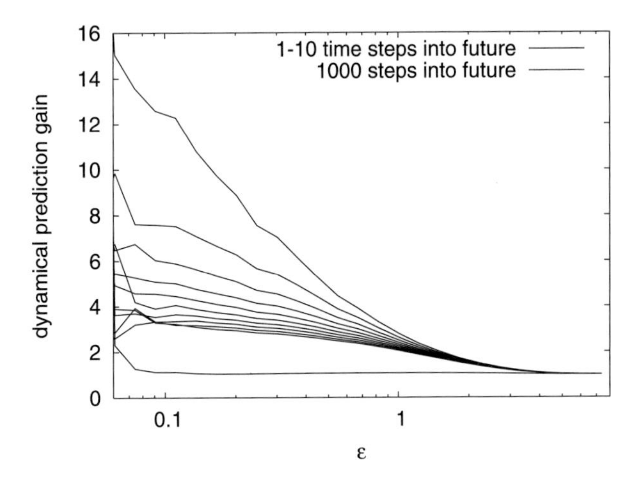

Fig. 4.4. The gain in predictability due to the dynamical part as compared to the static part. The *continuous curves* show the ratio of the width of the distribution of wind speeds divided by the width of the conditional distribution of the future wind speed, conditioned to the last ten measurements, and an assumed uncertainty in these given by ϵ. These curves are computed from the ϵ-entropies mentioned in Sect. 4.2.1. The curves show the predictability gain for predictions made 1–10 time steps into the future (from top to bottom), reflecting the trivial observation that predictions are worse the further into the future they apply to. The *dashed curve* is the numerical benchmark obtained for predictions 1000 steps into the future, where no dynamical predictability can be expected. This benchmark is, however, needed to confirm the significance of the results, since finite sample effects for small ϵ could result in a fake prediction gain. Hence, where this curve deviates from unity, the results for the other curves are doubtful as well

probabilities of the m-step-memory Markov chain, it is possible that its mean does not depend too significantly on the temporal pattern represented by the m-dimensional condition $\boldsymbol{s}$. For successful prediction of Xevents beyond the static average risk, it is necessary that the variance of this distribution depends on the conditioning state $\boldsymbol{s}$. If this is the case, then we are able to compute a time-dependent risk for a large increase of the wind occur.

Returning to the wind data, there are two indications that there are real precursors. One comes from estimating the distribution of wind speed fluctu-

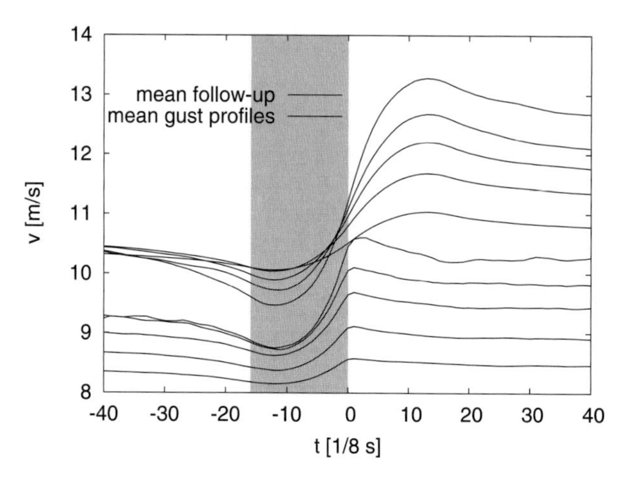

Fig. 4.5. The *continuous curves* shown the gust profiles for gusts of strength $g = 0.7, 1.4, \ldots, 3.5$. The *dashed curves* represent the means of all those data segments which are close to one of the vertically shifted gust profiles during the sample times $[-16 : 0]$ (*grey area*), whereas they are unconstrained outside this interval. In other words, they show the mean history and the mean future of profiles that contain the precursor of a gust. Since most of them are not gusts, the mean future does not look like a gust. However, the mean future velocity is systematically larger that the mean past velocity

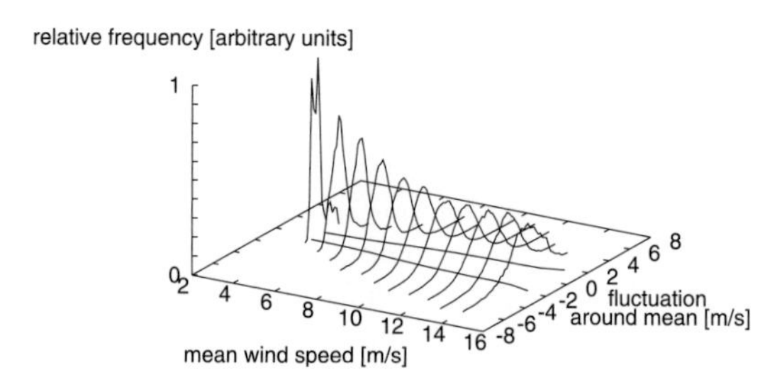

Fig. 4.6. Distributions of fluctuations of wind speeds around a one-minute moving mean value, as a function of this mean value. In the *bottom plane* the standard deviation of these distributions is shown by *dashed lines*, which increases roughly as 1/10 of the mean velocity

ations around a one-minute mean value, as a function of the latter. As one can see in Fig. 4.6, the standard deviation of these fluctuations increases with the one-minute mean $\bar{v}$ approximately as $\sigma(\bar{v}) \approx 0.1\bar{v}$. Hence, large fluctuations are much more likely if the actual wind speed is large. The minimal stochastic model that can generate such a finding is an AR-like process with *multiplicative* noise, for example, in the very simplest case, $x_{n+1} = ax_n + bx_n\xi_n$, where $0 < a < 1$, b is arbitrary, and ξ are independently drawn Gaussian random numbers.

The second indication of the existence of real precursors comes from Fig. 4.5. There we show the mean wind profiles when averaging over all gusts of strength g in a one-day data set. Additionally, we test for the specificity of these profiles: we accumulate all data segments over the 2 s that are similar to those in the (shifted in v) gust profiles on the 2 s interval right before the onset of the gust (causality). The sample averages of these are almost shifted copies of the gust profiles in the last 2 s before the gust (this similarity is imposed by the way in which the samples have been selected), and they have the freedom to exhibit any structure outside this window. Of course, they can do so only within the restrictions imposed by the (nonlinear) correlations in the data. We observe that the stronger the gust from which the precursor is taken, the larger the difference between the wind speed 5 s before the gust and 5 s after it. Also, we see a (very slight) increase in the wind speed at the sample times 1–3 s after the end of the conditioning window. It is beyond the scope of this article to discuss the additional investigations that would be needed to verify that these structures are really due to nonlinearities or higher order correlations. However, the observed increase in the standard deviations of fluctuations in Fig. 4.6 with the mean wind speed and the fact that the curves in Fig. 4.5 start from initial velocities that are the bigger the larger the gust increment g fit this idea nicely.

The prediction method described in Sect. 4.3 has been used to predict turbulent gusts in [19]. It is clear that the presence of precursors is automatically taken into account in this method once they lie inside the m-step memory (for $m \approx 10$ [1/8 s]). The conditional probability to be extracted from the data under the Markov chain hypothesis was the probability that a gust would follow, and hence it was a slight modification of the above probability $p(s'|\boldsymbol{s})$. As a result, at every instance one can predict the probability that the wind speed in the near future will exceed the current wind speed by more than g m/s. Verification of the predicted probabilities using a reliability analysis was successful. In order to test the success of this scheme for actual gust prediction, we introduced a threshold p_c for the predicted gust probability: if $p(s'|\boldsymbol{s}(t)) > p_c$ at time t then a gust warning was issued, whereas for $p(s'|\boldsymbol{s}(t)) < p_c$ no warning was given. Such a scheme should have an optimal hit rate, giving warnings when a gust is following, and it should have a low false alarm rate. Evidently, the lower the value of p_c, the more warnings are issued, and both the hit rate and the false alarm rate will be large, whereas

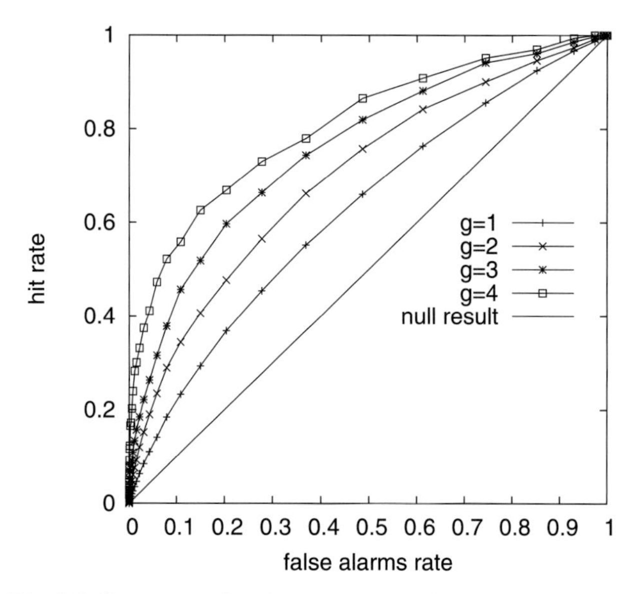

Fig. 4.7. The ROC statistics for the prediction of sudden wind gusts in surface wind. The *plot* shows hit rate versus false alarm rate under the variation of the threshold p_c for issuing a warning. Evidently, strong gusts (large g) have a higher hit rate than weak gusts

for high p_c both rates will be low. If warnings are issued randomly without any correlation to the real state of the system, hit rate and false alarm rate will be identical (both depending on p_c). In order to verify that the scheme possesses predictive power at all, one therefore studies the receiving operator characteristic (ROC) statistics [20]. Using p_c as a parameter and tuning it from 0 to 1, one plots the hit rate versus the false alarm rate. The scheme is only useful if the hit rate is always larger than the false alarm rate. In Fig. 4.7 we show these statistics for the case of wind speed data, where we consider different gust classes. Evidently, and this leads us back to the issue of Xevents, the stronger gusts (large g) possess a better predictability than the weaker ones.

4.4 Conclusions

The most relevant message we want to convey with this article is that systems which have to be assumed to contain stochastic components also allow for some degree of predictability. Although one can still make specific predictions about the future value of the observable in the sense of the most probable outcome or the outcome that yields the least average prediction error, it is much more reasonable to extract probabilistic information. This is information of the kind "an event of magnitude larger than some threshold is expected to occur with a given probability". We have mentioned that such a prediction can be verified using the reliability test, and that its usefulness can be tested using ROC statistics. Finally, in order to convert it into an action, the threshold for issuing a warning has to be chosen. This is done on the basis of the ROC curve together with a cost function: if one knows the costs for a false alarm and the costs of an unpredicted Xevent, then one can compute the costs corresponding to all points in the ROC curves. The minimum of this cost function then defines the optimal threshold p_c for warning. Of course, this implies a pragmatic point of view, which in some cases might appear inhuman: not to generate a warning if a big earthquake is predicted with some low probability might seem cynical, but when computing the real costs of an erroneous evacuation of a city with a million inhabitants it would soon become apparent that false alarms should be minimised.

We want to come back now to Xevents and discuss how their prediction differs from other prediction tasks. Dynamical predictability of a given phenomenon means that there are temporal correlations in the time series data which in principle allow us to make predictions with a given accuracy. In other words, the average prediction error is limited from below by the dynamics. An Xevent is usually of very large magnitude. Hence, a prediction error of a given size, which might be so big that it does not help with the prediction of "normal" events, might still be small compared to the magnitude of an Xevent and hence might not be prohibitive for its prediction. Let us illustrate this for the case of the wind speed data. If we want to predict whether the wind speed will increase or decrease (for example, in order to re-adjust the pitch angle of the rotor blades of a wind turbine), then for normal changes in wind speed (compare Fig. 4.3: 90% of all changes are by less than 0.5 m/s) the prediction error is too big to give any indication of the sign of the change. The predicted value is, even including the uncertainty due to the prediction error, significantly different from zero only when a really big change is to be expected. Of course, this reasoning requires that prediction errors are of about the same absolute size for Xevents as for normal events. Although there are indications that this is the case for wind speed data, this is not guaranteed, since, in particular, the KS-entropy refers to the *mean* prediction error averaged over all events according to their relative frequency. Since Xevents were assumed to be rare, a much larger than average prediction error in extreme cases would not affect the KS-entropy very much.

The dynamical origin of Xevents, which is clearly observed in the wind speed data (for instance, as the precursors shown in Fig. 4.5), leads also to a related issue. Xevents are usually characterised by extreme values of a given observable. The fact that the observable assumes an extreme value corresponds to a state vector of the system that lies inside a certain region in phase space. If we study the same system using a different observable, these "new" Xevents will correspond to a different set of state vectors. An example for weather is extreme temperature compared to extreme precipitation events. One does not imply the other, but both are a consequence of the dynamics of the atmosphere. Now the crucial question is whether extreme values of the observable and the corresponding dynamical states are in some abstract sense extreme states. For those phenomena where this is the case this might imply enhanced predictability of Xevents. Since the Xevent is usually a consequence of a great physical impact, we expect to identify an extreme state when the observable used in the time series is physically relevant and free from human-made thresholds. We thus see that there are both statistical (smaller relative error) and dynamical reasons to expect, at least in some systems, a higher predictability for Xevents when compared to typical events. Indeed we found an enhanced predictability for the more extreme turbulent gusts than for less extreme ones (see Fig. 4.7). Similar results have also been suggested in at least two more papers [21, 22]. Right now it is not clear to us how typical this is, and if it is typical, whether this is a dynamical or a statistical effect. Further work is needed for clarification.

In our discussion of dynamical structures we focused on nonlinear stochastic processes. Of course, if a phenomenon were perfectly deterministic, much better predictions would be possible. However, it is unrealistic to expect pure determinism in too many natural phenomena. In the stochastic setting, predictions are probabilistic themselves. As we have shown, probabilistic predictions are also of considerable benefit, since the probability of an Xevent occurring imminently can fluctuate considerably over the course of time, depending on the actual state of the system. The dynamical part of predictability can also be explored by searching for precursors or specific signatures before Xevents; for example, by searching for log-periodic structures [23]. The main advantage of the Markov chain model presented here is that it is a nonparametric method (in the sense that the parameters have no physical meaning) that can be applied to virtually any correlated time series, so we are not assuming any previous knowledge about the dynamics of the system or the existence of characteristic structures in the time series.

One might ask why predictability beyond the mere statistical probability of occurrence (static part) should be present in a real world system. In fact, we are aware of certain deterministic feedback loops in almost all systems (for example, in the stock market). The presence of these deterministic components enforces correlations in time even in the presence of strong stochastic components that drive the dynamics. From this point of view, non-zero dynamical predictability is the rule rather than the exception.

Acknowledgement. Figure 4.7 is a result of joint work with M. Ragwitz and N. Vitanov. E.G.A. acknowledges financial support from CAPES (Brazil) and DAAD (Germany).

References

1. D. Sornette, this book
2. J.P. Crutchfield, D.P. Feldman, *Regularities unseen, randomness observed: Levels of entropy convergence*, Chaos **13**, 25 (2003)
3. H.G. Schuster, *Deterministic chaos*, VCH Wiley (1997)
4. E.J. Gumbel, *Statistics of extremes*, Columbia University Press, New York (1958)
5. S. Coles, *An introduction to statistical modeling of extreme values*, Springer, Berlin Heidelberg New York (2004)
6. V.S. L'vov, A. Pomyalov, I. Procaccia, *Outliers, extreme events, and multiscaling*, Phys. Rev. E **63**, 056118 (2001)
7. A. Bunde, J.F. Eichner, S. Havlin, J.W. Kantelhardt, *The effect of long-term correlations on the return periods of rare events*, Physica A **330**, 1 (2003)
8. P. Gaspard, X.-J. Wang, Phys. Rep. **235**, 291 (1993)
9. H. Kantz, T. Schreiber, *Nonlinear time series analysis*, 2nd edn, Cambridge University Press, Cambridge, UK (2004)
10. F. Takens, *Detecting strange attractors in turbulence*, Lecture Notes in Math. 898, Springer, Berlin Heidelberg New York (1981)
11. T. Sauer, J. Yorke, M. Casdagli, *Embedology*, J. Stat. Phys. **65**, 579 (1991)
12. E.N. Lorenz, *Atmospheric predictability as revealed by naturally occurring analogues* J. Atmosph. Sci. **26**, 636 (1969)
13. J.D. Farmer, J.J. Sidorowich, *Predicting chaotic time series*, Phys. Rev. Lett. **59**, 845 (1987)
14. M. Ragwitz, H. Kantz, *Markov models from data by simple nonlinear time series predictors in delay embedding spaces*, Phys. Rev. E **65**, 056201 (2002)
15. A.D. Sabin, Z. Sen, *First-order Markov approach to wind speed modelling*, J. Wind Eng. Ind. Aerodynam. **89**, 262 (2001)
16. Main I. et al., *Is the reliable prediction of individual earthquakes a scientific goal?* Nature debates, 25 Feb. 1999 (see http://www.nature.com/nature/debates/earthquake/) (1999)
17. F. Paparella, A. Provenzale, L.A. Smith, C. Taricco, R. Vio, *Local random analogue prediction of nonlinear processes*, Phys. Rev. A **235**, 233 (1997)
18. P. Grassberger, I. Procaccia, *Characterization of strange attractors*, Phys. Rev. Lett. **50**, 346 (1983)
19. H. Kantz, D. Holstein, M. Ragwitz, N.K. Vitanov, *Markov chain model for turbulent wind speed data*, Physica A **342**, 315 (2004)
20. J.A. Hanley, B.J. McNeil, *The meaning and use of the area under the receiver operating characteristic (ROC) curve*, Radiology **143**, 29–36 (1982)
21. D. Sornette, *Predictability of catastrophic events: Material, rupture, earthquakes, turbulence, financial crashes, and human birth*, Proc. Natl. Acad. Sci. USA **99**, 2522 (2002)
22. D. Lamper, S.D. Howison, N.F. Johnson, *Predictability of large fluctuations in a competitive evolving population*, Phys. Rev. Lett. **88**, 017902 (2002)
23. N. Vandewalle, M. Ausloos, Ph. Boveroux, A. Minguet, *How the financial crash of October 1997 could have been predicted*, Eur. Phys. J. B **4**, 139 (1998)

5 Endogenous versus Exogenous Origins of Crises

Didier Sornette

Summary. Are large biological extinctions such as the Cretaceous/Tertiary KT boundary due to a meteorite, extreme volcanic activity or self-organized critical extinction cascades? Are commercial successes due to a progressive reputation cascade or the result of a well orchestrated advertisement? Determining the chain of causality for Xevents in complex systems requires disentangling interwoven exogenous and endogenous contributions with either no clear signature or too many signatures. Here, I review several efforts carried out with collaborators which suggest a general strategy for understanding the organizations of several complex systems under the dual effect of endogenous and exogenous fluctuations. The studied examples are: internet download shocks, book sale shocks, social shocks, financial volatility shocks, and financial crashes. Simple models are offered to quantitatively relate the endogenous organization to the exogenous response of the system. Suggestions for applications of these ideas to many other systems are offered.

5.1 Introduction

Xevents are pervasive in all natural and social systems: earthquakes, volcanic eruptions, hurricanes and tornadoes, landslides and avalanches, lightning strikes, magnetic storms, catastrophic events of environmental degradation, failure of engineering structures, crashes in the financial stock markets, social unrests leading to large-scale strikes and upheaval and perhaps to revolutions, economic drawdowns on national and global scales, regional and national power blackouts, traffic gridlocks, diseases and epidemics, and so on.

Can we forecast them, manage, mitigate or prevent them? The answer to these questions requires us to investigate their origin(s).

Self-organized criticality, and more generally, complex system theory contends that out-of-equilibrium slowly driven systems with threshold dynamics relax through a hierarchy of avalanches of all sizes. Accordingly, Xevents are seen to be endogenous [1, 2], in contrast with previous prevailing views. In addition, the preparation processes before large avalanches are almost undistinguishable from those before small avalanches, making the prediction of the former seemingly impossible (see [3] for a discussion). But how can one assert with 100% confidence that a given Xevent is really due to an endogenous self-organization of the system, rather than to the response to an external shock? Most natural and social systems are indeed continuously subjected to external stimulations, noises, shocks, solicitations, forcing, all of which can widely vary in amplitude. It is thus not clear a priori whether a given large event is due to a strong exogenous shock, to the internal dynamics of the system organizing in response to the continuous flow of small solicitations, or maybe to a combination of both. Addressing this question is fundamental to gaining an understanding of the relative importance of self-organization versus external forcing in complex systems and for the understanding and prediction of crises.

This leads to two questions:

1. Are there distinguishing properties that characterize endogenous versus exogenous shocks?
2. What are the relationships between endogenous and exogenous shocks?

Actually, the second question has a long tradition in physics. It is at the basis of the interrogations that scientists perform on the enormously varied systems they study. The idea is simple: subject the system to a perturbation, a "kick" of some sort, and measure its response as a function of time, of the nature of the solicitations and of the various environmental factors that can be controlled. In physical systems at thermodynamic equilibrium, the answer is known as the theorem of fluctuation-dissipation, sometimes also referred to as the theorem of fluctuation-susceptibility [4]. In a nutshell, this theorem relates quantitatively, in a very precise way, the response of the system to an instantaneous kick (exogeneous) to the correlation function of its spontaneous fluctuations (endogenous). An early example of this relationship is found in Einstein's relation between the diffusion coefficient D of a particle in a fluid subjected to the chaotic collisions of the fluid molecules and the coefficient η of viscosity of the fluid [5,6]. The coefficient η controls the drag; the response of the particle velocity when subjected to an exogenous force impulse. The coefficient D can be shown to be a direct measure of the (integral of the) correlation function of the spontaneous (endogenous) fluctuations of the particle velocity.

In out-of-equilibrium systems, the existence of a relationship between the response function to external kicks and spontaneous internal fluctuations has not been settled [7]. In many complex systems, this question amounts to distinguishing between endogeneity and exogeneity and is important for understanding the relative effects of self-organization versus external impacts. This is difficult in most physical systems because externally imposed perturbations may lie outside the complex attractor, which itself may exhibit bifurcations. Therefore, observable perturbations are often misclassified.

It is thus interesting to study other systems in which the dividing line between endogenous and exogenous shocks may be clearer in the hope that it will lead to insights into complex physical systems. The investigations of the two questions above may also bring a new understanding of these systems.

The systems to which the endogenous-exogenous question (which we will refer to as "endo-exo" for short) is relevant include the following:

- Biological extinctions, such as the Cretaceous/Tertiary KT boundary (meteorite versus extreme volcanic activity (Deccan traps) versus self-organized critical extinction cascades)
- Immune system deficiencies (external viral/bacterial infections versus internal cascades of regulatory breakdowns)
- Cognition and brain learning processes (role of external inputs versus internal self-organization and reinforcements)
- Discoveries (serendipity versus the outcome of slow endogenous maturation processes)
- Commercial successes (progressive reputation cascade versus the result of a well orchestrated advertisement)
- Financial crashes (external shocks versus self-organized instability)
- Intermittent bursts of financial volatility (external shocks versus cumulative effects of news in a long-memory system)
- The aviation industry recession (9/11/2001 terrorist attack versus structural endogenous problems)
- Social unrests (triggering factor or decay of social fabric)
- Recovery after wars (internally generated (civil wars) versus imported from the outside) and so on

It is interesting to mention that the question of exogenous versus endogenous forcing has been hotly debated in economics for decades. A prominent example is the theory of Schumpeter on the importance of technological discontinuities in economic history. Schumpeter argued that "evolution is lopsided, discontinuous, disharmonious by nature ... studded with violent outbursts and catastrophes ... more like a series of explosions than a gentle, though incessant, transformation" [8]. Endogeneity versus exogeneity is also paramount in economic growth theory [9]. Our analyses, reviewed below, suggest a subtle interplay between exogenous and endogenous shocks, which may cast a new light on this debate.

In the following, we review the works of the author with his collaborators, in which the endo-exo question is investigated in a variety of systems.

5.2 Exogenous and Endogenous Shocks in Social Networks

One defining characteristics of humans is their organization in social networks. It is probable that our large brains have been shaped by social interactions, and may have co-evolved with the size and complexity of social groups [10, 11]. A single individual may belong to several intertwined social networks, associated with different activities (work colleagues, college alumni societies, friends, family members, and so on). The formation and the evolution of social networks and their mutual entanglements control the hierarchy of interactions between humans, from the individual level to society and to culture. In this section, we review a few original probes of several social networks which unearth a remarkable universality: the distribution of human decision times in social networks seem to be described by a power law $1/t^{1+\theta}$ with $\theta = 0.3 \pm 0.1$. This constitutes an essential ingredient in models describing how the cascade of agent decisions leads to the bottom-up organization of the response of social systems. We first present such a model in terms of a simple epidemic process of word-of-mouth effects [12–14] and then discuss the different data sets.

5.2.1 A Simple Epidemic Cascade Model of Social Interactions

Let us consider an observable characterizing the activity of humans within a given social network of interactions. This activity can be the rate of visits or downloads on an internet website, the sales of a book or the number of newpaper articles on a given subject.

We envision that the instantaneous activity results from a combination of external forces such as news and advertisement, and from social influences in which each past active individual may prompt other individuals in her network of acquaintances to act. This impact of an active individual on other humans is not instantaneous, as people react on a variety of timescales. The time delays capture the time interval between social encounters, the maturation of the decision process, which can be influenced by mood, sentiments, and many other factors and the availability and capacity to implement the decision. We postulate that this latency can be described by a memory kernel $\phi(t - t_i)$, giving the probability that an action at time t_i leads to another action at a later time t by another person in direct contact with the first active individual. We consider the memory function $\phi(t - t_i)$ as a fundamental macroscopic description of how long it takes for a human to be triggered into action, following the interaction with an already active human.

Then, starting from an initial active individual (the "mother") who first acts (either from exogenous news or by chance), she may trigger actions by first-generation "daughters," which themselves prompt the actions of their own friends, who become second-generation active individuals, and so on. This cascade of generations can be shown to renormalize the memory kernel $\phi(t - t_i)$ into a dressed or renormalized memory kernel $K(t - t_i)$ [12, 13, 15], giving the probability that an action at time t_i leads to another action by another person at a later time t through any possible generation lineage. In physical terminology, the renormalized memory kernel $K(t)$ is nothing but the response function of the system to an impulse. This is captured by the following equations:

$$A(t) = s(t) + \int_{-\infty}^{t} d\tau \; A(\tau) \; \phi(t - \tau) = \int_{-\infty}^{t} d\tau \; s(\tau) \; K(t - \tau) \; . \qquad (5.1)$$

The meaning of these two equivalent formulations is as follows. The $s(t)$'s are the spontaneous exogenous activations. The integral $\int_{-\infty}^{t} d\tau\ A(\tau)\ \phi(t-\tau)$ gives the additional contribution due to past activities $A(\tau)$, whose influences on the present are mediated by the direct influence kernel ϕ of the first generation. The last integral $\int_{-\infty}^{t} d\tau\ s(\tau)\ K(t-\tau)$ expresses the fact that the present activity $A(t)$ can also be seen as resulting from all past exogenous sources $s(\tau)$ mediated to the present by the renormalized kernel K, which takes into account all of the generations of cascades of influences.

The following functional dependence is found to provide an accurate description, as we shall discuss below:

$$K(t) \sim 1/(t-t_c)^p\ , \qquad \text{with } \ p=1-\theta\ . \tag{5.2}$$

The dependence (5.2) implies that ([12,13,15]):

$$\phi(t) \sim 1/(t-t_c)^{1+\theta}\ . \tag{5.3}$$

We should stress that the renormalization from the usually (but not always) unobservable "bare" response function $\phi(t)$ with exponent $1+\theta$ in (5.3) to the observable "renormalized" response function $K(t)$ in (5.2) with exponent $1-\theta$ is obtained if the network is close to critical; in other words if the average branching ratio n is close to 1 (n is defined as the average number of daughters of the first generation per mother). In other words, there is on average approximately one triggered daughter per active mother. This condition of criticality ensures, in the language of branching processes, that avalanches of active people triggered by a given mother are self-similar (power law distributed). In contrast, for $n<1$, the cascade of triggered actions is "sub-critical" and avalanches die off more rapidly. It can be shown [12,13,15] that in this case there is a characteristic timescale

$$t^* \sim \frac{1}{(1-n)^{1/\theta}} \tag{5.4}$$

acting like a correlation time, which separates two regimes:

- for $t<t^*$, the renormalized response function $K(t)$ is indeed of the form (5.2);
- for $t>t^*$, the renormalized response function $K(t)$ crosses over to an asymptotic decay with exponent $1+\theta$, of the form of $\phi(t)$ in (5.3).

For $n>1$, the epidemic process is supercritical and has a finite probability of growing exponentially. We will not be concerned with this last regime, which does not seem relevant in the data discussed below.

In the absence of strong external influences, a peak in social activity can occur spontaneously due to the interplay between a continuous stochastic flow of small external news and the amplifying impact of the epidemic cascade of social influences. It can then be shown that, for n close to 1 or equivalently for $|t - t_c| < t^*$, the average growth of the social activity before such an "endogenous" peak and the relaxation after the peak are proportional to [13, 16]

$$\int_0^{+\infty} K(t - t_c + u)K(u)du \sim 1/|t - t_c|^{1-2\theta} , \quad (5.5)$$

where the right-hand-side of the expression holds for $K(t)$ of the form (5.2). The prediction that the relaxation following an exogeneous shock should happen faster (larger exponent $1 - \theta$) than for an endogeneous shock (with exponent $1 - 2\theta$) agrees with the intuition that an endogeneous shock should have inpregnated the network much more and should thus have a longer lived influence. In a nutshell, the mechanism producing the endogenous response function (5.5) is the constructive interference of accumulated small news cascading through the social influence network. In other words, the presence of a hierarchy of nested relaxations $K(t)$ given by (5.2), each one associated with each small news, creates the effective endogenous response (5.5).

Dodds and Watts have recently introduced a general contagion model which, by explicitly incorporating memories of past exposures to, for example, an infectious agent, a rumour, or a new product, includes the main features of existing contagion models and interpolates between them [17].

5.2.2 Internet Download Shocks

In [18], Johansen and Sornette report the following experiment. The authors were interviewed by a journalist from the leading Danish newspaper Jyllands Posten on a subject of rather broad interest, namely stock market crashes. The interview was published on April 14, 1999 in both the paper version of the newspaper as well as in the electronic version (with access restricted to subscribers) and included the URLs where the authors' research papers on the subject could be retrieved. It was hence possible to monitor the number of downloads of papers as a function of time since the publication date of the interview. The rate of downloads of the authors' papers as a function of time was found to obey a $1/t^p$ power law, with exponent $b = 0.58 \pm 0.03$, as shown in Fig. 5.1.

Within the model of epidemic word-of-mouth effect summarized in Sect. 5.2.1, the relaxation of the rate of downloads after the publication of the interview characterizes the response function $K(t)$ given by (5.2) with respect to an exogenous peak: prior to the publication of the interview, the rate of downloads was slightly less than one per day; it suddenly jumped to several tens of downloads per day in the first few days after the publication and then relaxed slowly according to (5.2). The reported power law with exponent $p \simeq 0.6$ is compatible with the form of (5.2) with $\theta = 0.4$, which is within the range of other values: $\theta = 0.3 \pm 0.1$.

Johansen [19] has reported another similar observation following another web interview on stock market crashes, which contained the URL of his articles on the subject. He again found a power law dependence (5.2), but with

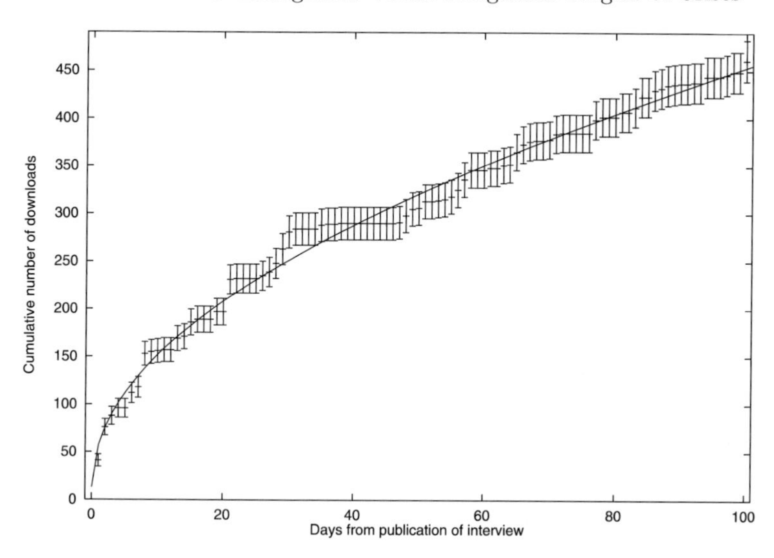

Fig. 5.1. Cumulative number of downloads N as a function of time t from the appearance of the interview on Wednesday 14th April 1999. The *fit* is $N(t) = \frac{a}{1-p}t^{1-p} + ct$ with $b \approx 0.58 \pm 0.03$. Reproduced from [18]

an exponent p close to 1, leading in the terminology of the model of epidemic word-of-mouth effect to $\theta \simeq 0$. Two interpretations are possible: (i) the exponent θ is non-universal; (ii) the social network is not always close to criticality ($n \simeq 1$) and the observable response function $K(t)$ is then expected to cross over smoothly from a power law with exponent $1 - \theta$ to another asymptotic power law with exponent $1 + \theta$. According to this second hypothesis, the exponent p of the relaxation kernel $K(t)$ may be found in the range $1 - \theta$ to $1 + \theta$, depending upon the range of investigated timescales and the proximity of $1 - n$ to criticality. We find hypothesis (ii) more attractive as it places the blame on the non-universal parameter n, which embodies the connectivity structure, static and dynamic, of social interactions at a given moment. It does not seem unrealistic to think that n may not always be at its critical value 1, due to many other possible social influences. In constrast, one could postulate that the power law (5.3) for the direct influence function $\phi(t)$ between two directly linked humans may reflect a more universal character. But, of course, only more empirical investigations will allow us to shed more light on this issue.

Eckmann, Moses and Sergi [20] also report on an original investigation probing the temporal dynamics of social networks using email networks in their universities. They find a distribution of response times for answering a message that seems to be a power law with an exponent of less than 1 for rapid response times (one hour) to another power law with an exponent larger than 1 at slower response times (days), which could be a direct evidence of the direct response function $\phi(t)$ defined in (5.3). The relationship between their investigation and the previous works using web downloads [18, 19] has been noted by Johansen [21].

5.2.3 Book Sale Shocks

Sornette, Deschatres, Gilbert and Ageon have used a database of sales from Amazon.com as a proxy for commercial growth and successes [14]. Figure 5.2 shows about 1.5 years of data for two books, Book A ("Strong Women Stay Young" by Dr. M. Nelson) and Book B ("Heaven and Earth (Three Sisters Island Trilogy)" by N. Roberts), which are illustrative of the two classes found in this study. On 5th June 2002, Book A jumped from a sales rank of over 2,000 to a rank of six in less than 12 hours. On 4th June 2002, the New York Times published an article crediting the "groundbreaking research done by Dr. Miriam Nelson" and advising the female reader, interested in having a youthful postmenopausal body, to buy the book and consult it directly [22]. This case is the archetype of an "exogenous" shock. In contrast, the sales rank of Book B peaked at the end of June 2002 after slow and continuous growth, with no such newspaper article, followed by a similar almost symmetrical decay, the entire process taking about four months. We will show below that the peak for Book B belongs to the class of endogenous shocks. This endogeneous growth is well explained qualitatively in [23] by taking the example of the book "Divine Secrets of the Ya-Ya Sisterhood" by R. Wells, which became a bestseller two years after publication, with no major advertising campaign. After reading this (originally) small budget book, "Women began forming *Ya-Ya* Sisterhood groups of their own [...]. The word about *Ya-Ya* was spreading [...] from reading group to reading group, from *Ya-Ya* group to *Ya-Ya* group" [23]. Generally, the popularity of a book is based on whether the information associated with that book will be able to propagate far enough into the network of potential buyers.

Another dramatic example of exogenous shocks is shown in Fig. 5.3. Here, the personal trainer of Oprah Winfrey had his book presented seven or eight times during the Oprah Winfrey Show, leading to dramatic overnight jumps in sales.

The declines in the sales of about 140 books that reached the top 50 in the Amazon.com ranking system have been analysed and shown to fall into two categories: relaxations described by a power law with an exponent close to $0.7 = 1 - \theta$, and relaxations described by a power law with an exponent close to $0.4 = 1 - 2\theta$, for $\theta \simeq 0.3$. Examples of these fits for the two books shown in Fig. 5.2 are presented in Fig. 5.4. In addition, Sornette et al. [14] checked that an overwhelming majority of those sale peaks classified as exogenous from the value of their exponent $\simeq 0.7 = 1 - \theta$ were preceded by an abrupt jump, in agreement with the epidemic cascade model of social interactions described in Sect. 5.2.1. In contrast, those sale peaks that fell into the endogenous class

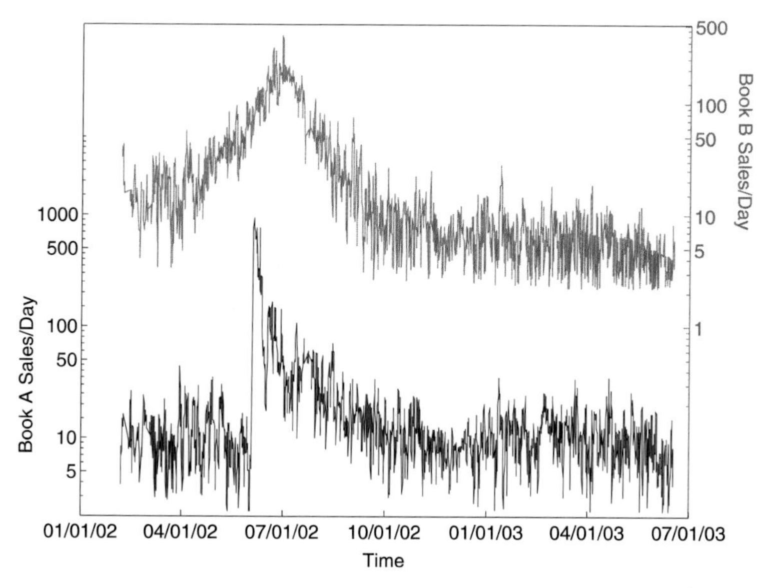

Fig. 5.2. Time evolution over a year and a half of the sales per day of two books: Book A (*bottom, blue, left scale*) is "Strong Women Stay Young" by Dr. M. Nelson and Book B (*top, green, right scale*) is "Heaven and Earth (Three Sisters Island Trilogy)" by N. Roberts. The difference in the patterns is striking, Book A undergoing an exogenous peak on 5th June 2002, and Book B endogenously reaching a maximum on 29th June 2002. Reproduced from [14]

according to the exponent $\simeq 0.4 = 1 - 2\theta$ of their relaxation after the peak were found to be preceded by approximately symmetric growth described by a power law with the same exponent, as predicted by (5.5). An example is shown also for Book B in Fig. 5.4.

The small values of the exponents (close to $1 - \theta$ and $1 - 2\theta$) for both exogenous and endogenous relaxations imply that the sales dynamics are dominated by cascades involving higher-order generations rather than by interactions that stop after first-generation buy triggering. Indeed, if buys were initiated mostly due to news or advertisements, and not much by triggering cascades in the acquaintance network, the cascade model predicts that we should then measure an exponent $1 + \theta$ given by the "bare" memory kernel $\phi(t)$, as already said. This implies that the average number n (the average branching ratio in the language of branching models) of prompted buyers per initial buyer in the social epidemic model is on average very close to the critical value 1, because the renormalization from $\phi(t)$ to $K(t)$ given by (5.2) only operates close to criticality, as characterized by the occurrence of large cascades of buys. Reciprocally, a value of the exponent p that is larger than 1 suggests that the associated social network is far from critical. Such instances can actually be observed. Examples of crossovers from the renormalized re-

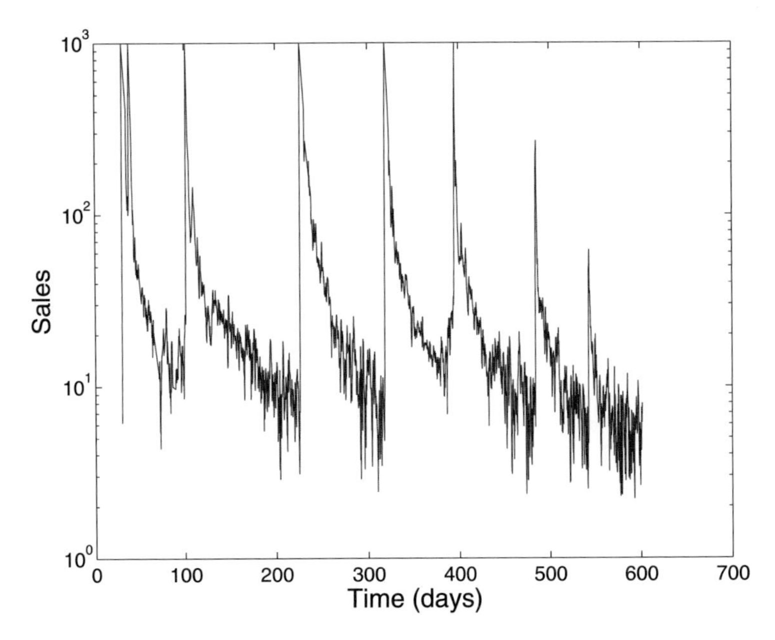

Fig. 5.3. Time evolution of the book entitled "Get with the Program." Each time the book appeared on the Oprah Winfrey Show (B. Greene was Oprah Winfrey's personal trainer), the sales jumped overnight

sponse function $K(t)$ (5.2) to $\phi(t)$ in (5.3) with an asymptotic decay with exponent $1+\theta$ have been documented ([14], Deschatres, F. and D. Sornette, in preparation). Note that it is possible to give an analytical description of this crossover exhibited by $K(t)$ as a function of n [12], thus allowing us, in principle, to invert for n for a given data set. This opens up the tantalizing possibility of measuring the dynamical connectivity of the social network, and possibly of monitoring it as a function of time.

These findings open up other interesting avenues of research. While this first investigation has emphasised the distinction between exogenous and endogenous peaks, setting the fundamentals for a general study, repeating peaks as well as peaks that may not be pure members of a single class are also frequent. In a sense, there are no real "endogenous" peaks, one could argue, because there is always a source or a string of news impacting upon the network of buyers. What Sornette et al. [14] have done is to distinguish between two extremes, the very large news impact and the structureless flow of small news amplified by the cascade effect within the network. One can imagine and actually observe a continuum between these two extremes, with feedbacks between the development of endogeneous peaks and the increased interest of the media as a consequence, feeding back and providing a kind of exogenous boost, and so on. In those and in more complicated cases, the

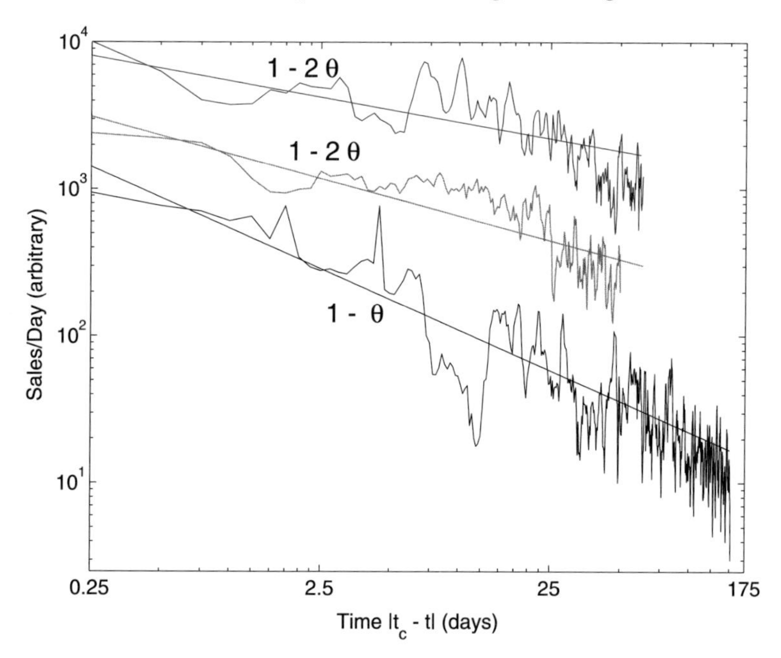

Fig. 5.4. The *bottom curve* (*blue*) shows the relaxation in the sales of Book A after the sales peak at t_c = 5th June 2002 as a function of the time $t - t_c$ from the time of the peak. A least squares best fit with a power law gives a slope of ≈ -0.7. Since this peak is identified as exogenous with a theoretical slope of $1 - \theta$, we obtain the estimate $\theta = 0.3 \pm 0.1$. *The curve in the middle* (*green*, shifted up by a factor 6 compared to the bottom curve) shows the relaxation in sales of Book B after the peak at t_c = 29th June 2002 as a function of the time $t - t_c$ from the time of the peak. The least squares fit gives a slope of ≈ -0.4, which provides the independent estimate $\theta = 0.3 \pm 0.1$ from the theoretical endogenous exponent $1 - 2\theta$. The *top curve* (*red*, shifted up by a factor 25 with respect to the bottom curve) shows the acceleration in sales of Book B leading to the same peak at t_c = 29th June 2002 as a function of the time $t_c - t$ to the time of the peak. The time on the x-axis has been reversed to compare the precursory acceleration with the aftershock relaxation. The least squares slope is ≈ -0.3, not far from the predicted $1-2\theta$ of the cascade model, with $\theta = 0.3 \pm 0.1$

epidemic model of word-of-mouth effects should provide a starting platform for predicting the sales dynamics as a function of an arbitrary set of external sources. By dynamically tracking the connectivity $n(t)$ of each social network relevant to a given product, it should also be possible to target the most favourable times, corresponding to the largest $n(t)$, for promoting or sustaining the sales of a given product, with obvious consequences for marketing and advertisement strategies. An additional extension includes the possible feedback of the marketing strategy into the control parameter $n(t)$, which could be manipulated so as to keep the system critical, an ideal situation from the point of view of marketers and firms. Quantifying this effect requires us to extend the simple epidemic model in the spirit of mechanisms leading to self-organized criticality by positive feedbacks of the order parameter onto the control parameter [24, 25]. The results of Sornette et al suggest that social networks have evolved to converge very close to criticality. As Andreas S. Weigend, chief scientist of Amazon.com (2002–2004) wrote on his webpage: "Amazon.com might be the world's largest laboratory to study human behaviour and decision making." I share this viewpoint.

Actually, I envision that an extension of the study of Sornette et al to a broad database of sales from all products sold by e-retailers like Amazon.com could give access to the equivalent of the "social climate" of a country like the USA and its evolution as a function of time under the various exogenous and endogenous factors at work. Indeed, Amazon.com categorizes its products into different (tradable) compartments of possible interest, such as

- Books, Music, DVD,
- Electronics (audio and video, cameras and photography, software, computers and video games, cell phones...)
- Office
- Children and Babies
- Home and Garden (which includes pets)
- Gifts, Registries, Jewellery and Watches
- Apparel and Accessories
- Food
- Health, Personal Care, Beauty
- Sports and Outdoors
- Services (movies, restaurants, travel, cars, ...)
- Arts and Hobbies
- Friends and Favourites

with many subcategories. Monitoring and analysing the sales as a function of time in these different categories is like getting the temperature, wind velocity, humidity in meteorology in many different locations. The flow of interest of society at large and of subgroups could in principle tell us how society is responding in its spending habits to large scale influences. As an illustrative example, it has been shown that, during bullish periods characterized by strong stock market gains (bubble regimes), the number of books written and sold related to financial investments soar [26, 27].

Another potentially fruitful application is the music industry and the impact upon sales of internet piracy, the quality of performers (endogenous effect on the network of potential buyers who can promote a CD by word-of-mouth in the network of potential buyers), as well as the promotion campaigns of short-lived performers and their one-hit wonders [28]. Indeed, according to an internal study performed by one of the big companies that dominate the production and distribution of music, the drop in sales in America may have less to do with internet piracy than with other factors, among them the decreasing quality of music itself. The days of watching a band develop slowly over time with live performances are over, according to some professionals. Even Wall Street analysts are questioning quality. If CD sales have shrunk, one reason could be that people are less excited by the industry's product. A poll by Rolling Stone magazine found that fans believe that relatively few "great" albums have been produced in recent years [28]. This is clearly an endo-exo question that can be analysed with databases available on the Internet.

5.2.4 Social Shocks

Roehner, Sornette and Andersen [29] have used the concept of exogeneous shocks to propose a general method for quantifying the response function in order to advance the social sciences. By using a database of newspaper articles called Lexis-Nexis, which is available in many departments of political science or sociology, they have quantified the response to shocks, such as the following:

- On 31st October 1984, the Prime Minister of India, Indira Gandhi, was assassinated by two of her Sikh bodyguards. This event triggered a wave of retaliations against Sikh people and Sikh property, not only in India (particularly in New Delhi), but in many other countries as well.
- In the early hours of 6th December 1992, thousands of Hindus converged on the holy city of Ayodhya in northern India and began to destroy the Babri mosque which was said to be built on the birthplace of Lord Rama. The old brick walls came down fairly easily and soon the three domes of the mosque crashed to the ground. This event triggered a wave of protestations and retaliations which swept the whole world from Bangladesh to Pakistan, to England and the Netherlands. In all of these countries, Hindu people were assaulted and Hindu temples were firebombed, damaged or destroyed.
- On 11th September 2001, two planes crashed into the twin towers of the World Trade Center in New York. This event triggered a wave of reactions against Islamic people and property, not just in the United States.

For these different events, Roehner et al. [29] show that different quantitative measures of social responses exhibit an approximately universal behaviour, again characterised by a power law, as shown in Fig. 5.5. This figure gives the time evolution after 11th September 2001 of newspaper articles, anti-Arab incidents and the Dow Jones Industrial Average, which are approximated by a power law $\sim 1/t^p$. Due to the coarseness of the measures, the exponent p is not well-constrained: $p = -1.8 \pm 0.7$ (newspaper articles), $p = -1.4 \pm 0.5$ (anti-Arab incidents) and $p = -2.2 \pm 1.6$ (DJI). Comparing the reaction to 11th September 2001 in different countries such as Canada, Great Britain and the Netherlands, Roehner et al. [29] have suggested that

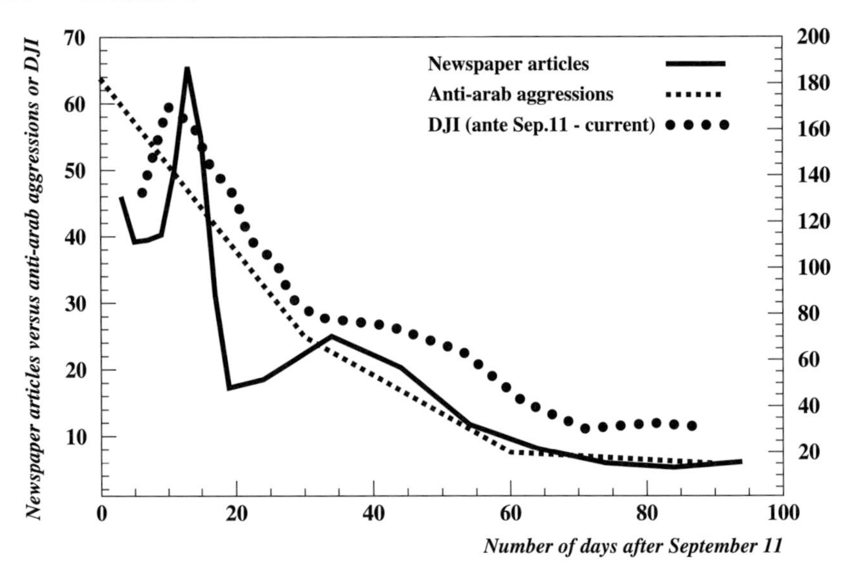

Fig. 5.5. Relaxation of three different social variables after the events of 11th September 2001. The *solid line curve* is the number of articles reporting on the destruction of mosques after the event; the *broken line* (scale on the right-hand side) shows the number of anti-Arab incidents in California in the three months after 11th September; the *dotted line* shows the changes in the level of the Dow Jones Index with respect to its pre-9/11 level, as given by the difference DJI(pre-9/11)−DJI(current). Source: California's Attorney General Office; published in the San Jose Mercury News, 11th March 2002. Reproduced from [29]

the response function actually expresses information on "cracks" that pre-existed in the social networks of the corresponding countries. For instance, the number of attacks on Mosques was larger in the Netherlands, which is in line with other information on the concern expressed at high political levels (private communication to the authors) about the integrity of the social fabric of the Netherlands, a fact illustrated more recently on the political scene by the rapid rise and then assassination of the rightist politician Fortuyn in May 2002. This line of evidence can be quantified within the epidemic model of social influence by different values of the connectivity parameter n in different countries.

Burch, Emery and Fuerst [30] have used also the unique opportunity offered by the 9/11 terrorist attack to clearly confirm the hypothesis that closed-end mutual fund discounts from fund net asset values reflect small investor sentiment. Carter and Simkins [31] investigated the reaction of airline stock prices to the 9/11 terrorist attack and found that the market was concerned about the increased likelihood of bankruptcy in the wake of the attacks and distinguished between airlines based on their ability to cover short-term obligations (liquidity).

5.3 Exogenous and Endogenous Shocks in Financial Markets

5.3.1 Volatility Shocks

Standard economic theory maintains that the complex trajectory of stock market prices is the faithful reflection of the continuous flow of news that is interpreted and digested by an army of analysts and traders. Accordingly, large shocks should result from really bad surprises. It is a fact that exogenous shocks exist, as epitomized by the recent events of 11th September 2001, and there is no doubt about the existence of utterly exogenous bad news that moves stock market prices and creates strong bursts of volatility. One case that cannot be refuted is the market turmoil observed in Japan following the Kobe earthquake of 17th January 1995, the estimated cost of which was around $200 billion dollars. Indeed, so longinfancy, destructive earthquakes cannot be not endogenized in advance in stock market prices by rational agents ignorant of seismological processes. One may also argue that the invasion of Kuwait by Iraq on 2nd August 1990 and the coup against Gorbachev on 19th August 1991 were strong exogenous shocks. However, some could also argue that precursory fingerprints of these events were known to some insiders, suggesting the possibility that the action of these informed agents may have been reflected in part in stock markets prices. Even more difficult is the classification (endogenous versus exogenous) of the hierarchy of volatility bursts that continuously shake stock markets. While it is a common practice to associate the large market movements and strong bursts of volatility with external economic, political or natural events [32], there is no convincing evidence to support this.

Perhaps the most robust observation in financial stock markets is that volatility is serially correlated with long-term dependence (approximately power law-like). Volatility autocorrelation is typically modelled using autoregressive conditional heteroskedasticity (ARCH) [33], generalized ARCH [34], stochastic volatility [35], Markov switching [36, 37], nonparametric [38] and extensions of these models (see [39] for comparisons). Recent powerful extensions include the Multifractal Random Walk model (MRW) introduced by Muzy, Bacri and Delour [40, 41], which belongs to the class of stochastic volatility models. Using the MRW, Sornette, Malevergne and Muzy [42] have shown that it is possible to distinguish between an endogenous and an exogenous originated volatility shock. Tests on the October 1987 crash on a hierarchy of volatility shocks and on a few of the obvious exogenous shocks have validated the concept. This study shows that the relaxation with time of a burst of volatility is distinctly different after a strong exogenous shock compared with the relaxation of volatility after a peak with no identifiable exogenous sources. This study does not explain the origin of volatility correlation. But it identifies the "natural" response function of the system to an external shock, from which the stationary long-term dependence structure of the volatility and its intermittent bursts derive automatically. In other words, the study of Sornette et al. leads to the view that the properties of the volatility can be largely understood from a single characteristic, which is the response of the agents to a new piece of news. This response function must ultimately be derived from the behaviour of financial agents, for instance taking into account their sensitivity to changes in wealth, their loss aversion as well as their finite-time memory of past losses that may impact their future decisions [44].

The multifractal random walk is an autoregressive process with a long-range memory decaying as $t^{-1/2}$, which is defined using the logarithm of the volatility. Using the MRW model for the dependence structure of the volatility, Sornette et al. predict that exogenous volatility shocks will be followed by a universal relaxation

$$\simeq \lambda/t^{1/2} , \tag{5.6}$$

where λ is the multifractal parameter, while endogenous volatility shocks relax according to a power law

$$\simeq 1/t^{p(V_0)} , \qquad \text{with } p(V_0) \simeq \lambda^2 \ \ln(V_0) , \tag{5.7}$$

with an exponent $p(V_0)$ which is a linear function of the logarithm $\ln(V_0)$ of the shock of volatility V_0. The difference between these behaviours and those reported above modelled by the epidemic process with long-term memory stems from the fact that the stock market returns $r_{\Delta t}(t)$ at timescale Δt at a given time t can be accurately described by the following process [40,41]:

$$r_{\Delta t}(t) = \epsilon(t) \cdot \sigma_{\Delta t}(t) = \epsilon(t) \cdot e^{\omega_{\Delta t}(t)} , \tag{5.8}$$

where $\epsilon(t)$ is a standardized Gaussian white noise independent of $\omega_{\Delta t}(t)$, and $\omega_{\Delta t}(t)$ is a near-Gaussian process with mean and covariance

$$\mu_{\Delta t} = \frac{1}{2} \ln(\sigma^2 \Delta t) - C_{\Delta} t(0) \tag{5.9}$$

$$C_{\Delta} t(\tau) = Cov[\omega_{\Delta t}(t), \omega_{\Delta t}(t+\tau)] = \lambda^2 \ln \left(\frac{T}{|\tau| + e^{-3/2} \Delta t} \right) . \tag{5.10}$$

where $\sigma^2 \Delta t$ is the return variance at scale Δt and T represents an "integral" (correlation) timescale. λ is called the multifractal parameter: when it vanishes, the MRW reduces to a standard Wiener process (standard continuous random walk). Such a logarithmic decay of the log-volatility covariance at different timescales has been shown empirically in [40,41]. Typical values for T and λ^2 are respectively one year and 0.04.

The MRW model can be expressed in a more familiar form, in which the log-volatility $\omega_{\Delta} t(t)$ obeys an auto-regressive equation whose solution reads

$$\omega_{\Delta} t(t) = \mu_{\Delta} t + \int_{-\infty}^{t} d\tau \ \eta(\tau) \ K_{\Delta} t(t-\tau) , \tag{5.11}$$

where $\eta(t)$ denotes a standardized Gaussian white noise and the memory kernel $K_{\Delta}t(\cdot)$ is a causal function, ensuring that the system is not anticipative. The process $\eta(t)$ can be seen as the information flow. Thus $\omega(t)$ represents the response of the market to incoming information up to the date t. At time t, the distribution of $\omega_{\Delta}t(t)$ is Gaussian with mean $\mu_{\Delta}t$ and variance $V_{\Delta}t = \int_0^{\infty} d\tau\ K_{\Delta}^2t(\tau) = \lambda^2 \ln\left(\frac{Te^{3/2}}{\Delta t}\right)$. Its covariance, which entirely specifies the random process, is given by

$$C_{\Delta}t(\tau) = \int_0^{\infty} dt\ K_{\Delta}t(t)K_{\Delta}t(t+|\tau|)\ . \tag{5.12}$$

Performing a Fourier transform, we obtain

$$\hat{K}_{\Delta}t(f)^2 = \hat{C}_{\Delta}t(f) = 2\lambda^2\ f^{-1}\left[\int_0^{Tf} \frac{\sin(t)}{t} dt + O\left(f\Delta t \ln(f\Delta t)\right)\right]\ , \tag{5.13}$$

which shows, using (5.10), that for a small enough τ,

$$K_{\Delta}t(\tau) \sim K_0\sqrt{\frac{\lambda^2 T}{\tau}} \qquad \text{for} \quad \Delta t \ll \tau \ll T\ , \tag{5.14}$$

which is the previously stated exogenous response function (5.6). The slow power law decay (5.14) of the memory kernel in (5.11) ensures the long-range dependence and multifractality of the stochastic volatility process (5.8).

The main difference between the MRW model and the previous class of epidemic process is that the long-term memory appears in the logarithm of the variable in the former, as shown from (5.11). As a consequence, the MRW basically describes a variable which is the exponential of a long-memory process. It is the interplay between this strongly nonlinear exponentiation and the long-memory which gives multifractal properties to the MRW and, as a consequence, the shock amplitude dependence of the exponents $p(r)$ of the relaxation of the volatility following endogenous shocks. In contrast, the linear long-term memory structure (5.1) of the epidemic processes of Sect. 5.2.1 ensures universal exponents that are independent of the shock amplitudes (but not of the endo-exo nature). In the epidemic process (5.1), the relationship between exogenous and endogenous relaxations is expressed by the exponents of the power laws $\sim 1/t^{1-\theta}$ (exo) versus $\sim 1/t^{1-2\theta}$ (endo). In the MRW, notice that the relationship between exogenous (5.6) and endogenous relaxations (5.7) is through the multifractal parameter λ: the fact that an amplitude of the exogenous response function impacts the power law exponent of the endogenous relaxation is again a signature of the exponential structure of the multifractal model. The MRW extends the realm of possible relationships between endogenous and exogenous responses discussed until now.

5.3.2 Financial Crashes

The endo-exo question also appears to be crucial for understanding financial crashes. In contrast with the previous examples, the strongest distinction is not in the relaxation or recovery after the shock but rather in the precursory behaviour before the crash. An endogenous crash might be expected to end a period of strong price gains, due to speculative herding for instance. In contrast, an exogenous crash would be the response of the financial system to a very strong adverse piece of information.

Indeed, according to standard economic theory, the complex trajectory of stock market prices is the faithful reflection of the continuous flow of news that are interpreted and digested by an army of analysts and traders [45]. Accordingly, large market losses should result only from really bad surprises. It is indeed a fact that exogenous shocks exist, as epitomized by the recent events of 11th September 2001 and the coup in the Soviet Union on 19th August 1991, which move stock market prices and create strong bursts of volatility [42], as discussed above. However, is this always the case? A key question is whether large losses and gains are indeed slaved to exogenous shocks, or whether they may result from endogenous origins in the dynamics of that particular stock market. The former possibility requires the risk manager to closely monitor the world of economics, business, political, social, environmental news for possible instabilities. This approach is associated with standard "fundamental" analysis. The latter endogenous scenario requires an investigation of the signs of instabilities to be found in the market dynamics itself, and it could, in part, rationalize so-called "technical" analysis (see [43] and references therein).

Johansen and Sornette [47] have carried out a systematic investigation of crashes to clarify this question. They have proceeded in several steps:

1. They have developed a methodology to identify crashes as objectively and unambiguously as possible. Specifically, they have studied the distributions of drawdowns (runs of losses) in several markets: the two leading exchange markets (US dollar against the Deutsch and against the Yen), the major world stock markets, the U.S. and Japanese bond market and the gold market. By introducing and varying a certain degree of fuzziness in the definition of drawdowns, they have tested the robustness of the empirical distributions of drawdowns.
2. By carefully analysing these distributions, they have shown that the extreme tail belongs to a different population than the bulk (typically the top 1% (most extreme) drawdowns occur 10–100 times more often than would be predicted by an extrapolation of the distribution of the other 99% of the drawdowns).
3. The Xevents which seem to belong to a different population have been called "outliers" [46,48–50]. Others have referred to such events as "kings" or "black swans." Johansen and Sornette [47] have taken these kings to be

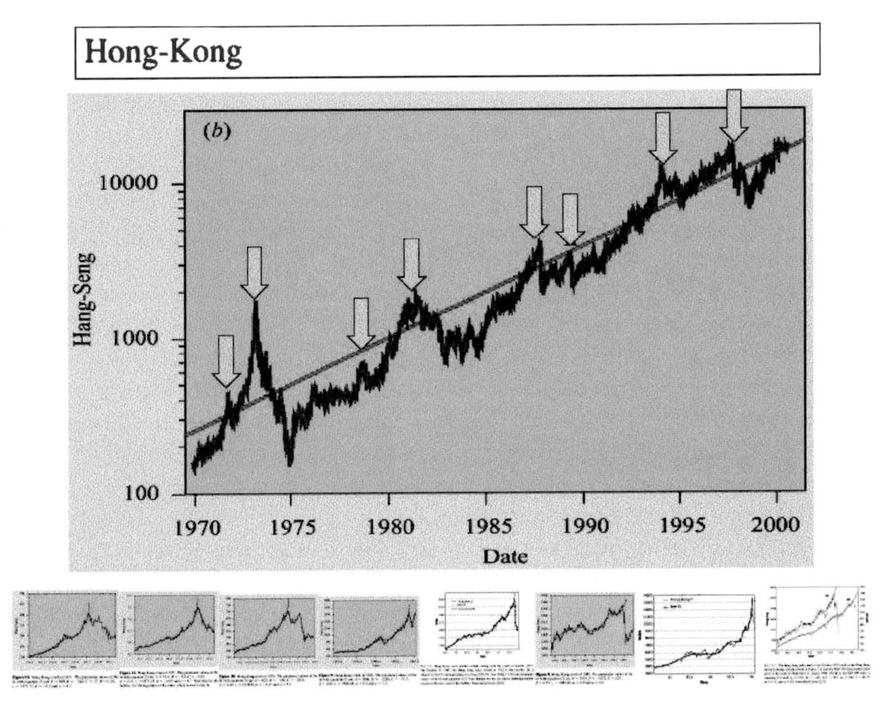

Fig. 5.6. The Hang-Seng composite index of the Hong Kong stock market from November 1969 to September 1999. Note the logarithmic scale on the *vertical axis*. The peaks of the bubbles followed by strong crashes are indicated by the *arrows* and correspond to the times Oct. 1971, Feb. 1973, Sept. 1978, Oct. 1980, Oct. 1987, April 1989, Jan. 1994 and Oct. 1997. This figure shows that the Hang-Sing index has grown exponentially on average at the rate of $\approx 13.6\%$ per year, represented by the *straight line* corresponding to the best exponential fit to the data. Eight large bubbles (five of which are very large) can be observed as upward accelerating deviations from the average exponential growth, and are characterized by LPPL signatures ending in a crash, here defined as a drop of more than 15% in less than two weeks. The eight small panels at the *bottom* are given to show the LPPL price trajectory over a period of six months preceding each of these eight crashes. Constructed from [46] and other papers from the author

the crashes that need to be explained. Note that this procedure ensures that the definition of a crash is relative to the specific market rather than obeying such an arbitrary absolute rule.

4. Then, for each identified king, Johansen and Sornette [47] checked whether a specific market structure, called log-periodic power law (LPPL), is present in the price trajectory *preceding* the occurrence of the draw-down king. The rational for this approach was based on their previous works [46, 51–53], in which they documented the existence of such log-periodic power law signatures associated with speculative bubbles before

crashes. The work [47] is in this respect an out-of-sample test of the LPPL bubble-crash hypothesis applied to a population of financial time series selected according to a criterion (outlier test in the distribution of drawdowns) which is unrelated to the LPPL structure itself.

5. In this test, Johansen and Sornette [47] take the existence of a LPPL as the qualifying signature for an endogenous crash: a drawdown outlier is seen as the end of a speculative unsustainable accelerating bubble generated endogenously.
6. With these criteria fixed, Johansen and Sornette [47] identify two classes of crashes. Those that are not preceded by a LPPL price trajectory are classified as exogenous. For those, it was possible to identify what seems to have been the relevant historical event (a new piece of information of such magnitude and impact that it is reasonable to attribute the crash to it, following the standard view of the efficient market hypothesis). Such drawdown outliers are classified as having an exogenous origin.
7. The second class, characterized by LPPL price trajectories, is called endogenous. Figure 5.6 illustrates a series of endogenous crashes preceded by LPPL bubble trajectories on the Heng-Seng composite index of the Hong-Kong stock market, perhaps one of the most speculative markets in the world. All of the events shown belong to the endogenous class.
8. Globally over all of the markets analysed, Johansen and Sornette [47] identified 49 outliers, of which 25 were classified as endogenous, 22 as exogenous and two as associated with the Japanese "anti-bubble" that started in January 1990. Restricting to the world market indices, they found 31 outliers, of which 19 are endogenous, ten are exogenous and two are associated with the Japanese anti-bubble.

The combination of the two proposed detection techniques, one for drawdown outliers and the second for LPPL signatures, provides a novel and systematic taxonomy of crashes, further substantiating the importance of LPPL (see also [54–58] for reviews and extensions).

A more microscopic approach, formulated in terms of agent-based models has also allowed some mechanisms to be identified with the occurrence of Xevents, such as excess bias on nodes in the de Bruijn diagram of active agent strategies [59], or the decoupling of strategies which become transiently independent from the recent past [60].

5.4 Concluding Remarks

Let us end with a discussion of other domains of applications.

While the idea is not yet well developed, I think that beyond the products sold by e-retailers discussed above, which are proxies of reputation and commercial successes, the endo-exo question is relevant to understanding the characteristics of Initial Public Offerings (IPO) [62] and the movie industry [63]. In the latter, the mechanism of information cascade derives from the fact that agents can observe box office revenues and communicate via word-of-mouth about the quality of the movies they have seen.

Earthquakes are now thought to be caused by a mixture of spontaneous occurrences driven by plate tectonics and triggering by previous earthquakes. Within such a picture [12], which rationalizes much of the phenomenology of seismic catalogs, Helmstetter and Sornette have shown that there is a fundamental limit to earthquake predictability resulting from the "exogenous" class of earthquakes that are not triggered by other earthquakes [61]. Furthermore, the rate of foreshocks preceding mainshocks can be understood from the idea that mainshocks may result from endogenous triggering by previous events, as developed above in Sect. 5.2.1. The time dependence of the seismic rate of foreshocks is predicted and observed to follow (5.5). The memory kernels $\phi(t)$ given by (5.3), and $K(t)$ given by (5.2), correspond respectively in the present case to the bare and renormalized Omori law [64] for triggered aftershocks [12, 15].

The weather and the climate also involve extremely complex processes, which are often too difficult to disentangle. This leads to major uncertainties about the important mechanisms that need to be taken into account, for instance, to forecast the future global warming of the earth due to anthropogenic activity coupled with natural variability. 9/11 has again offered a unique window. Travis and Carleton [65] noted the following: "Three days after suicide airplane hijackers toppled the World Trade Center in New York and slammed into the Pentagon in Washington, D.C., the station crew noted an obvious absence of airborne jetliners from their perch 240 miles (384 kilometers) above Earth. "I'll tell you one thing that's really strange: Normally when we go over the U.S., the sky is like a spider web of contrails", U.S. astronaut and outpost commander Frank Culbertson told flight controllers at NASA's Mission Control Center in Houston. "And now the sky is just about completely empty. There are no contrails in the sky," he added. "It's very, very weird." "I hadn't thought of that perspective," fellow astronaut Cady Coleman replied." Travis and Carleton [65] showed that there was a significant elevation of the average diurnal temperature of the US in the three days following 9/11, when most jetliners were grounded and no contrails were present. This is the archetype of an exogenous response. It remains to be seen if the endo-exo viewpoint will offer new fruitful perspectives that will allow us to make progress in understanding and in forecasting the weather and the climate.

Finally, from a theoretical viewpoint, another potentially interesting domain of research is to extend the concept of the response function to nonlinear systems [66,67] and to study its relationship with the internal fluctuations [7].

Acknowledgement. I am grateful to my collaborators and colleagues who helped shape these ideas, among them, Y. Ageon, J. Andersen, R. Crane, D. Darcet, F. Deschatres, T. Gilbert, S. Gluzman, A. Helmstetter, A. Johansen, Y. Malevergne, J.-F. Muzy, V.F. Pisarenko, B. Roehner and W.-X. Zhou.

References

1. Bak, P., How Nature Works: the Science of Self-organized Criticality (Copernicus, New York, 1996)
2. Bak, P. and M. Paczuski, Complexity, contingency, and criticality, Proc. Natl. Acad. Sci. USA, 92, 6689–6696 (1995)
3. Sornette, D., Predictability of catastrophic events: material rupture, earthquakes, turbulence, financial crashes and human birth, Proc. Natl. Acad. Sci. USA, 99 S1, 2522–2529 (2002)
4. Stratonovich, R.L., Nonlinear Nonequilibrium Thermodynamics I: Linear and Nonlinear Fluctuation-Dissipation Theorems (Springer, Berlin Heidelberg New York, 1992)
5. Einstein, A., Über die von der molekularkinetischen Theorie der Wärme geforderte Bewegung von in ruhenden Flüssigkeiten suspendierten Teilchen, Ann. Phys., 17, 549 (1905)
6. Einstein, A., Investigations on the Theory of Brownian Movement (Dover, New York, 1956)
7. Ruelle, D., Conversations on nonequilibrium physics with an extraterrestrial, Physics Today, 57(5), 48–53 (2004)
8. Schumpeter, J.A., Business Cycles: A Theoretical, Historical and Statistical Analysis of the Capitalist Process (McGraw-Hill, New York, 1939)
9. Romer, D., Advanced Macroeconomics (McGraw-Hill, New York, 1996)
10. Dunbar, R.I.M., The social brain hypothesis, Evol. Anthrop., 6, 178–190 (1998)
11. Zhou, W.-X., D. Sornette, R.A. Hill and R.I.M. Dunbar, Discrete hierarchical organization of social group sizes, Proc. Royal Soc. London, 272, 439–444 (2005) doi:10.1098/rspb.2004.2970
12. Helmstetter, A. and Sornette, D., Sub-critical and supercritical regimes in epidemic models of earthquake aftershocks, J. Geophys. Res., 107, B10, 2237, doi:10.1029/2001JB001580 (2002)
13. Sornette, D. and A. Helmstetter, Endogeneous versus exogeneous shocks in systems with memory, Physica A, 318, 577 (2003)
14. Sornette, D., F. Deschatres, T. Gilbert and Y. Ageon, Endogenous versus exogenous shocks in complex networks: an empirical test using book sale ranking, Phys. Rev. Letts., 93 (22), 228701 (2004)
15. Sornette, A. and D. Sornette, Renormalization of earthquake aftershocks, Geophys. Res. Lett., 6, N13, 1981–1984 (1999)
16. Helmstetter, A., D. Sornette and J.-R. Grasso, Mainshocks are aftershocks of conditional foreshocks: How do foreshock statistical properties emerge from aftershock laws, J. Geophys. Res., 108 (B10), 2046, doi:10.1029/2002JB001991 (2003)
17. Dodds, P.S. and D.J. Watts, Universal behavior in a generalized model of contagion, Phys. Rev. Lett., 92, 218701 (2004)
18. Johansen, A. and D. Sornette, Download relaxation dynamics on the WWW following newspaper publication of URL, Physica A, 276(1-2), 338–345 (2000)
19. Johansen A., Response time of internauts, Physica A, 296(3-4), 539–546 (2001)
20. Eckmann, J.P., E. Moses and D. Sergi, Entropy of dialogues creates coherent structures in e-mail traffic, Proc. Nat. Acad. Sci. USA, 101(40), 14333–14337 (2004)

21. Johansen, A., Probing human response times, Physica A, 338(1-2), 286–291 (2004)
22. Brody, J., Push up the weights, and roll back the years, The New York Times, F 7 (June 4, 2002)
23. Gladwell, M., The Tipping Point: How Little Things Can Make a Big Difference (Back Bay Books, Boston, MA, 2002)
24. Sornette, A. Johansen and I. Dornic, Mapping self-organized criticality onto criticality, J. Phys. I France, 5, 325–335 (1995)
25. Gil, L. and D. Sornette, Landau-Ginzburg theory of self-organized criticality, Phys. Rev. Lett., 76, 3991–3994 (1996)
26. Roehner, B.M. and D. Sornette, "Thermometers" of speculative frenzy, Eur. Phys. J., B 16, 729–739 (2000)
27. Roehner, B.M., Patterns of Speculation: A Study in Observational Econophysics (Cambridge University Press, Cambridge, UK, 1st edition, 2002)
28. The Economist, Music's brighter future: The music industry, Business Special, The Economist, Friday 12th November (2004)
29. Roehner, B.M., D. Sornette and J.V. Andersen, Response functions to critical shocks in social sciences: An empirical and numerical study, Int. J. Mod. Phys., C 15 (6), 809–834 (2004)
30. Burch, T.R., D.R. Emery and M.E. Fuerst, What can "Nine-Eleven" tell us about closed-end fund discounts and investor sentiment, Financial Review, 38 (4), (2003)
31. Carter, D.A. and B.J. Simkins, Do Markets React Rationally? The Effect of the September 11th Tragedy on Airline Stock Returns, Working Paper (2002), see http://papers.ssrn.com/paper.taf?abstract_id=306133
32. White E.N., Stock market crashes and speculative manias. In: Capie F.H., ed, The International Library of Macroeconomic and Financial History 13 (Edward Elgar, Brookfield, US, 1996)
33. Engle, R., Autoregressive conditional heteroskedasticity with estimates of the variance of United Kingdom inflation, Econometrica, 50, 987–1008 (1982)
34. Bollerslev, T., Generalized autoregressive conditional heteroskedasticity, J Econometrics, 31, 307–327 (1986)
35. Anderson, T., Stochastic autoregressive volatility, Mathematical Finance, 4, 75–102 (1994)
36. Hamilton, J., Rational-expectations econometric analysis of changes of regimes: an investigation of the term structure of interest rates, J Econometric Dynamics Control, 12, 385–423 (1988)
37. Hamilton, J., A new approach to the economic analysis of nonstationary time series and the business cycle, Econometrica, 57, 357–384 (1989)
38. Pagan, A. and A. Ullah, The econometric analysis of models with risk terms, J Applied Econometrics, 3, 87–105 (1988)
39. Pagan, A. and G.W. Schwert, Alternative models for conditional stock volatility, J Econometrics, 45, 267–290 (1990)
40. Bacry, E., J. Delour and J.-F. Muzy, Multifractal random walk, Phys. Rev. E, 64, 026103 (2001)
41. Muzy, J.-F., J. Delour and E. Bacry, Modelling fluctuations of financial time series: from cascade process to stochastic volatility model, Eur. Phys. J. B, 17, 537–548 (2000)

42. Sornette, D., Y. Malevergne and J.-F. Muzy, What causes crashes? Risk 16 (2), 67–71 (2003)
43. Andersen, J.V., S. Gluzman and D. Sornette, Fundamental framework for technical analysis, Eur. Phys. J. B, 14, 579–601 (2000)
44. McQueen, G. and K. Vorkink, Whence GARCH? A preference-based explanation for conditional volatility, Rev. Financ. Stud., 17, 915–949 (2004)
45. Cutler, D., J. Poterba and L. Summers, What moves stock prices? J. Portfolio Manag., Spring, 4–12 (1989)
46. Sornette, D. and A. Johansen, Significance of log-periodic precursors to financial crashes, Quant. Finance, 1, 452–471 (2001)
47. Johansen, A. and D. Sornette, Endogenous versus Exogenous Crashes in Financial Markets, In: Columbus F., ed, Contemporary Issues in International Finance, in press, (Nova Science, New York, 2004) (http://arXiv.org/abs/cond-mat/0210509)
48. Johansen, A. and D. Sornette, Stock market crashes are outliers, Eur. Phys. J. B 1, 141–143 (1998)
49. Johansen, A. and D. Sornette, Large stock market price drawdowns are outliers, J. Risk, 4(2), 69–110 (2001/02)
50. Johansen, A., Comment on "Are financial crashes predictable?", Eur. Phys. Lett., 60(5), 809–810 (2002)
51. Johansen, A. and D. Sornette, Critical crashes, RISK, 12 (1), 91–94 (1999)
52. Johansen, A., D. Sornette and O. Ledoit, Predicting financial crashes using discrete scale invariance, J. Risk, 1 (4), 5–32 (1999)
53. Johansen, A., O. Ledoit and D. Sornette, Crashes as critical points, Int. J. Theor. Appl. Finance, 3 (2), 219–255 (2000)
54. Sornette, D., Why Stock Markets Crash (Critical Events in Complex Financial Systems) (Princeton University Press, Princeton, NJ, 2003)
55. Sornette, D., Critical market crashes, Phys. Rep., 378 (1), 1–98 (2003)
56. Zhou, W.-X. and D. Sornette, Non-parametric analyses of log-periodic precursors to financial crashes, Int. J. Mod. Phys. C, 14 (8), 1107–1126 (2003)
57. Sornette, D. and W.-X. Zhou, Evidence of fueling of the 2000 new economy bubble by foreign capital inflow: implications for the future of the US economy and its stock market, Physica A, 332, 412–440 (2004)
58. Sornette, D. and W.-X. Zhou, Predictability of large future changes in complex systems, Int. J. Forecasting, in press (2004) (http://arXiv.org/abs/cond-mat/0304601)
59. Johnson, N.F., P. Jefferies and P. Ming Hui, Financial Market Complexity (Oxford Univ. Press, Oxford, UK, 2003)
60. Andersen, J.V. and D. Sornette, A mechanism for pockets of predictability in complex adaptive systems, Europhys. Lett., 70 (5), 697–703 (2005)
61. Helmstetter and D. Sornette, Predictability in the ETAS model of interacting triggered seismicity, J. Geophys. Res., 108, 2482, 10.1029/2003JB002485 (2003)
62. Jenkinson, T. & Ljungqvist, A., Going Public: The Theory and Evidence on How Companies Raise Equity Finance (Oxford Univ. Press, Oxford, UK, 2nd edition 2001)
63. De Vany, A. and Lee, C., Quality signals in information cascades and the dynamics of the distribution of motion picture box office revenues. J. Econ. Dyn. Control, 25, 593–614 (2001)

64. Omori, F., On the aftershocks of earthquakes, J. Coll. Sci. Imp. Uni., 7, 111 (1894)
65. Travis, D. J., A.M. Carleton and R.G. Lauritsen, Contrails reduce daily temperature range, Nature, 418, 601 (2002)
66. Potter, S.M., Nonlinear impulse response functions, J. Econ. Dynam. Control, 24(10), 1425–1446 (2000)
67. Dellago, C. and S. Mukamel, Nonlinear response of classical dynamical systems to short pulses, Bull. Korean Chem. Soc., 24(8), 1107–1110 (2003)

Part II

Scenarios

6 Epilepsy: Extreme Events in the Human Brain

Klaus Lehnertz

Summary. The analysis of Xevents arising in dynamical systems with many degrees of freedom represents a challenge for many scientific fields. This is especially true for the open, dissipative, and adaptive system known as the *human brain.* Due to its complex structure, its immense functionality, and – as in the case of epilepsy – due to the coexistence of normal and abnormal functions, the brain can be regarded as one of the most complex and fascinating systems in nature. Data gathered so far show that the epileptic process exhibits a high spatial and temporal variability. Small, specific, regions of the brain are responsible for the generation of focal epileptic seizures, and the amount of time a patient spends actually having seizures is only a small fraction of his/her lifetime. In between these Xevents large parts of the brain exhibit normal functioning. Since the occurrence of seizures usually can not be explained by exogenous factors, and since the brain recovers its normal state after a seizure in the majority of cases, this might indicate that endogenous nonlinear (deterministic and/or stochastic) properties are involved in the control of these Xevents. In fact, converging evidence now indicates that (particularly) nonlinear approaches to the analysis of brain activity allow us to define precursors which, provided sufficient sensitivity and specificity can be obtained, might lead to the development of patient-specific seizure anticipation and seizure prevention strategies.

6.1 Introduction

Xevents are critical determinants of the evolution and character of a vulnerable system. Xevents are usually considered to be rare and unpredictable events and/or events that strongly deviate from *normality.* However, objective criteria that can be used to define Xevents are yet to be defined. Thus, an Xevent might not simply be characterized by features such as *intensity* and *rareness.* Rareness demands some characteristic scales or some temporal and spatial boundaries, while intensity should reflect an event's potential to cause a large change. However, both intenseness and rareness are derived from the human perception of consequences, which in turn reflects the character of the affected system.

Xevents are inherently contextual and relational. Events with disastrous natural, social or financial consequences, such as floods, earthquakes, heavy storms, meltdown of nuclear power plants, or financial crashes are certainly extreme. The extremeness of such events commands broad attention and demands both comprehension and action. In contrast, epileptic seizures might be considered less extreme, mainly due to limited public awareness. But they are doubtlessly extreme for those affected.

Epilepsy affects more than 50 million individuals worldwide – approximately 1% of the world's population (see [1] for a comprehensive overview). The disease is characterized by a recurrent and sudden malfunction of the brain that is termed a *seizure*. Epileptic seizures are the clinical manifestation of excessive and hyper-synchronous neuron activity in the brain. Depending on the extent of involvement of other brain areas during the course of the seizure, epilepsies can be divided into two main classes. While primary generalized seizures involve almost the entire brain, focal (or partial) seizures originate from a circumscribed region of the brain (the epileptic focus) and remain restricted to this region. Epileptic seizures may be accompanied by an impairment or loss of consciousness, psychic, autonomic or sensory symptoms or motor phenomena. With today's available antiepileptic drugs, seizures can be controlled satisfactorily in about two thirds of affected individuals; another 8% may profit from epilepsy surgery. The remaining 25% of epilepsy patients can not be adequately treated by any available therapy.

The fact that seizures strike *like a bolt from the blue* in the majority of cases is one of the most disabling aspects of epilepsy. The recurrent and sudden incidence of seizures as well as the disturbance of consciousness and sudden loss of motor control can lead to dangerous and possibly life-threatening situations. The ability to forecast epileptic seizures would dramatically change therapeutic possibilities. One might envisage a simple warning system that could eventually decrease the risk of injury, patients' anxiety and the feeling of helplessness resulting from the seemingly unpredictable occurrence of seizures. Long-term treatment with antiepileptic drugs, which can cause cognitive or other neurological deficits, could be diminished to an on-demand application of a short-acting drug during the pre-seizure period. Together with other suitable prevention techniques, this would reduce morbidity and mortality and would greatly improve the quality of life for epilepsy patients. In addition, the identification of a pre-seizure period could aid investigations of the pathophysiological mechanisms causing seizures in humans.

6.2 Basic Mechanisms

Knowledge about the basic mechanisms leading to seizures is mainly derived from animal experiments, and must still be regarded as fragmentary. Focal seizures are assumed to be initiated by abnormally discharging neurons (so-called *bursters*; see Fig. 6.1 [2–4]) that recruit and entrain neighboring neurons into a *critical mass*.

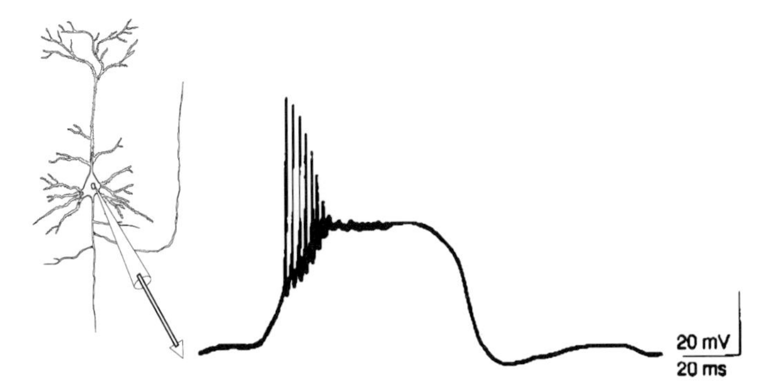

Fig. 6.1. Membrane potential changes (paroxysmal depolarization shift, PDS) of an impaled neuron during focal epileptiform activity. PDS represents a shift of the resting membrane potential which is accompanied by a rise in intracellular calcium and a massive burst of action potentials (500–800 per second). PDSs originating from a larger cortical region are usually associated with steep field potentials (known as spikes) recorded in the scalp EEG

This process manifests itself as an increasing synchronization of neuronal activity, accompanied by a loss of inhibition which is usually maintained by surrounding neural networks. The build-up of such a critical mass might be mediated by facilitating processes in the sense of nonspecific predisposing factors that permit seizure emergence by lowering the threshold [5]. In this context the term *critical mass* should not be interpreted as a highly localized mass phenomenon that would be easily accessible for conventional simultaneously recorded electroencephalogram analyses, which fail to detect it. Instead, the interactions between neurons that play a crucial role in seizure generation probably take place on different spatial and temporal scales and are known to be nonlinear in nature.

The phase of transition to the seizure state (*pre-ictal state*) is thought to be related to a breakdown in local inhibitory mechanisms that are mediated by different synaptic and non-synaptic processes. Epilepsy involves multiple neurotransmitter systems, where glutamate/aspartate (especially N-methyl-D-aspartate) represents the major excitatory and gamma aminobutyric acid the major inhibitory neurotransmitter. The assumption that an imbalance of these two substances causes epilepsy, however, is too simplistic and does not take into account the enormous complexity involved with the biochemical regulation of paroxysmal activity. Changes in intra-/extracellular ion concentrations (sodium, potassium, calcium, and magnesium) are also assumed to contribute to the depression of inhibitory mechanisms.

The spreading of epileptic activity to remote brain areas is mainly mediated by neuronal transmission, while passive volume conduction plays a less significant role. Target regions may be divided into *low threshold* or *high threshold* areas with different propensities for epileptic responses. However, the locations and spatial extents of these areas may differ from case to case.

The exact mechanisms underlying seizure termination are as uncertain as the mechanisms underlying seizure initiation and spread. Early hypotheses suggested that seizure termination was caused by depletion of metabolic supplies and increasing hypoxia. At present, neuronal processes like diminishing neuronal synchrony or active inhibition are assumed to be responsible. The significance of the latter is stressed by a transient phase of seizure refractiveness that is often observed after a seizure. During this *post-ictal state* specific neurological deficits are often accompanied by an electrographical *silence* in brain activity that may reflect interference phenomena between inhibitory processes and the return to physiological neuronal functioning.

6.3 EEG and Epilepsy

In 1875 the British physiologist Richard Caton published his observations on electrical activity in animal brains, but it was not until 1929 that Hans Berger, a psychiatrist working in Jena, Germany, first reported on the electroencephalogram (EEG) of humans [6]. EEG signals reflect the dynamics of electrical activity in populations of neurons. Although it is commonly accepted that postsynaptic potentials represent the neurophysiological basis of the EEG [7], the mechanisms underlying the generation of rhythms in the EEG are not yet fully understood.

In recent years, technical advances such as digital video-EEG monitoring systems and increased computational performance have led to highly sophisticated clinical epilepsy monitoring systems that allow huge amounts of data to be processed in real-time. In addition, chronically implanted intracranial electrodes allow continuous recording of brain electrical activity from the surface of the brain and/or from within specific brain structures at a high spatial resolution. Due to its high temporal resolution and its close relationship to physiological and pathological functions of the brain, and particularly since it directly measures the pathophysiological substrate of the Xevent seizure (see Fig. 6.2), electroencephalography is regarded as indispensable for clinical practice despite the rapid development of imaging technologies such as magnetic resonance tomography and positron emission tomography.

The extraction of information from EEG recordings relevant for diagnostic purposes may be divided into two classes. Visual EEG evaluation involves measuring the frequencies, amplitudes, and morphologies of waves using special rulers. More complex patterns and their association with normal or pathological conditions may also be recognized by visual inspection. However, it is sometimes extremely difficult to define reliable criteria when rating transient phenomena as spikes, sharp waves or other patterns associated with

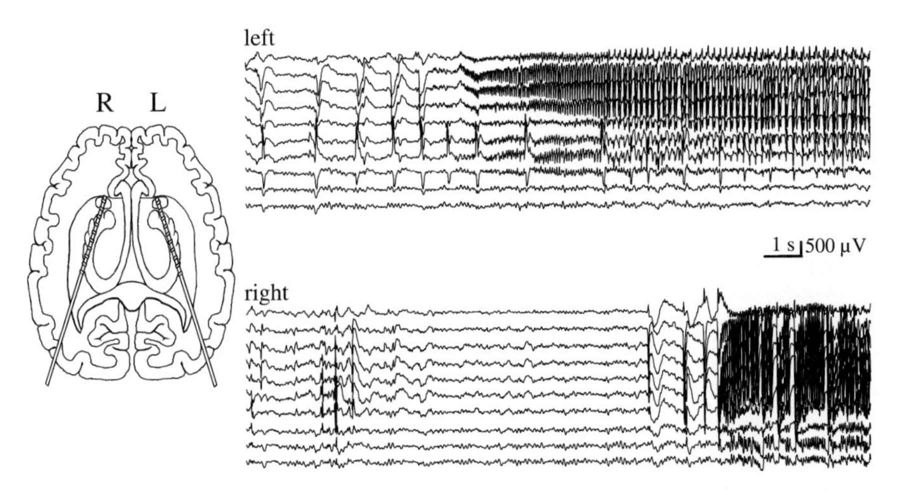

Fig. 6.2. Intracranial EEG recording of a seizure originating from the left hippocampus of a patient suffering from mesial temporal lobe epilepsy. Seizure activity involves the right hippocampus after approximately 6 s. Signals were recorded from intrahippocampal depth electrodes that were implanted stereotaxically along the longitudinal axis of the hippocampal formation, with the amygdala as the target for the most anterior electrode. Each catheter-like, 1 mm-thick silastic electrode contained ten cylindrical contacts of a nickel-chromium alloy (2.5 mm) every 4 mm

pathological states. The limitations of such methods become particularly obvious in clinical problems when large amounts of EEG data must be evaluated and rather complex questions are being asked. This particularly holds true in epileptology, where, as Engel pointed out, *the clinical interpretation of EEG recordings and application of EEG findings to the diagnosis and treatment of epilepsy remains more of an art than a science* [5].

The second class comprises a variety of time series analysis techniques, which are usually applied to long-lasting multichannel recordings in a moving-window fashion. The time length of the window is chosen in such a way that it represents a reasonable trade-off between approximate stationarity of the system but still allows a sufficient number of samples to be recorded to achieve a statistically reliable estimate. However, most analyses cannot be applied in a strict mathematical sense because the necessary theoretical conditions cannot be met in practice – a common problem that applies to any analysis of finite (and noisy) data segments or non-stationary systems.

Classical linear EEG analysis techniques can be divided into two main categories (see [8] for a comprehensive overview). Nonparametric methods comprise analysis techniques such as evaluations of amplitude, interval or period distributions, estimations of auto- and cross-correlation functions, as well as analysis in the frequency domain, such as power spectral estimates and cross-spectral functions. Parametric methods include, among others, AR (autoregressive) and ARMA (autoregressive moving average) models, inverse

AR-filtering and segmentation analysis. These main branches are accompanied by pattern recognition methods involving either a mixture of the techniques mentioned above or, more recently, the wavelet transform. The reason for applying these methods is the classical view that brain rhythms may be described as linear resonance phenomena of neuronal networks. Despite their intrinsic limitations, classical EEG analysis techniques have significantly contributed to and are still advancing our understanding of the physiological and pathophysiological mechanisms of the brain.

Nonlinear time series analysis techniques [9, 10] have been developed to analyze and characterize deterministic nonlinear dynamical systems that exhibit apparently irregular behavior – a distinctive feature also found in the EEG. The traditional linear time series approaches mentioned above insufficiently detect or explain a wide range of EEG phenomena such as bursting, amplitude-dependent frequencies, or frequency doubling, all thought to reflect nonlinear processes. Thus, during the last 10–15 years a variety of nonlinear analysis techniques has been applied to EEG recordings during physiological and pathological conditions and they have been shown to offer new information about complex brain dynamics. Usually, the well known nonlinear behavior of individual neurons ("all-or-none" firing) and synapses, and with it the expectation that neuronal networks behave in a similar way, is assumed when applying these methods. This assumption, however, is still matter of debate [11, 12]. The fact that a system contains nonlinear components does not prove that this nonlinearity is also reflected in a specific signal measured from that system. Although there is ample evidence for nonlinearity, in particular, in small assemblies of neurons [13], it is now commonly accepted that the existence of a low-dimensional deterministic or even chaotic structure in the EEG is difficult if not impossible to prove. Nevertheless, approaches that seek to correlate values of operationally defined nonlinear measures with disease states, in space or time, have generated new, clinically relevant measures as well as new ways of interpreting brain functioning, particularly with regard to epileptic brain states [14–17].

6.4 Nonlinear EEG Analysis

There is increasing evidence that a number of key conceptual features of nonlinear dynamical systems have particular relevance to improving our understanding of the spatio-temporal dynamics of the seizure generating process.

The basic principle of almost all nonlinear time series analysis techniques is to reconstruct the system dynamics observed in so-called state space. Although an unknown system may well be dependent on a large (and, for the EEG, often unknown) number of variables, the theorem of Takens [18] states that under certain genericity assumptions the system's behavior in state space can be approximated using only a single observed variable (such as the EEG).

If the system is governed by some nonlinear laws, a simple cause-effect relationship should not be expected. Rather, nonlinear systems are characterized by a rich variety of dynamics including bifurcations that indicate abrupt state transition or intermittent behavior. EEG phenomena like spike-burst suppression patterns, epileptiform activity such as spikes, or the interictal-ictal state transition point to nonlinearity. Abrupt state transitions from highly complex, irregular to less complex, almost periodic dynamics appear to be a characteristic feature of many dynamical disorders [19] including epilepsy, and are one of the most compelling reasons for the notion of *complexity loss*. Moreover, it is the very occurrence of periodicities and highly-structured patterns that allows identification and classification of many pathological phenomena.

Due to the sensitive dependence on initial conditions (the *butterfly effect*) of a deterministic chaotic system, its long-term behavior is very difficult to predict. On the other hand, a nonlinear system has – under certain conditions – an inherent ability to "self-organize" in the sense that it evolves towards an ordered temporal and spatial structure. This concept might explain the well-organized, self-sustained oscillations in EEG recordings during seizure activity. This dualism of chaos and order is the key feature of nonlinear dynamics.

Generally, the initial conditions and the rules that govern a system like the epileptic brain are unknown. However, a variety of new concepts and measures have been developed that allow us to characterize the dynamical behavior of an unknown system: Lyapunov exponents characterize the system's stability under small perturbations and are therefore a measure of how "chaotic" a system behaves; dimension estimates are closely related to the number of degrees of freedom of a system; entropies measure the degree of order/disorder. Dimensions and entropies can thus be regarded as an estimate of the system's complexity. These *univariate* nonlinear analysis techniques quantify certain properties of the EEG signal, thus possibly reflecting the state of a certain region of the brain. Recently, these techniques have been supplemented by *bivariate* approaches that quantify the amount of interaction between different areas of the brain and provide information about spatial synchronization phenomena, which are considered to play a crucial role in seizure generation. Due to the large number of influencing factors, however, the limitations of the techniques have to be taken into consideration and results must be interpreted carefully. This has led to the current point of view that it is advisable to use operationally defined or relative measures, thereby focusing on the existence of a change and not necessarily on the nature of this change.

6.4.1 State Space Reconstruction

The time evolution of a dynamical system (such as the brain) in a state space $\Gamma \subset \mathbf{R}$ can be expressed in discrete time $t = n\Delta t$ (where Δt is the sampling interval of some observable, such as the EEG) by maps of the form

$$\boldsymbol{x}_{n+1} = \mathbf{f}(\boldsymbol{x}_n) \tag{6.1}$$

A time series can then be thought of as a sequence of observations $v_n = g(\boldsymbol{x}_n)$ (where $n = 1, \ldots, N$) performed with some measurement function $g(\cdot)$. Since the (usually scalar) time series $\{v_n\}$ may not properly represent the (high-dimensional) state space of the dynamical system, a reconstruction technique must be employed in order to unfold the high-dimensional structure using the data available. The most important and widely used state space reconstruction technique is the *method of delays* [18]. State space vectors in an m-dimensional embedding space are formed from time delayed values of the scalar measurements:

$$\boldsymbol{v}_i = (v_i\,, v_{i-\tau}\,, \ldots\,, v_{i-(m-1)\tau}) \tag{6.2}$$

with $i = 1\,, \ldots\,, M = N - m\tau$. Under certain genericity assumptions [18,20], the time delay embedding provides a one-to-one image of the original set $\{\boldsymbol{x}\}$, provided m is large enough. The *optimal* choice of the parameters embedding dimension m and time delay τ largely depends on the application.

6.4.2 Measures Based on the Correlation Sum

The correlation sum [21] is an estimate of the local probability density in state space. It counts the number of pairs of vectors in state space that are closer than a given hypersphere radius ϵ:

$$C(m, \epsilon) = \frac{2}{(M-W)(M-W-1)} \sum_{i=1}^{M} \sum_{j=i+W}^{M} \Theta(\epsilon - |\boldsymbol{v}_i - \boldsymbol{v}_j|) \tag{6.3}$$

where Θ is the Heaviside step function. The exclusion of pairs closer in time than the length of the so-called Theiler window W is essential to reduce the unwanted influence of temporal correlations on $C(m, \epsilon)$ [22].

The majority of approaches based on the use of measures derived from the correlation sum to detect precursors of epileptic seizures in the EEG assume that neuronal networks involved in the seizure generating process exhibit a decreased level or loss of *complexity*; in other words, a reduced number of active degrees of freedom. For deterministic dynamics, the correlation dimension D_2 [21] is related the number of active degrees of freedom. Using the local slope of the correlation sum $d(\epsilon) = \frac{d \ln C(\epsilon)}{d \ln(\epsilon)}$, the correlation dimension is defined as $D_2 = \lim_{N\to\infty} \lim_{\epsilon\to 0} d(\epsilon)$. From the limits it follows that the calculation of D_2 would require an infinite length N and an unlimited accuracy for the time series. However, an estimate for an *effective correlation dimension* [23] can be obtained if an almost constant value of $d(\epsilon)$ is found for at least a range of ϵ values, the quasi-scaling region.

In our investigations [24, 25] we calculate $d(\epsilon)$ for embedding dimensions of $m = 1$ and $m = 25$ using a fixed time delay ($\tau = \Delta t$) and Theiler window ($W = 5\Delta t$). The range of ϵ is chosen to match the resolution of the analog-to-digital converter. We define a *quasi-scaling region* $[\epsilon_l, \epsilon_u]$ by

$$\begin{aligned} \epsilon_u &= \max\left\{\epsilon |\ d(\epsilon)_{|m=1} > 0.975\right\} \\ \epsilon_l &= \min\left\{\epsilon |\ d(\epsilon_u)_{|m=25} - d(\epsilon)_{|m=25} \leq 0.05 \cdot d(\epsilon)_{|m=25} \wedge \epsilon < \epsilon_u\right\} \end{aligned} \tag{6.4}$$

If ϵ_u and ϵ_l exist and the number n_ϵ of values in $[\epsilon_l, \epsilon_u]$ is greater than 4, the estimate

$$D^* = \frac{1}{n_\epsilon} \sum_{\epsilon=\epsilon_l}^{\epsilon_u} d(\epsilon)_{|m=25} \tag{6.5}$$

is computed. If no quasi-scaling behavior is found for $d(\epsilon)$ or if $D^* \geq 9.5$, an arbitrary but fixed value of $D^* = 10$ is set (see [26]). We emphasize that we do *not* interpret D^* as an estimate of the correlation dimension D_2. Our considerations are aimed solely at maximizing its discriminative power.

A similar approach is based on the *correlation density* [27]. In order to limit the use of computational resources, this measure is estimated by computing $C(m, \epsilon)$ for some fixed $\epsilon = \epsilon_0$. It should be noted, however, that a proper choice of ϵ_0 is mandatory in order to achieve a meaningful estimate of the correlation density (see [28]).

The *correlation entropy* h_2 [29, 30] is a lower bound of the Kolmogorov–Sinai entropy, which describes the level of uncertainty about the future state of the system, and therefore relates to predictability. Provided a scaling region exists, h_2 can be estimated from the correlation sum as

$$h_2 \approx \ln \frac{C_m(\epsilon)}{C_{m+1}(\epsilon)}, \tag{6.6}$$

using an extrapolation to large m. Alternatively, an entropy estimate can be derived from the sum of the positive Lyapunov exponents [31].

The *dynamical similarity index* [32] is based on an extension of the (auto–) correlation sum (6.3) to the cross-correlation sum [33]. Rather than measuring the lengths of state space vectors generated from the same EEG time series, this method computes nonlinear characteristics of a *reference* EEG window which is then compared to a similar scanning window that is moved forward in time towards known seizure onsets.

6.4.3 Lyapunov Exponents

The exponential divergence of nearby trajectories in state space is conceptually the most basic indicator of deterministic chaos. This exponential instability is characterized by the spectrum of Lyapunov exponents. The largest Lyapunov exponent λ_{max} can be determined without the explicit construction of a model for the time series. In EEG analysis, λ_{max} is used, for example, to characterize phase transitions from *order to chaos* in the epileptic brain (see [34,35] and references therein). The algorithm most widely used to compute λ_{max} from a time series [36] suffers from severe drawbacks that occur particularly with short and noisy time series, strongly depends on parameters used for the state space reconstruction, and is highly computationally intensive. In order to avoid these shortcomings we use a combination of improved algorithms [37,38] according to which λ_{max} can be estimated from

$$\delta_j(i) \approx C_j e^{\lambda_{max} \cdot i \cdot \Delta t} \tag{6.7}$$

where $\delta_j(i)$ denotes the average divergence between two trajectory segments at time t_i. C_j with $j = 1, \ldots, M$ is a constant that is given by the initial separation of a reference vector $\boldsymbol{z}_j$ in state space from its nearest neighbor. In order to improve statistics we follow [38] and search for all neighbors starting within a hypersphere of radius ϵ around $\boldsymbol{z}_j$ using a box-assisted algorithm [39]. Taking the logarithm of (6.7), λ_{max} is then calculated using a least-squares fit to an average line defined by $y(i) = \frac{1}{\Delta t} \langle \ln \delta_j(i) \rangle_j$. For our analyses, we estimate Λ_{max} using an embedding dimension of $m = 7$ and a fixed time delay of $\tau = 5 \cdot \Delta t$. In order to reduce the unwanted influence of temporal correlations we follow [37] and choose a Theiler window with a length given by the reciprocal of the mean frequency of the power spectrum.

6.4.4 Synchronization and Interdependencies

In almost all of the theories on seizure generation commonly accepted today, pathological neuronal synchronization is considered to play a crucial role. Since synchronization phenomena can manifest themselves in many different ways, a unifying framework for synchronization in chaotic dynamical systems is still missing. Instead various concepts for its description have been offered including, among others, phase synchronization and generalized synchronization (see [40] for an overview).

The classical concept of phase synchronization was extended from linear to nonlinear or even chaotic systems for cases where the definition of a phase variable is possible for the analyzed systems. Traditionally, phase synchronization is defined as the locking of the phases ϕ of two oscillating systems a and b:

$$\phi_a(t) - \phi_b(t) = const. \tag{6.8}$$

As a measure to quantify the degree of phase synchronization between two EEG time series, we have introduced the so-called *mean phase coherence* R [41], defined as

$$R = \left| \frac{1}{K} \sum_{j=0}^{K} e^{i[\phi_a(j\Delta t) - \phi_b(j\Delta t)]} \right| = 1 - V \tag{6.9}$$

where $1/\Delta t$ is the sampling rate of the discrete time series of length K, and V denotes the circular variance of an angular distribution obtained by transforming the differences in phase onto the unit circle in the complex plane [42]. By definition R is confined to the interval [0,1] where $R = 1$ ($V = 0$) indicates fully synchronized systems (see [43] for alternative measures of phase synchronization).

In order to determine the phases $\phi_a(t)$ and $\phi_b(t)$ of two EEG signals $s_a(t)$ and $s_b(t)$, we followed the analytic signal approach, which renders an unambiguous definition of the so-called instantaneous phase for an arbitrary signal $s(t)$:

$$\phi(t) = \arctan \frac{\tilde{s}(t)}{s(t)} \tag{6.10}$$

where

$$\tilde{s}(t) = \frac{1}{\pi} p.v. \int_{-\infty}^{+\infty} \frac{s(\tau)}{t - \tau} d\tau \tag{6.11}$$

is the Hilbert Transform of the signal ($p.v.$ denoting the Cauchy principal value). Alternatively, the phase variable can be obtained from the wavelet transform [44]:

$$\phi(t) = \arctan \frac{\mathrm{ImW(t)}}{\mathrm{ReW(t)}} \tag{6.12}$$

using the wavelet coefficients

$$W(t) = \int_{-\infty}^{\infty} \Psi(t - \tilde{t}) s(\tilde{t}) d\tilde{t} \tag{6.13}$$

of a complex Morlet wavelet.

The *nonlinear interdependence* S [45] quantifies the degree to which similarity of states of one (sub-)system (for example brain region A) implies similarity of simultaneous states of the other (sub-)system (such as brain region B) and is closely related to other attempts to detect *generalized synchronization*. In contrast to commonly used measures like cross-correlation, coherence and mutual information, S is non-symmetric and provides information about the direction of interdependence. Let $\boldsymbol{v}_i$ and $\boldsymbol{w}_i$ denote state space trajectories reconstructed from two EEG time series recorded simultaneously at different sites. Let $\alpha_{i,j}$ and $\beta_{i,j}$, $j = 1, \ldots, k$ denote the time indices of the k nearest neighbors of $\boldsymbol{v}_i$ and $\boldsymbol{w}_i$, respectively. For each $\boldsymbol{v}_i$ the mean squared Euclidean distance to its k nearest neighbors is given by

$$R_i^{(k)}(\boldsymbol{v}) = \frac{1}{k} \sum_{j=1}^{k} (\boldsymbol{v}_i - \boldsymbol{v}_{\alpha_{i,j}})^2 \tag{6.14}$$

while the w-*conditioned* mean squared Euclidean distance is constructed by replacing the nearest neighbors by the equal time partners of the closest neighbors of $\boldsymbol{w}_i$:

$$R_i^{(k)}(\boldsymbol{v}|\boldsymbol{w}) = \frac{1}{k}\sum_{j=1}^{k}(\boldsymbol{v}_i - \boldsymbol{v}_{\beta_{i,j}})^2 \tag{6.15}$$

$R_i^{(k)}(\boldsymbol{w})$ and $R_i^{(k)}(\boldsymbol{w}|\boldsymbol{v})$ are defined accordingly. The local and global interdependence measures $S_i^{(k)}(\boldsymbol{v}|\boldsymbol{w})$ and $S^{(k)}(\boldsymbol{v}|\boldsymbol{w})$ are defined as

$$S_i^{(k)}(\boldsymbol{v}|\boldsymbol{w}) \equiv \frac{R_i^{(k)}(\boldsymbol{v})}{R_i^{(k)}(\boldsymbol{v}|\boldsymbol{w})} \tag{6.16}$$

and

$$S^{(k)}(\boldsymbol{v}|\boldsymbol{w}) \equiv \frac{1}{M}\sum_{i=1}^{M} S_i^{(k)}(\boldsymbol{v}|\boldsymbol{w}) = \frac{1}{M}\sum_{i=1}^{M}\frac{R_i^{(k)}(\boldsymbol{v})}{R_i^{(k)}(\boldsymbol{v}|\boldsymbol{w})} \tag{6.17}$$

Since $R_i^{(k)}(\boldsymbol{v}|\boldsymbol{w}) \geq R_i^{(k)}(\boldsymbol{v})$ by construction, we have

$$0 < S^{(k)}(\boldsymbol{v}|\boldsymbol{w}) \leq 1\ . \tag{6.18}$$

If $S^{(k)}(\boldsymbol{v}|\boldsymbol{w}) \approx (k/M)^{2/D} \ll 1$, where D denotes an effective dimension, then obviously $\boldsymbol{v}$ and $\boldsymbol{w}$ are independent within the limits of accuracy. If, however, $S^{(k)}(\boldsymbol{v}|\boldsymbol{w}) \gg (k/M)^{2/D}$, we say that $\boldsymbol{v}$ depends on $\boldsymbol{w}$, thereby *not* implying any causal relationship. This dependence becomes maximum when $S^{(k)}(\boldsymbol{v}|\boldsymbol{w}) \to 1$.

The opposite dependences $S_n^{(k)}(\boldsymbol{w}|\boldsymbol{v})$ and $S^{(k)}(\boldsymbol{w}|\boldsymbol{v})$ are defined in complete analogy. They are in general *not* equal to $S_n^{(k)}(\boldsymbol{v}|\boldsymbol{w})$ and $S^{(k)}(\boldsymbol{v}|\boldsymbol{w})$. Both $S^{(k)}(\boldsymbol{v}|\boldsymbol{w})$ and $S^{(k)}(\boldsymbol{w}|\boldsymbol{v})$ may be of the order of 1. Therefore $\boldsymbol{v}$ can depend on $\boldsymbol{w}$, and at the same time can $\boldsymbol{w}$ depend on $\boldsymbol{v}$. If $S^{(k)}(\boldsymbol{v}|\boldsymbol{w}) > S^{(k)}(\boldsymbol{w}|\boldsymbol{v})$, in other words if $\boldsymbol{v}$ depends more on $\boldsymbol{w}$ than vice versa, we say that $\boldsymbol{w}$ is more "active" than $\boldsymbol{v}$. Again we do *not* imply this to have any causal meaning, a priori.

6.4.5 Testing for Nonlinearity

Almost all of the measures mentioned above share a common property. Their probability distribution on finite data sets is not known analytically. In order to derive confidence limits or probability distributions of nonlinear statistics, it is therefore highly preferable to use a Monte Carlo resampling technique. The method of surrogate time series (see [46] for a comprehensive overview) allows us to test a specified null hypothesis about the dynamics underlying a given time series. For this purpose, an ensemble of surrogate time series is constructed from the original EEG time series in such a way that the surrogates have all properties included in the null hypothesis in common with the original, but are otherwise random. Then a certain measure (for example a dimension), which has to be sensitive to at least one property that is not included in the null hypothesis, such as nonlinearity, is calculated for the original and the surrogates. If the result for the original time series deviates significantly from the distribution of the surrogates, the null hypothesis can be rejected. The probability of false rejections (the nominal size of the test) is adjustable via the number of surrogates. In our investigations we applied a technique for generating so-called iterative amplitude-adjusted surrogates. These surrogates allow us to test the null hypothesis that the EEG time series was measured from a Gaussian linear stochastic and stationary dynamics by means of a static and invertible but possibly nonlinear measurement function. Starting from a random permutation of the original amplitudes of the EEG time series, the surrogates are constructed by an iterative algorithm that alternately adjusts the power spectrum and the amplitude distribution to the original values, resulting in a deviation in the other quantity. After a sufficient number of iterations (typically 20–50), deviations of both quantities from values of the original EEG time series will be reduced to negligibly small values.

6.5 Can Epileptic Seizures Be Anticipated?

Seizure prediction or anticipation or forecasting (despite their different meanings these terms are currently used interchangeably), is a field of great interest to the clinical and basic neuroscience communities, not only due to its potential clinical applications in warning and therapeutic devices, but because it holds great promise for increasing our understanding of the mechanisms underlying epilepsy and seizure generation.

In some patients, seizures appear to occur unpredictably, with no discernible patterns. In others, seizures appear to occur in cycles. In some cases, the cycling patterns have been attributed to other biological rhythms such as the menstrual cycle. Clustering patterns, where one seizure appears to increase or decrease the likelihood of subsequent seizures, are a common clinical observation. Analyses of long-term seizure patterns are usually based on seizure diaries [47–52]. While some authors have concluded that the timing of seizure recurrence is random and follows a homogeneous Poisson distribution, others have observed significant deviations from a homogeneous Poisson process and hypothesized that seizures occur in a probabilistically nonlinear fashion.

Because of this inconsistency, the transition to a seizure is generally believed to be an abrupt phenomenon, occurring without warning. Once a seizure has occurred, it is trivial to retrospectively postulate the existence of a transitional pre-seizure phase. Its unequivocal a priori definition and prospective detection (and with it the possibility to actually forecast an impending seizure) is, however, obviously far from being trivial. Nevertheless, a variety of clinical observations support the notion that at least some seizures can be anticipated.

Several seizure-facilitating factors are known. In the context of his reservoir theory, Lennox [53] has defined seizure facilitation as the input of sensory, metabolic, emotional, or other yet unknown factors that fill up some reservoir until it overflows, which in turn results in a seizure. Among others, levels of consciousness, sleep deprivation, tension states, disturbances of water and acid-base balance, sensory and drug stimulation are regarded as potential influencing factors [54]. However, apart from the rare exception of sensory-evoked or reflex epilepsies, these factors are rather unspecific and highly variant since they depend on the habits and daily activities of the patient. It is clinically undisputed from many descriptions of close relatives that long-lasting behavioral and/or prodromal changes in the autonomous nervous system exist in certain patients prior to seizures [55]. These alterations include depressive mood changes, irritability, sleep problems, nausea, and headache. Few reports indicate the possibility of seizure self-abatement [56]. Moreover, when asked more thoroughly, certain patients declared that they had developed their own seizure prevention strategies that are used with a varying degree of success. Although these strategies often appear extremely complicated, they can nevertheless be considered specific in the sense that patients attempt to prevent neuronal networks from being recruited into the epileptic process by forcing them into some physiological processing.

In EEG analysis, the search for hidden information that may be used to predict an impending seizure has a long history. As early as 1975, researchers considered analysis techniques such as pattern recognition, analytic procedures of spectral data [57], or autoregressive modeling of EEG data [58, 59] for predicting epileptic seizures. Findings indicated that EEG changes characteristic of pre-seizure states may be detectable, a few seconds before the actual seizure onset at the most. None of these techniques have been implemented clinically. Apart from applying signal analysis techniques, the relevance of steep, high-amplitude epileptiform potentials (spikes, the hallmark of the epileptic brain) were investigated in a number of clinical studies [60–63]. While some authors reported a decrease or even total cessation of spikes before seizures, re-examination in a larger sample did not confirm this phenomenon.

The earliest attempts to use nonlinear time series analysis were performed in the 1990s using the largest Lyapunov exponent (see Sect. 6.4.3) to describe changes in brain dynamics [34]. The first studies to describe characteristic changes in the EEG shortly before an impending seizure in a larger group of patients used the correlation dimension [21] as an estimate for neuronal complexity [24, 26, 64, 65] and the correlation density [27] (see Sect. 6.4.2). These studies were followed by others using measures such as dynamic similarity [32, 66–68], entropy [30], predictability [69], or certain signal patterns ("bursts") and changes in signal energy [70,71]. More recently, bivariate measures (see Sect. 6.4.4), such as nonlinear interdependence [45], measures for phase synchronization and cross-correlation [41,72,73], the difference between the largest Lyapunov exponents of two channels [35], as well as a multivariate approach based on a fusion of multiple features with neural networks [70] and on simulated neuronal cell models [74] have been shown to be capable of defining a pre-seizure period (see [75–77] for an overview).

Summarizing these studies, pre-ictal states ranging from several minutes up to hours in duration could be observed (see Fig. 6.3).

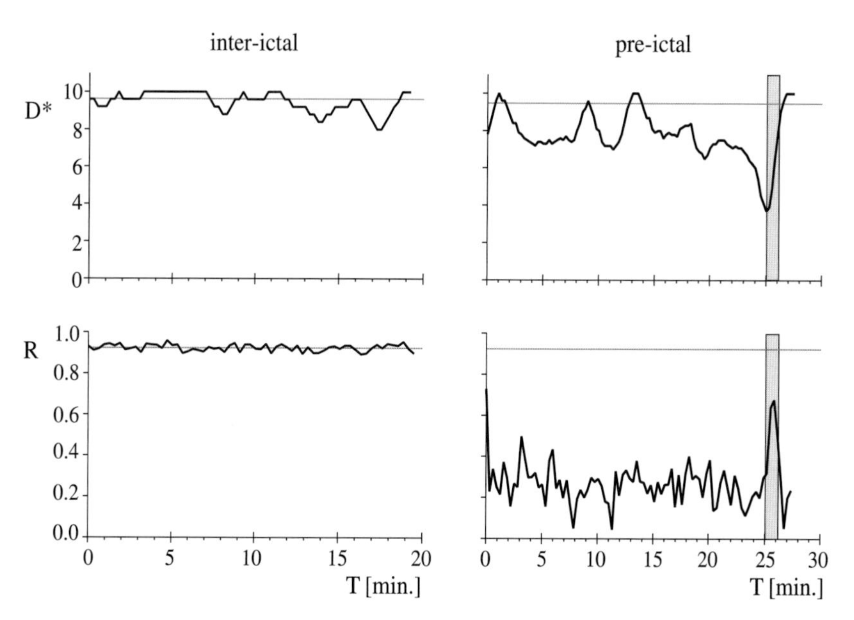

Fig. 6.3. Representative examples of discrimination between the inter-ictal states (temporally far away from any seizure) and pre-ictal states in one patient. An estimate of an effective correlation dimension D^* (see (6.5)) is shown in the *upper row*, and the mean phase coherence R (see (6.9)), a measure of phase synchronization, is shown in the *lower row*. Note that changes in D^* characteristic of the pre-seizure state occur about 12 min before seizure onset, whereas those for R occur at least 25 min ahead of the seizure. *Gray vertical lines* indicate the mean inter-ictal value for each measure. The *gray shaded area* indicates the ictal state, from onset to end of electrical seizure activity

Long-lasting pre-ictal states might possibly reflect non-specific, widespread changes that decrease the threshold for seizure activity. Short-lasting states may indicate critical recruitment phenomena within the epileptic focus and its surroundings, where hypersynchronous behavior is gradually intensified by the aforementioned generalized changes.

Although these studies have brought the possibility of anticipating epileptic seizures into sharp focus, it is not yet clear whether sensitivity and specificity of analysis techniques are sufficient [25]. This can mainly be attributed to the fact that there are currently no accepted methods for assessing performance and validating the statistical significance of seizure anticipation algorithms, although recent attempts aim to bridge this gap [78–81]. In addition, the current impact of this topic is highlighted by recent controversies about the relevance of nonlinear approaches for the anticipation of epileptic seizures [28, 73] and by studies raising doubts about the reproducibility of reported claims [79, 82, 83].

However, it is worth noting that a study on the predictability of seizures, in its most promising form, requires large amounts of high quality, continuous intracranial EEG data, which are very difficult to acquire in a busy, noisy clinical environment. In addition, despite excellent work in the field, convincing evidence demonstrating unequivocal seizure anticipation in blinded, prospective, randomized clinical trials has been elusive. There are a number of reasons for this, in addition to the challenge of developing algorithms to detect the unknown patterns associated with seizure generation: (1) there has been no accepted test data set; (2) standardized methods and nomenclature for marking continuous EEG data are missing; (3) even a clear definition of exactly what constitutes seizure onset and seizure anticipation is difficult to obtain. Recent efforts aim at minimizing these shortcomings and try to move the field forward from "proof of principle" experiments into validated, well-understood methods that can be applied in basic-science and clinical applications.

At present, it is hard to judge which seizure anticipation technique is the best. All methods seem to provide important insights into seizure generation, and they probably constitute different ways of viewing the same phenomenon. At present, there appears to be a tendency that bivariate approaches perform better than univariate ones. The combined use of different techniques along with appropriate classification schemes will probably be required to carry out reliable seizure anticipation tailored to individual patients. As a consequence, real progress requires interdisciplinary research and collaborations.

6.6 Can Epileptic Seizures Be Controlled?

Many patients with epilepsy ($\sim 30\%$) remain inadequately controlled despite optimal use of antiepileptic medication. These patients have refractory epilepsy. The administration of a new antiepileptic drug results in seizure freedom in only a low percentage of these patients. Epilepsy surgery is the treatment of choice for medically refractory patients in whom the seizure onset zone, which is responsible for the generation of habitual seizures, can be identified and subsequently resected. Epilepsy surgery results in seizure freedom in 50–85% of cases. At least 50% of the presurgical candidates will not ultimately undergo resective surgery, because a single seizure onset zone could not be identified or because it was located in functional brain tissue.

Since no adequate therapy is currently available for these patients, the possibility of anticipating seizures with a sufficient sensitivity and specificity in real-time would afford new therapeutic possibilities. One might consider miniaturized, possibly implantable, seizure anticipation and prevention devices (see Fig. 6.4), similar to cardiac devices such as pacemakers and implantable defibrillators, or to devices already in use in patients with movement disorders such as Parkinson's disease. In its most basic form a prevention device could deliver a warning to take self-protective action. One might also consider control of patient-specific prevention strategies such as biofeedback operant conditioning by applying neuropsychological or behavioral tools (sensory processing, motor task, or memory processing).

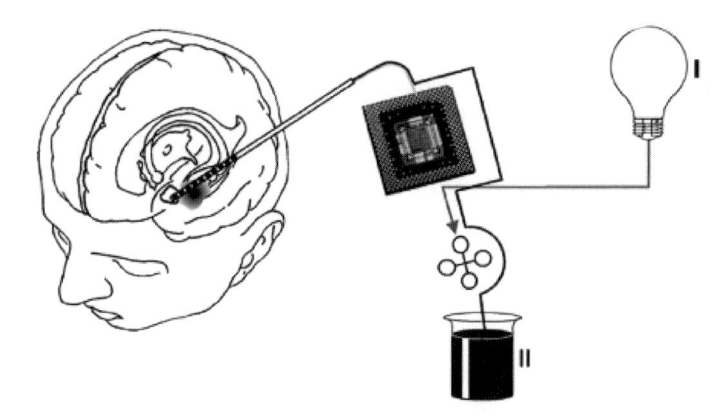

Fig. 6.4. Scheme of a miniaturized device for seizure anticipation and prevention. The EEG is recorded from electrodes implanted near the epileptic focus and it is fed to an analysis system. This hypothetical system should be powerful enough to allow both real-time extraction of features that can be used to predict an impending seizure and to enforce suitable prevention techniques (such as a *simple* warning system (I) or an *on-demand* infusion of short-acting drugs in the area of the epileptic focus (II))

Other intervention strategies in the brain aim to arrest or prevent seizures and these are guided by two major physiological paradigms: (a) modulation of abnormal cortical activity by exciting or inhibiting central structures (such as the thalamus and brainstem), and (b) intervening directly in the region of the epileptic focus.

Research into new implantable devices for treating epilepsy is expanding rapidly [84,85]. Vagus nerve stimulation is an alternative treatment that has recently become available that reduces seizure frequency by at least 50% in one third of patients, with only minor side effects. Deep brain stimulation or direct electrical stimulation of specific brain areas could be another alternative neurostimulation modality. In the past, different structures of the central nervous system have been chosen as stimulation targets in different types of epilepsy in humans, resulting in variable seizure control. Other techniques like focal cooling or localized drug infusion have been demonstrated to stop seizures where they appear to begin.

Despite the many prevention techniques currently available, convincing evidence demonstrating improved seizure control through combined use with seizure anticipation techniques has been elusive. This deficiency can partly be attributed to the fact that further optimization of analysis techniques and development of a miniaturized analysis system are necessary. Although the optimization of algorithms underlying the computation of specific characterizing measures has already allowed us to continuously track the temporal behavior of these measures in real-time, at present these applications still require the use of powerful computer systems. Nevertheless, taking into account the technologies currently available, miniaturized systems can be expected to be realized within the next few years.

In addition to these technical requirements, it is important to increase both the sensitivities and the specificities of seizure anticipation methods. Adjusting the parameters of an anticipation method to achieve higher sensitivity typically results in an increase in the rate of false alarms (decreases specificity), and vice versa. However, too many false alarms may cause patients to ignore a warning system or may lead to possible side effects from unnecessary interventions, causing physiological impairment. A clinical application achieving high sensitivity at the expense of low specificity is questionable with respect to the quality of life of patients. Frequent false anticipations may even immobilize the patients' coping processes and contribute to the patients' helplessness and depression.

6.7 Conclusions

We are currently witnessing a rapid increase in knowledge concerning the generation, predictability, and management of epileptic seizures. This can mainly be attributed to the development of refined mathematical and physical theories and analysis methods, the development of refined microscopic models of interconnected individual neurons [86], macroscopic, neurophysiologically relevant models of the EEG [87], and most importantly to the increasing willingness and acceptance of interdisciplinary research. For almost 40 years, neuroscientists thought that epileptic seizures began abruptly, just a few seconds before clinical attacks. There is now mounting evidence that seizures gradually develop minutes to hours before clinical onset. This change in thinking is based on the possibility of quantitatively study long-lasting electroencephalographic (EEG) recordings from patients being evaluated for epilepsy surgery. Advances in seizure anticipation promise to give rise to implantable devices able to warn of impending seizures and to trigger therapy to prevent clinical epileptic attacks.

Doubtlessly, research into Xevents arising in other scientific areas will fertilize the field of seizure anticipation and prevention and (hopefully) vice versa.

Acknowledgement. I gratefully acknowledge contributions by Anton Chernihovskyi, Christian E. Elger, and Florian Mormann. This work was supported by the Deutsche Forschungsgemeinschaft.

References

1. J. Engel Jr., T.A. Pedley: *Epilepsy: a comprehensive text-book.* (Lippincott-Raven, Philadelphia, PA 1997)
2. R.D. Traub, R.K. Wong: Science **216**, 745 (1982)
3. E.R.G. Sanabria, H. Su, Y. Yaari: J. Physiol. **532**, 205 (2001)
4. Y. Yaari, H. Beck: Brain Pathol. **12**, 234 (2002)
5. J. Engel Jr.: *Seizures and epilepsy.* (F.A. Davis Co., Philadelphia, PA 1989)
6. H. Berger: Archiv für Psychiatrie **87**, 35 (1929)
7. E.J. Speckmann, C.E. Elger: In *Electroencephalography*, ed. E. Niedermeyer, F.H. Lopes da Silva (Williams and Wilkins, Baltimore, MD 1999), pp 15
8. E. Niedermeyer, F.H. Lopes da Silva: *Electroencephalography.* (Williams and Wilkins, Baltimore 1999)
9. E. Ott: *Chaos in dynamical systems* (Cambridge University Press, Cambridge, UK 1993)
10. H. Kantz, T. Schreiber: *Nonlinear time series analysis* (Cambridge University Press, Cambridge, UK 2003)
11. T. Schreiber: Is nonlinearity evident in time series of brain electrical activity? In *Chaos in Brain?*, eds. K. Lehnertz, J. Arnhold, P. Grassberger, C.E. Elger (World Scientific, Singapore 2000) pp 13–22
12. F.H. Lopes da Silva et al..: Rhythms of the brain: between randomness and determinism. In *Chaos in Brain?*, eds. K. Lehnertz, J. Arnhold, P. Grassberger, C.E. Elger (World Scientific, Singapore 2000) pp 63–76
13. R.G. Andrzejak et al..: Phys. Rev. E **64**, 061907 (2001)
14. E. Başar: *Chaos in Brain Function* (Springer, Berlin Heidelberg New York 1990)
15. D. Duke, W. Pritchard: *Measuring chaos in the human brain* (World Scientific, Singapore 1991)
16. B.H. Jansen, M.E. Brandt: *Nonlinear dynamical analysis of the EEG* (World Scientific, Singapore 1993)
17. K. Lehnertz, J. Arnhold, P. Grassberger, C.E. Elger: *Chaos in Brain?* (World Scientific, Singapore 2000)
18. F. Takens: Detecting strange attractors in turbulence. In *Lecture Notes in Mathematics 898*, eds. D.A. Rand, L.S. Young (Springer, Berlin Heidelberg New York 1981) pp 366–381
19. L. Glass: Nature **410**, 277 (2001)
20. T. Sauer, J. Yorke, M. Casdagli: J. Stat. Phys. **65**, 579 (1991)
21. P. Grassberger, I. Procaccia: Phys. Rev. Lett. **50**, 346 (1983)
22. J. Theiler: Phys. Rev. A **34**, 2427 (1986)

23. P. Grassberger, T. Schreiber, C. Schaffrath: Int. J. Bifurcation Chaos **1**, 521 (1991)
24. K. Lehnertz, C.E. Elger: Phys. Rev. Lett. **80**, 5019 (1998)
25. F. Mormann et al..: Clin. Neurophysiol. (in press)
26. K. Lehnertz, C.E. Elger: Electroencephalogr. Clin. Neurophysiol. **95**, 108 (1995)
27. J. Martinerie et al.: Nature Med. **4**, 1173 (1998)
28. P.E. McSharry, L.E. Smith, L. Tarassenko: Nature Med. **9**, 241 (2003)
29. P. Grassberger, I. Procaccia: Phys. Rev. A **28**, 2591 (1983)
30. W. van Drongelen et al.: Pediatr. Neurol. **129**, 207 (2003)
31. H.R. Moser et al.: Pre-ictal changes and EEG analysis within the framework of Lyapunov theory
32. M. Le Van Quyen et al.: Neuroreport **10**, 2149 (1999) In *Chaos in Brain?*, eds. K. Lehnertz, J. Arnhold, P. Grassberger, C.E. Elger (World Scientific, Singapore 2000) pp 96–111
33. H. Kantz: Phys. Rev. E **49**, 5091 (1994)
34. L.D. Iasemidis et al.: Brain Topogr. **2**, 187 (1990)
35. L.D. Iasemidis et al.: J. Comb. Optimization **5**, 9 (2001)
36. A. Wolf, J.B. Swift, L. Swinney, A. Vastano A: Physica D **16**, 285 (1985)
37. M.T. Rosenstein, J.J. Collins, C.J. De Luca: Physica D **65**, 117 (1993)
38. H. Kantz: Phys. Lett. A **185**, 77 (1994)
39. T. Schreiber: Int. J. Bifurc. Chaos **5**, 349 (1995)
40. A.S. Pikovsky, M.G. Rosenblum, J. Kurths: *Synchronization – A universal concept in nonlinear sciences* (Cambridge University Press, Cambridge, UK 2001)
41. F. Mormann et al.: Physica D **144**, 358 (2000)
42. K.V. Mardia: *Probability and mathematical statistics: Statistics of directional data* (Academy, London 1972)
43. P. Tass et al.: Phys. Rev. Lett. **81**, 3291 (1998)
44. J.P. Lachaux et al..: Hum. Brain Mapp. **8**, 194 (1999)
45. J. Arnhold et al.: Physica D **134**, 419 (1999)
46. T. Schreiber, A. Schmitz: Physica D **142**, 346 (2000)
47. C.D. Binnie et al.: Electroencephalogr. Clin. Neurophysiol. **58**, 498 (1984)
48. J.G. Milton et al.: Epilepsia **28**, 471 (1987)
49. M. Balish, P.S. Albert, W.H. Theodore: Epilepsia **32**, 642 (1991)
50. E. Tauboll, A. Lundervold, L. Gjerstad: Epilepsy Res. **8**, 153 (1991)
51. L.D. Iasemidis et al.: Epilepsy Res. **17**, 81 (1994)
52. J. Bauer, W. Burr: Seizure **10**, 239 (2001)
53. W.G. Lennox: *Science and seizures.* (Harper, New York 1946)
54. R.B. Aird: Epilepsia **24**, 567 (1983)
55. P. Rajna et al.: Seizure **6**, 361 (1997)
56. P.B. Prichard, V.L. Holstrom, J. Giacinto: Ann. Neurol. **18**, 265 (1985)
57. S.S. Viglione, G.O. Walsh: Electroencephalogr. Clin. Neurophysiol. **39**, 435 (1975)
58. Z. Rogowski, I. Gath, E. Bental: Biol. Cybern. **42**, 9 (1981)
59. R.B. Duckrow, S.S. Spencer: Electroencephalogr. Clin. Neurophysiol. **82**, 415 (1992)
60. H.H. Lange et al.: Electroencephalogr. Clin. Neurophysiol. **56**, 543 (1983)
61. J. Gotman, M.G. Marciani: Ann. Neurol **17**, 597 (1985)

62. J. Gotman, D.J. Koffler: Electroencephalogr. Clin. Neurophysiol. **72**, 7 (1989)
63. A. Katz et al.: Electroencephalogr. Clin. Neurophysiol. **79**, 153 (1991)
64. C.E. Elger, K. Lehnertz: Ictogenesis and chaos. In *Epileptic seizures and syndromes*, ed. P. Wolf (J. Libbey & Co, London 1994) pp 547–552
65. C.E. Elger, K. Lehnertz: Eur. J. Neurosci. **10**, 786 (1998)
66. M. Le Van Quyen et al.: Eur. J. Neurosci. **12**, 2124 (2000)
67. M. Le Van Quyen et al.: Lancet **357**, 183 (2001)
68. V. Navarro et al.: Brain **125**, 640 (2002)
69. I. Drury et al.: Exp. Neurol. **184**, S9, (2003)
70. B. Litt et al.: Neuron **30**, 51 (2001)
71. S. Gigola et al.: J. Neurosci. Meth. **138**, 107 (2004)
72. F. Mormann et al.: Epilepsy Res. **53**, 173 (2003)
73. F. Mormann et al.: Phys. Rev. E **67**, 021912 (2003)
74. K. Schindler et al.: Clin. Neurophysiol **113**, 604 (2002)
75. J. Clin. Neurophysiol. ("Special Issue on Seizure Prediction") **18**, 191 (2001)
76. B. Litt, K. Lehnertz: Curr. Opin. Neurol. **15**, 173 (2002)
77. H. Witte, L.D. Iasemidis, B. Litt. IEEE Trans. Biomed. Eng. ("Special Issue on Epileptic Seizure Prediction") **50**, 537 (2003)
78. R.G. Andrzejak et al.: Phys. Rev. E **67**, 010901 (2003)
79. R. Aschenbrenner-Scheibe et al.: Brain **126**, 2616 (2003)
80. M. Winterhalder et al.: Epilepsy Behav. **4**, 318 (2003)
81. T. Kreuz et al.: Phys. Rev. E **69**, 061915 (2004)
82. Y.C. Lai et al.: Phys. Rev. Lett. **91**, 068102 (2003)
83. W. de Clerq et al.: Lancet **361**, 970 (2003)
84. B. Litt: Epilepsia **44**(Suppl. 7), 30 (2003)
85. W.H. Theodore, R.S. Fisher: Lancet Neurol. **3**, 111 (2004)
86. R.D. Traub, J.G.R. Jefferys, M.A. Whittington: J. Comput. Neurosci. **4**, 141 (1997)
87. F. Wendling et al.: Biol. Cybern. **83**, 367 (2000)

7 Extreme Events in the Geological Past

Jürgen Herget

Summary. Many Xevents in the geological past exceeded the strengths and intensities observed for modern-day natural events. The number of extraordinary events that occurred in the geological past is of course much larger than the number we witness today because the geological timescale covers millions of years. This contribution focuses on these Xevents from earth's geological history, including selected examples from plate tectonics, earth magnetism, ice age cycles, volcanism, earthquakes, meteorite impacts and floods. Events related to these processes occur on different timescales. For example, drastic modifications of atmospheric and oceanic circulation due to continental shift (which creates new mountain ranges and reshapes land masses and oceans) take millions of years, while meteorite impacts happen within seconds. However, any these processes can be the trigger for dramatic consequences, like mass extinctions of life, or global glaciations. An overview of a research program that considers historic and prehistoric flood events is given. Based on the water levels observed during floods, the palaeodischarge can be determined and used to improve the reliability of flood predictions. Investigations of Pleistocene ice-dammed lake outburst floods (the largest flood events in the Earth's history) are useful when developing new methods and techniques that can be applied to younger events of a smaller scale in other environments.

7.1 Introduction

This contribution focuses on extraordinary events that occurred on the Earth in prehistoric times. Although this work concentrates on the prehistoric timescale, extraordinary events have of course occurred in more recent times and they will continue to occur, as illustrated by other contributions in this book, such as the one by Hense on the climatic system. Events in modern times are frequently characterised as being catastrophes due to the impact on human life and infrastructure. However, as we will see from the selected examples used in this chapter, even the most extraordinary volcanic eruptions, earthquakes or floods observed in recent times were significantly exceeded in scale many times in the prehistoric past. Therefore, while recent Xevents might be extraordinarily strong compared with other recent events, if we consider much longer timescales reaching far back into Earth's history these events may not seem so extreme after all. The relatively high number of Xevents that have occurred over thousands of millions of years (the geological timescale) leads us to the question of the frequencies of various Xevents (see Table 7.1).

The important issue here is to find out the cause of the event, be it a unique mechanism based on specific conditions or some other plausible explanation for the stochastic nature of the individual phenomenon. In the context of this contribution, Xevents are seen as prehistoric events that were strong enough to leave evidence of their occurrence that has survived until today. Recently, mankind has begun to have an influence on the natural environment, and especially its dynamics [1], that will probably outlast our civilisation. The word "anthropocene" has been coined, characterising the period of time that humans have made a dominant impact on the Earth (since the eighteenth century), so the geology of mankind is now being written [2].

In the following survey, selected Xevents in the geological past are presented and briefly reviewed along with key references for further reading. We also touch on the background of any scientific debates about the Xevents. The order that the examples are presented is roughly based on the durations of the Xevents. These Xevents may last from millions of years (the folding of mountain ranges until their final decay by denudation), which might still be seen as relatively short compared to the age of our planet, down to seconds (meteoritic impacts). In keeping with the general theme of this book, we close the chapter by discussing approaches to prognosing and forecasting Xevents over geological timescales.

7.2 Extreme Events in the Geological Past

7.2.1 Events Driven by Plate Tectonics

The concept of plate tectonics, which explains the gradual creation and erosion of mountain ranges and the slowly changing shapes and locations of the continents and oceans by invoking continental plate drift, was first mentioned in the 1920s by Alfred Wegener, a German meteorologist. It became an accepted theory among geologists in the 1960s, when evidence for it was gained via ocean drilling projects. The current generally accepted explanation for continental drift is that circulation cells exist within the liquid part of the earth's interior, upon which the Earth's crust sits. The flow within these cells drags the relatively thin crust sitting on top of them [3–5]. Typical drift velocities are on the order of cm/a, which is very fast on the geological timescale.

The positions and shapes of the continents and oceans have been changing continuously since the "supercontinent" Pangea was split apart (Fig. 7.1), about 200 Ma years ago. At the beginning of the Cretaceous, about 135 Ma ago, North America and Europe/Asia were still connected, while India and Antarctica/Australia became separated from the Gondwana supercontinent and started to move in a north-easterly direction. At the end of this geological period, North-America and Eurasia began to split, resulting in the formation

Table 7.1. Generalised geological timescale (compiled from various sources). The geological timescale is classified into eras, periods and epochs. The given names are derived from key locations of geological research or separate distinct units of more or less homogenous fauna and flora. Note that time periods typically extend for millions of years. Hence, human observations are negligible in the context of the timescale of the geological past, as the first traces of mankind developed about 1.5 Ma ago (Ma = million years ago) and written history began just a few thousand years ago. Periods of mountain folding are explained in the context of plate tectonics in the text, while global glaciations are discussed in the section on ice ages

ERA Period Epoch	**Millions of years (since start of time period)**	**Periods of mountain folding**	**Global glaciations**
CENOZOIC			
Quaternary			
Holocene	0.01		
Pleistocene	2.6		numerous
Tertiary			
Pliocene	5.3		
Miocene	24		
Oligocene	37		
Eocene	58	Alpine orogenesis (Alps, Pyrenees, Himalayas, ...)	
Palaeocene	65		
MESOZOIC			
Cretaceous	135		
Jurassic	205		
Triassic	250		
PALAEOZOIC			
Permian	290		several (350–250 Ma)
Carboniferous	355	Variscan orogenesis (central and west European mountain ranges, Appalachians)	
Devonian	405		
Silurian	435	Caledonian orogenesis (Scandinavia, Scotland, ...)	
Ordovician	510		
Cambrian	545		

Table 7.1. (continued)

ERA Period Epoch	**Millions of years (since start of time period)**	**Periods of mountain folding**	**Global glaciations**
Precambrian	> 545		several (800–550 Ma)
	3900	Oldest rocks	
	4450	Cooling of the liquid Earth	
	4600	Earth is formed	

of the Atlantic Ocean, while the continental plates of Africa and Eurasia collided. Parts of the Tethys Ocean closed and the folding of the seafloor began to generate the Alps. Then the Indian plate collided with the Asian plate causing the Himalayas to rise in the collision zone. The Australian-Antarctica plate split into several fragments, while South America linked with North America, closing the connection between the Atlantic and the Pacific. Based on modern drift velocities, it has been possible to forecast the movements and positions of the continents in the future. In about 30 Ma the African plate will move further to the west and split at its eastern margin at the so-called Afar Triangle. It is also expected that the Persian Gulf will close and the land bridges between North and South America and in South-East Asia will change.

Even though the timescales of events related to continental drift are measured in millions of years, such changes in alignments of the continental plates might be classed as events, since the age of our planet in measured in billions of years (Table 7.1). Sealing oceans from each other has a huge influence on the currents within the oceans, while the creation of high mountain ranges changes atmospheric circulation patterns. New land bridges are also highly relevant to events associated with biological evolution and species distribution (consider, for example, the species endemic to Australia).

Beyond the effects mentioned above, that occur on the geological timescale, plate tectonics also produce earthquakes. The movements (and therefore collisions) of continental plates are not smooth and continuous on shorter timescales (for example, years). Frictional stress is released abruptly and results in earthquakes (covered in more depth later).

The movement of continental plates away from each other is accompanied by the generation of new rock of magmatic origin on the seafloor between the continental plates. This magmatic rock can be used to derive important information on the history of the magnetic field of this planet.

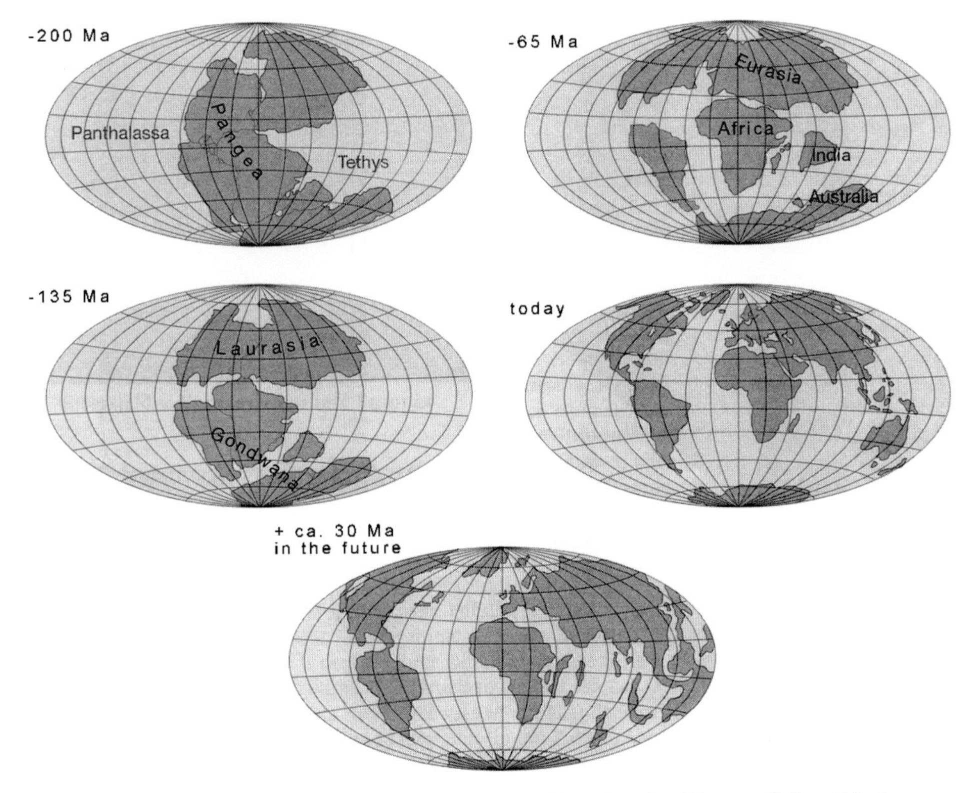

Fig. 7.1. Continental drift from prehistory to modern day (and beyond) (modified from [6])

7.2.2 Changes in the Earth's Magnetic Field

The fact that the Earth generates a magnetic field was first discovered in China at around 1000 AD, although magnetism itself was known to the ancient Greeks [7]. This magnetic field is generated by the circulation of material within the Earth's liquid metal core [3]. Most people are aware that the Earth's magnetic pole is close to but not in exactly the same place as geographic north, due to the recent movement of the magnetic north pole. In 2004 the magnetic north pole was located around 82.3° N 113.4° W, although it wanders in a well-defined oval (with a length of about 85 km) each day [8]. Over the last century the magnetic north pole moved about 1100 km towards its current position from a location further south, and it is expected to continue its movement across the geographic north pole to northern Siberia over the next 50 years [8].

On the geological timescale, the Earth's magnetic field has often changed its orientation – in other words, the geographic north became the magnetic south pole and vice versa. The last reversal occurred about 780000 years

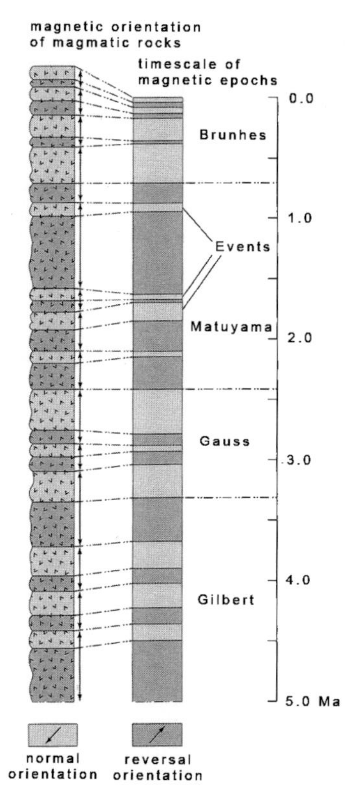

Fig. 7.2. Changes of the orientation of the magnetic field of the Earth over the last 5 Ma years (modified from [9])

ago (Fig. 7.2). Typically, the timescale between these switches is measured in millions of years. The long epochs of one orientation (each of which are named after scientists involved with magnetic research) are interrupted by "events" where the orientation briefly flips back. Hence, the orientation of the magnetic field changes roughly every 200000 years, although there is no obvious periodicity.

Investigations into the prehistoric magnetism of the earth have been made possible because the orientation of the magnetic field is documented in magmatic rocks [10]. As they cool, the magnetic minerals in the liquid lava align according to the current orientation and strength of the magnetic field. By measuring this alignment and isotopically dating the rocks, it is possible to investigate the geomagnetism at the time the rocks cooled. For example, the magmatic rock between drifting continental plates, provides useful geomagnetic information. This is illustrated by the characteristic symmetrical pattern of magnetism present on the seafloor of the Atlantic Ocean parallel to the central oceanic ridge, which documents the changes in the Earth's magnetism since the separation of Europe and North America. For older rock samples the time resolution decreases and data must be collected and correlated from different places around the world, and the continental drift and the rotations of the plates must be considered.

If we consider them on the geological timescale, changes of magnetic epoch can be regarded as "events", as can the brief flips in orientation that occur within epochs.

The magnetic field and the atmosphere of the Earth protect its surface from the dangers of the solar wind, which produces the aurora borealis (the "Northern Lights") seen in polar regions. Changes in geomagnetism (and solar wind strength) influence radio transmission, flow through pipelines, electrical power supplies and animal migrations. We should note that the strength of the magnetic field of the Earth dropped at various times in prehistory. A statistical analysis of magnetic reversal appears infeasible since this phenomenon occurs aperiodically. However, extrapolating the current decay in field strength into the future, the next "flip" could take place in about 1500 years. It is assumed that this flip would take about 5000 years, during which field strength would drop to about 20% of the current intensity and the geometry of the pole would fluctuate [11]. This would lead to disruption of most electric power and radio transmission networks – a serious event in our modern technology-based civilisation.

7.2.3 Periods and Cycles of Ice Ages

Several periods with extended glaciation, typically covering about 30% of the surface, have been discovered when investigating the prehistory of the Earth (Table 7.1). Typical glacial deposits (like moraines) have been found on almost every continent, providing evidence for various periods of glaciation that have occurred since Precambrian times [12]. We should bear in mind that for times this far back, different oceans, continents and especially mountain ranges were present compared to those around today (due to continental drift) [13]. This means that it is difficult to compare these older glaciation periods with more recent (and better understood) glaciations.

The Pleistocene epoch of the Quaternary period is characterised by temporary glaciation periods lasting from 2.6 Ma until the current warm period, called the Holocene, starting about 10000 years ago (10 ka). Scandinavia, Northern America and Siberia were all covered with vast ice fields, and extended valley glaciers in nearly all of the mountain ranges worldwide document the temporary global climatic shift that occurred during the Pleistocene [14,15]. Figure 7.3 illustrates the extent of the glaciation in the northern hemisphere 18000 years ago, during the peak of the last ice age.

The Pleistocene epoch is characterised by cyclic warm and cold periods. Depending on local conditions, glacial deposits and glacigenic landforms usually derive from the last three or four glaciations, because previous deposits will have been mantled. Parameters like stable oxygen isotope ratios can be used to evaluate climate conditions associated with rocks (see [17]). Based on

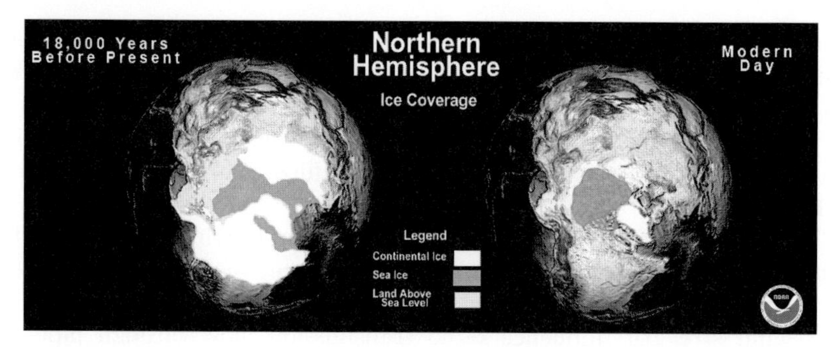

Fig. 7.3. Ice coverage of the northern hemisphere 18 ka ago and today, indicating the extent of glaciation during the last ice age (figure from [16], used with the kind permission of NOAA)

this indirect indicator, 51 warm and 52 cold periods have been found for the Pleistocene [18]. Considering this timescale, we are currently living in a warm period. For the last 800000 years, the typical duration of the cold period has been about 100000 years, while warm periods have typically lasted for about 10000 years. The mechanism of global climatic change resulting in extended ice coverage seems to be triggered by periodic overlaps between the Earth's orbital cycles. These cyclic changes modify the insolation (incoming solar radiation) for different parts of the Earth, which eventually results in global cooling. It is not fully understood why the glaciations cycle changed around 800000 ka [19], but energy exchange via ocean currents may be a significant factor [20]. Glaciations in Palaeozoic times appear to be more closely related to the positions of the continents, as the continental plates were located in polar positions (like Antarctica is today) during glaciations. However, new analysis tools are increasing our understanding of the connection between cyclic orbital parameters and the geological climate, even for the earliest times [21].

Obviously, ice ages can be seen as events on the geological timescale. These glaciated landscapes were similar to the ice shields of Greenland or Antarctica today. These climatic changes were naturally driven and much stronger than any scenario developed for current climate trends. Habitats for animals and plants were mechanically destroyed by the moving ice flow and glaciers and species distributions were modified by the changes in climate [19, 22]. The global hydrological cycle changed due to the storage of huge amounts of water in the ice. This produced a drop in sea levels, which opened land bridges between islands and continents, resulting in, for example, the migration of man from eastern Siberia and Alaska to North America.

Based on the evidence left over from the Pleistocene epoch, one might expect that the next ice age is imminent, especially when viewed against the background of the geological timescale. The current retreat of mountain glaciers represents a short-term climatic change. Recent modelling has revealed that the anthropogenic enrichment of greenhouse gases in the atmosphere has even decelerated the natural cooling trend of the global climate [23]. The so-called Little Ice Age was a recent global climate fluctuation within the current warm stage of the Holocene that lead to a dramatic increase in mountain glaciation [24], but one should bear in mind that the the mountain glaciers of the Alps are usually smaller than they are currently [25]. Previous glaciation periods started with intensive climate fluctuations, but current anthropogenic influences on the climatic system could modify this pattern [19, 26].

7.2.4 Volcanism

Numerous volcanic eruptions and explosions have been documented throughout history, most of which have had a tremendous influence on local and regional living conditions. Volcanic eruptions are undoubtedly (small-scale) Xevents: the landscape is modified due to new layers of solid rock, which is then weathered intensively over many years until new soil is created that can be used for plants and cultivation. Eruptions in historic times destroyed towns and villages, frequently killing thousands of people – the historic eruptions of Thera/Santorini (around 1500 BC ended the Minoan culture, Vesuvius/Somma destroyed Pompeii in 79 AD and the events at Tambora in 1815, Krakatoa in 1887 or Pinatubo in 1991 are well-known examples of extraordinary volcanic events. These eruptions were obviously strong enough to influence the global climate [27], even if we do not fully understand the detailed mechanisms involved [28].

On the geological timescale, volcanic eruptions have been identified that were much stronger and more intense than any of the events observed in historic times. Large areas of all of the continents are covered by lava sheets (plateau basalts) that are related to "hot spots" (Fig. 7.4). Hot spots are fixed locations where magma from the Earth's interior has been transported towards the surface over long geological periods [3]. As the continental plates drift across these locations, chains of volcanoes are generated on the surface. The islands of Hawaii are a famous example of this phenomenon: the volcanoes on the youngest island, Hawaii, are still active, while islands like Oahu and Kauai further to the west are inactive. The continental drift explains why the plateau basalts are now located several thousand kilometres away from their related hot spots (Fig. 7.4).

Some scientists have found a remarkable correlation between the ages of the extended plateau basalts and periods of mass extinction (Table 7.2) (reviews by [30,31]). If we take the example of the Deccan Trapp in India, which were generated at the same time as the lava sheets in the north Atlantic, it has been argued that these volcanic events triggered the mass extinction at the end of the Cretaceous Period (resulting in the extinction of dinosaurs) [30]. Originally the Deccan Trap covered an area of more than 2000000 km^2 and was largely generated over a period of less than 500000 years (as shown

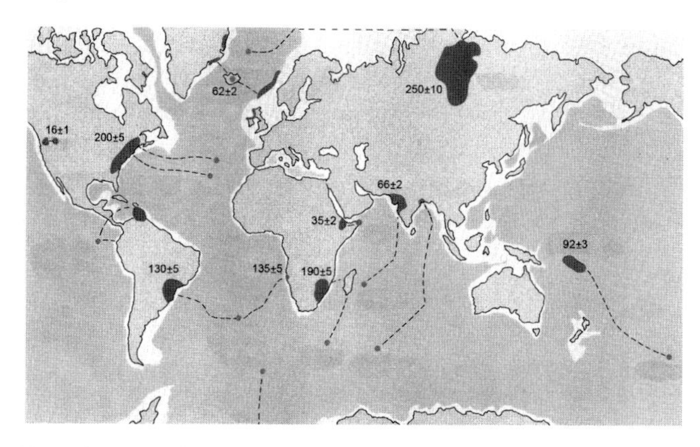

Fig. 7.4. Distribution of extended plateau basalts (with their ages in Ma) and related hot spots (modified from [29])

by dating the lava by isotope methods and performing a detailed analysis of the orientation of the Earth's magnetic field). Courtillot argues that the huge amount of dust that was blown into the atmosphere during the volcanic eruption significantly reduced insolation, which reduced plant photosynthesis. Food resources were therefore significantly reduced. This final effect was intensified by acid precipitation caused by gases released from the eruption. Careful review of the data on the extinction of species 65 Ma ago and related sediments indicate that it was a fast and abrupt event on the geological timescale, but lasted several hundreds of thousands of years. This is the main reason why these types of huge volcanic eruptions could be the trigger for the extinction of species at various geological times (Table 7.2). However, we should note that another mechanism has been proposed for mass extinctions: meteoritic impacts, which we discuss later.

It is worth mentioning, that even modern-day volcanic eruptions are sustainable; typically, single eruptions last for days to weeks and modify the environment drastically on local and regional scales over very short timescales. However, we should compare this with the fact that there were volcanic events on the geological timescale that had a significant global impact.

7.2.5 Earthquakes

Movements of the earth's crust related to plate tectonics, pressure releases on continental plates and volcanic eruptions are the main causes of earthquakes. These dynamics mean that millions of earthquakes occur on an annual basis, while thousands occur daily, as observed by seismic observation networks covering the entire planet. Threshold values are applied to separate large from ordinary events. By magnitude alone, the largest earthquake in history occurred in 1960 in Chile, with a magnitude of 9.5, while the most destructive one took place in China in 1556 AD, causing about 830000 deaths [32].

Table 7.2. Mass extinctions of species, eruptions of plateau basalts, and related hot spots (after [29])

Mass extinction at (time in Ma)	Eruption of plateau basalts at (Ma)	Related hot spot
14 ± 3	Columbia River (NW USA) 16 ± 1	Yellowstone
36 ± 2	Ethiopia 35 ± 2	Afrar
65 ± 1	North Atlantic/Arctic 62 ± 2 Deccan Trap (India) 66 ± 2	Iceland Reunion
91 ± 2	West-Pacific 92 ± 3	Central Pacific
137 ± 7	Parana (Brazil) 130 ± 5 Namibia 135 ± 5	Tristan da Cunha (mid-Atlantic)
191 ± 3	SE Africa 190 ± 5	South Atlantic Ocean/ SW Indic Ocean
211 ± 8	Eastern USA 200 ± 5	Azores
249 ± 4	Siberia 250 ± 10	Jan Mayen (Arctic Ocean)

There seems to be no reason to expect significant changes in the frequency of earthquakes on a global scale in the geological past. However, the effects of plate tectonics and the displacements of collision and spreading zones must be considered. Hence, regions with relatively high earthquake frequencies were located in different places in geological times. Earthquakes in the prehistoric past left traces in the sedimentary record. In most depositional environments, earthquakes cause disturbances in the regular steady accumulation of sediments. Comparing the disturbances within the deposits with the influences of modern earthquakes in similar environments has lead to estimations for the magnitudes of palaeoearthquakes [33] and prehistoric earthquake frequencies (see [34]).

7.2.6 Meteoritic Impacts

Although meteorites commonly reach the Earth's atmosphere and surface, most of them burn up completely on their way through the atmosphere (producing a "shooting star" in the night sky roughly every ten minutes), while the vast majority of those that do make it to the surface are relatively small. Most meteorites consist of rock while a smaller fraction are iron or an iron/rock mixture. Annually, several thousand tons of meteorites reach the Earth's surface [35–37]. Most of the > 20000 meteorites that have been recovered were found in Antarctica, due to the limited weathering that occurs under polar climate conditions and the ease with which they can be spotted in the ice. The largest iron-based meteorite found has a weight of 54 tons and was found in 1920 in Namibia, while the largest rock meteorite (1.8 tons) was detected in China.

Meteoritic impacts have been noted throughout history. Descriptions of falling fireballs and strange light phenomena that hit buildings and someimes even set them alight were handed down [38]. One of the more spectacular recent events was the impact that occurred on the 30th of June 1908 in the valley of the River Tunguska, Siberia. An eye-witness who was more than 100 km away from the impact area, observed a flash of light which became a fireball, and felt a wave of heat before he was knocked by an explosion. The change in air pressure was also measured thousands of kilometres away in England using barographs. The earth tremor from the impact was felt for several hundreds of kilometres around the impact site and was measured across the world. Clouds of dust rose up to 20 km into the atmosphere and were later precipitated in rain. Due to the remote location, it took almost 20 years before the impact area was investigated scientifically. All of the forest was burned out to about 65 km from the impact area, while no impact crater or remnants of the meteorite itself could be found. Subsequent theoretical calculations have estimated that the mass of the meteorite was about 1000000 tons, and it is believed that the meteorite exploded in the air before reaching the surface [39]. Hundreds of impact craters with diameters of 0.015–300 km have been found that date from prehistoric times, and most of them are still visible as craters [39]. Figure 7.5 shows the Canyon Diabolo/Barringer Crater in Arizona, generated by a meteoritic impact that occurred 49 ± 3 ka ago; this crater is currently 1200 metres in diameter with a depth of 180 m (note that erosion has reduced the size of the crater over time).

A large meteoritic impact that occurred during the transition from the Cretaceous to the Tertiary period has recently received a great deal of attention, since it has been linked to one of the largest mass extinctions to occur in prehistory: including all f thedinosaurs and half of all species. The impact occurred north of Yucatan, Mexico [41–43]. The Chicxulub Crater

Fig. 7.5. Aerial photograph of Diabolo Canyon/Barringer meteorite impact crater in Arizona (photo provided by [40])

associated with the impact has a diameter of 170–200 km, and was generated 64.98 ± 0.05 Ma ago; it now lies buried beneath 1100 m of limestone on the seafloor of the Gulf of Mexico. It is assumed that the meteorite involved had a diameter of about 10 km. During the impact, the meteorite itself, together with billions of tons of displaced rock, evaporated and generated a global cloud of dust, causing darkness for between one and six months, stopping any photosynthesis for about two to twelve months. The darkness seriously disrupted the food chains of marine and terrestrial life. Evidence for this cloud of dust is provided by a layer of clay deposited around the world during this time, which is enriched in iridium, an element that is rarely found in rocks on Earth but common in meteorites. While the existence of a catastrophic meteorite impact is a fact, its relation to the mass extinction is still the topic of debate, since it can also be related to a period of intense volcanism (see above). The iridium enrichment and the specific structure of the quartz found around the impact location and within the separation layer between Cretaceous and Tertiary deposits can also be explained in the context of volcanism [30]. The scientific debate on the cause of the mass extinction is therefore still open.

7.2.7 Floods

Floods occur frequently in all climate conditions. Caused by precipitation, melting snow or ice in spring, exceptional water levels and discharges can be reached in very short times in rivers. Frequently, the damage that results is unexpected, even when similar floods have been observed and documented for the same region throughout history [44,45]. Statistical analysis of water level measurements have been found to be inadequate for predicting floods, due to the limited time of observation and the stochastic generation of extrapolated values. Indirect methods of estimating flood discharges that are documented in the sedimentary record from prehistoric times have been developed, and an increasing importance has been placed on them in relation to the prediction of climatic/hydrologic changes in the future [46,47]. Extreme flood events can be generated by the failure of dammed natural lakes, like those generated by subglacial meltwater storage of volcanic eruptions in Iceland or where landslides have blocked rivers, causing temporary lakes [48].

The largest floods observed in geological times occurred during the last glaciation period about 20000 years ago in different parts of the world, due to outburst floods from ice-dammed lakes. Extraordinary large in this context means that these floods were orders of magnitude larger than any flood caused by precipitation observed in historic times, like that of the River Rhine (12600 m^3/s) in 1926, the River Amazon (370000 m^3/s) in 1953, or even the flood related to a subglacial volcanic eruption in Iceland (1500000 m^3/s) in 1918 [48,49]. Ice-dammed lakes are typically generated by advancing glaciers and ice sheets that block the course of a river, damming temporary lakes upstream. Due to the pressure of the stored water or the melting of the ice dam by the liquid water behind, sooner or later the ice dams fail and give way to the dammed water [50, 51].

One well investigated example of this is the Glacial Lake Missoula, generated in the north-western part of the USA. The advanced valley glaciers of the Rocky Mountains blocked the Clark Fork River and generated Lake Missoula, with a maximum volume of 2184 km^3 and a maximum depth of 635 m [52]. At least 25 outburst floods occurred in this area between 19000 and 13000 years ago [53], with peak discharges of more than 17000000 m^3/s. These outburst floods formed the unique feature of the Channeled Scabland. Outburst floods within the plateau basalts of the Columbia Plateau in eastern Washington generated numerous channels which are dry valleys and waterfalls with heights of more than 100 m today. After the lake had drained, the glacier tongue advanced again, blocked the river and generating the lake again. This cyclic behaviour explains the repeated outburst floods that were discovered due to the related deposits that were obviously separated from each other.

Similar events took place in the Altai Mountains, the source of the River Ob in southern Siberia, near the border to Mongolia. Again, valley glaciers blocked the course of the River Chuja, one of the sources of the River Ob, and generated an extended ice-dammed lake in intra-mountainous basins upstream with a volume of 603 km^3 and a maximum depth of 650 m near the ice-dam [49]. During repeated outburst floods – evidence for at least three events has been found in the sedimentary record in the field – the flood (with peak discharges of 10000000 m^3/s) was canalised within the mountain valleys of the Rivers Chuja and Katun, and left deposits indicating a flow depth of 250–400 m that remain today (Fig. 7.6).

Fig. 7.6. Surface deposits related to the ice-dammed lake outburst flood in the Altai Mountains, Siberia (view downstream). The deposits indicate a depth of flow of 310 m in this location. Run-up sediments deposited 124 m above the bar surface in the foreground are visible in front of the valley obstruction beyond the edge of the bedrock on the right side (photograph by J. Herget)

Outburst floods from large (continental-scale) ice-dammed lakes have been found to influence global climate conditions. Lake Agassiz, located in southern Canada in Pleistocene times, covered an area of about 150,000 km^2 and contained a volume of more than 163,000 km^3 [54]. Overflow through spillways influenced discharges of the Mackenzie and Mississippi Rivers, while subglacial outburst floods were even strong enough to block currents in the Atlantic Ocean. Outburst floods below the continental ice sheet covering most parts of northern Canada in these times can be related to global cold events that occurred during the final stage of the Pleistocene and even during the early Holocene [55, 56].

7.3 Predictions and Forecasts on the Geological Timescale

The key to predicting and forecasting the events presented here is the timescale. The events mentioned above require prognosis to protect man from related catastrophes in the near future, but in terms of the natural dynamics of the entire planet they are not of much real relevance. Earthquakes and local volcanic eruptions illustrate this difference in perspective towards the phenomenon: people might be victims of earthquakes and volcanic eruptions – thousands might be killed – but those events are not necessarily extreme when compared with others on the geological timescale from a global perspective.

Different approaches have been applied to predict the previously mentioned events:

- Based on the currently observed drift velocities, it has been possible to forecast the movements of continents and oceans, as shown in Fig. 7.1. Even though the movement of continental plates is very fast on the geological timescale, the opening or closing of ocean connections or the folding of new mountain ranges happens over periods of millions of years.
- Earthquakes are typically related to shifts of continental plates. They are observed using global seismic networks. The sedimentary record of palaeoearthquakes has been investigated in detail in order to increase our knowledge of the frequency and magnitude of earthquakes in specific areas.
- The intensity of the magnetic field of the Earth fluctuates with an apparently random frequency. As mentioned above, the current decrease in its strength cannot be extrapolated into the future with any real assurance, so the dynamics of the field in the future remain unsure and cannot be predicted at the current stage of knowledge.
- Global climate changes from warm periods to ice ages have occurred cyclically over the last 800000 years. Based on a statistical approach, the next cold period should be expected to occur within the next few centuries.

However, increasing levels of anthropogenic greenhouse gases in the atmosphere will probably modify the natural dynamics of this cycle. This anthropogenic aspect complicates our understanding of natural global climate changes and prevent reliable forecasts on the longer timescale. This aside, the analysis of previous climatic changes in prehistory, documented in various natural archives, provides a unique way to understand the processes related to climatic change. Without knowledge of these natural processes and former modes of change, even the most ambitious quantitative models and simulations will fail due to missing calibration data.

- Active volcanoes are carefully observed if they are located close the inhabited areas. New periods of activity are usually announced by specific tremors caused by the magma rising inside and below the volcano. The periods of increased volcanic activity, especially those lasting thousands of years, have gained importance in relation to mass extinctions only after studying them on the geological timescale. Against this background it is obvious that these events cannot be predicted at the time but noted afterwards.
- Meteorites frequently reach the Earth's atmosphere and even its surface. The environs of our planet is carefully watched by astronomers, who are yet to find a large meteorite that could collide with our planet in the future. Statistical analysis of the frequency of meteorite impacts indicates that events comparable with that which formed the Chicxulub Crater occur about once every 50–100 million years, although impacts that modify the global climate are expected to occur every 2–3 million years [39].
- Extrapolating from measured events via statistical methods is a common approach taken to predicting the magnitudes and frequencies of floods. Some approaches have extended their datasets by taking into account both historic and prehistoric events. Ice-dammed lake outburst floods (and their magnitudes in particular) are difficult to predict due to uncertainties with the outburst mechanisms. For modern events, a black-box regression of drained lake volume versus peak discharge is used and found to be of remarkable accuracy, even if local differences cannot be considered (reviews in [49, 57]). For large events like those that occurred in Pleistocene times, the database for statistical analysis is too limited [49].

The misinterpretation of events and phenomena is a familiar problem in the earth sciences [58]. Cause and consequence and the interaction with the geological system on a suitable spatial and temporal scale must be carefully considered. The problem of integrating extraordinary Xevents into the concept of the uniformitarianism of the dynamics of the Earth system has been recognised, and the special term "cataclysmic process" has been given to such events [59].

Although the Xevents that we have described in the geological past are very rare and thus might appear of academic relevance only, if we consider the issue in more depth – for example in relation to radioactive nuclear waste repositories [60] – it becomes clear that Xevents that occur on the geological timescale still require serious attention.

7.4 Research Perspectives

"The past is the key for understanding recent processes in earth sciences – hence, the fundament for predictions." This statement expresses the basis for earth sciences and explains the focus of many working groups in different disciplines on the geological past. The idea is to estimate magnitudes and frequencies of Xevents from the natural archives in order to gauge the probability of reoccurrence. This approach works quite well over the longer geological timescale. New methods and techniques are being developed by specialists in related disciplines to help us to read and understand the information about Xevents in the geological past documented in rocks and ice.

For shorter time periods, this attempt at prediction occasionally fails, which is illustrated by flood predictions based on reoccurrence intervals of observed flood events in modern times. Measured floods are statistically analysed by stochastic approaches to predict frequencies of Xevents (such as the once in hundred or thousand years flood). Modern large floods modify the stochastic magnitude–frequency relationship in such a way that extraordinary events become usual: a reoccurrence interval of once in hundred years is reduced to once in ten years. Obviously, this approach depends on the range of events considered, which depends on duration of measurements and stationary conditions of flood frequencies. This problem has been recognized [61,62] and illustrated by case studies in various places (e.g. [63–65]), but conclusions are yet to be drawn.

One way to approach this problem is to include events that took place before the era of quantitative observations. It is possible to gauge extreme flood events from historical documents and flood marks on old buildings. Using this approach, the frequencies of floods can be determined back to late mediaeval times in some locations, as illustrated by the flood events observed at the River Rhine near Bale (Fig. 7.7).

This qualitative information can be converted into quantitative discharge data for given water levels via hydraulic calculations [46, 49]. Hence, additional flood magnitudes can be estimated and added to recent measured data. This approach allows us to include Xevents that took place before the scientific era, increasing the amount of data, particularly for the prediction of large flood events. One of these approaches is to estimate the mean flow velocity using the empirical Manning equation (see [68]). Using this and the given water level (which limits the cross-sectional area), it is possible to calculate the discharge.

$$Q = AR^{2/3}S^{1/2}n^{-1} \tag{7.1}$$

Fig. 7.7. Frequency of extreme and strong floods of the River Rhine near Bale since 1496 (modified after [66]). The key is to derive quantitative discharge data from qualitative data through the application of hydraulic equations. Data can even be obtained for prehistoric events documented by fluvial sediments (for a review of the methods involved, see [49, 67])

where

- Q: discharge
- A: cross-sectional area (derived from the observed water level and channel topography)
- R: hydraulic radius $R = A/P$ (P: wetted perimeter, derived from channel topography and observed water level)
- S: slope of energy line (more or less identical to the slope of the water surface, derived from observation)
- n: hydraulic roughness (tabulated values, see [68])

One challenge for this kind of research is to estimate values for the hydraulic roughness of settled floodplains. Historical flood marks have been given for specific areas but the related hydraulic roughness has not yet been systematically investigated. Flume experiments can help to estimate related data if the density of settlement and other characteristics of the floodplain in historic times are known from maps and other sources. Extraordinary events, like the extreme flood of 1342 observed in numerous catchments and locations in Central Europe (see [69]), can be investigated in more detail and could provide insights into plausible flood events under slightly modified climate conditions, which are vital for predicting conditions in the near future. Palaeohydrological and palaeohydraulic investigations have been carried out under different conditions in several locations across the world, but have been less frequently applied in Europe [53, 67, 70].

Large flood events, like the extraordinary ice-dammed lake outburst floods in Pleistocene times mentioned above, can help us to understand the specific hydraulic conditions of other Xevents, and they provide concepts for new methods like the interpretation of run-up sediments, described for the first time in relation to the outburst floods in the Altai Mountains [49]. This kind of sediment is deposited in front of local obstructions, but at higher elevations than the usual water level due to the energy transfer as the velocity is abruptly reduced and flow energy is transferred into potential energy. The difference between the mean water level (indicated by other flood deposits) and the level of the run-up sediment quantifies the velocity head, which can then be easily and successfully converted into a mean flow velocity. According to the Bernoulli energy equation, the total energy of flow is constant along the channel. The total energy of flow H consists of three elements:

$$H = (y + z) + (\rho/(pg)) + (v^2/(2g)) \tag{7.2}$$

where

- $(y+z)$: potential energy, where y is the depth of flow and z is the elevation above datum,
- $(\rho/(pg))$: pressure energy, where ρ is the density of water, p is the pressure and g is the acceleration of gravity,
- $(v^2/(2g))$: kinetic energy, where v is the velocity of flow and g is the acceleration under gravity.

Due to the fact that all of these elements are expressed by a unit of length, they are also called the elevation head, the pressure head and the velocity head, and the result is called the energy head. For free surface flow, the pressure energy is zero, which simplifies the equation to

$$H = (y + z) + (v^2/(2g)) \tag{7.3}$$

This equation illustrates that in the case of a sudden drop of velocity, all kinetic energy is transferred into potential energy, so the water level rises and the depth of flow y increases.

Due to the non-uniform distribution of velocities within open channel flow, an energy coefficient α is introduced to calculate the change of height of the water level from the total loss of kinetic energy [68]. This coefficient is also called the velocity head coefficient or the Coriolis coefficient. The equation of the velocity head v_h can therefore be written as

$$v_h = \alpha v^2/(2g) \tag{7.4}$$

Values for the energy coefficient α are always larger than unity, except in the case of uniform flow. For a known velocity distribution within a channel, α can be calculated, otherwise it can be derived from experience (see [49] for details).

Assuming that the suspended sediment load is homogenously distributed throughout the cross-sectional area of the flow, the difference in elevation between usual water level indicators like bar surfaces and run-up sediments represents the velocity head v_h of flow. As this value can be measured in the field, the flow velocity v and finally the discharge ($Q = vA$) can be calculated after a simple algebraic transformation:

$$v = (2gv_h/\alpha)^{1/2} \tag{7.5}$$

For mega floods like the ice-dammed lake outburst floods in the Altai Mountains, the application of this approach results in a discharge of comparable magnitude to values derived by other attempts [49]. This concept of the run-up effect should be transferred to other environments like settled floodplains or narrow bridges, as obstruction by buildings (where the flood marks are fixed!) may lead to water levels that are too high, resulting in overestimations for the discharges.

Other large-scale flood-related features investigated in the Altai Mountains cannot be interpreted successfully at the current state of knowledge. Large boulders with diameters of 13 m were displaced by the outburst floods in the Altai Mountains. Obviously, the size distribution of transported boulders is related to the shear stress resulting from the high energy currents, indicating what is known as the competence of flow. The problem is that hydraulic interpretation is only possible for smaller boulders with diameters of up to 1–2 m. The macroturbulence phenomenon [71] results in a different vortex system with significant lifting forces beyond a boulder size of 1–2 m. This threshold system has not yet been thoroughly investigated.

Another challenge is the application of flow simulation software to extreme flood events. While simple one-dimensional models can be applied successfully to steady flow conditions [72], unsteady flow modelling results in the numerical instability of the previously applied model. More sophisticated two- and three-dimensional models have only occasionally been applied in palaeohydraulic investigations [73], but, alongwith a user-friendly interface, are needed for flood investigations.

The example of the run-up sediments illustrates the value and importance of the investigation of extreme flood events, even if they occurred thousands of years ago in remote places: unknown phenomenona can be detected and new methods developed that may help to predict flood magnitudes that could occur in settled areas in the near future. Those floods might be caused by intensive precipitation or abrupt snowmelting, but the failure of manmade dams can result in outburst floods that could be comparable with those that occurred during the last ice age.

References

1. Goudie, A. (ed.): *The human impact reader – readings and case studies*, (Blackwell, Oxford 1997)
2. Crutzen, P.J.: Geology of mankind. Nature **415**, (2002), p 23
3. Press, F., R. Siever: *Earth*, 4th rev. edn, (Freeman, San Francisco, CA 1985)
4. Sullivan, W.: *Continents in motion – the new earth debate*, (McGraw-Hill, New York 1974)
5. Summerfield, M. (ed.): *Geomorphology and global tectonics*, (Wiley, Chichester, UK 2000)
6. Bauer, J., W. Englert, U. Meier, F. Morgeneyer, W. Waldeck: *Physische Geographie kompakt*, (Spektrum, Heidelberg 2001), Abb. 36.1
7. Stern, D.P.: A millennium of geomagnetism, *Rev. Geophys.* **40**, (2002), pp 1–30

8. GSC (Geological Survey of Canada) (ed.) (2005): Geomagnetism webpage. http://gsc.nrcan.gc.ca/geomag/index_e.php (23. Oct. 2005).
9. Press, F., R. Siever: *Allgemeine Geologie – Einführung in das System Erde*, 3rd edn (Spektrum, Heidelberg 2003), Abb. 19.15
10. Butler, R.F.: *Paleomagnetism*, (Blackwell, Boston 1992)
11. Soffel H.C.: Feldumkehr. In: *Lexikon der Geowissenschaften*, vol **2**, ed by Landscape GmbH, Heidelberg (Spektrum, Heidelberg 2000) p 142
12. Deynoux, M., J.M.G. Miller, E.W. Domack, N. Eyles, I.J. Fairchild, G.M. Young (eds.): *Earth's glacial record*, (Cambridge University Press, Cambridge 1994)
13. Crowell, J.C.: Pre-Mesozoic ice ages: their bearing on understanding the climate system, *Geol. Soc. Am. Mem.* **192**, pp 1–112 (1999)
14. Benn, D.I., D.J.A. Evans: *Glaciers and glaciation*, (Arnold, London 1998)
15. Sibrava, V., D.Q. Bowen, G.M. Richmond (eds.): Quaternary glaciations in the northern hemisphere, *Quaternary Sci. Rev.* **5**, (1986), pp 1–510
16. McCaffrey M.: Figure showing ice coverage of northern hemisphere today and during the last ice age, accessed at http://www.ncdc.noaa.gov/paleo/slides/slideset/11/11_177_slide.html (23. Oct. 2005)
17. Bradley, R.S.: *Paleoclimatology – reconstructing climates of the Quaternary*, 2nd edn (Academic, San Diego, CA 1999)
18. Herget, J.: Klimaänderungen in Mitteleuropa seit dem Tertiär. *Petermanns Geog. Mitteil.*, **144**, (2000), pp 56–65
19. Wilson, R.C.L., S.A. Dury, J.L. Chapman: *The great ice age – climate change and life*, (Routledge/The Open University, London/New York 2000)
20. Rahmstorf, St.: On the freshwater forcing and transport of the Atlantic thermohaline circulation, *Clim Dynam*, **12**, (1996), pp 799–811
21. Hinnov, L.A.: New perspectives on orbitally forced stratigraphy, *Annu. Rev. Earth Pl. Sc.*, **28**, (2000), pp 419–475
22. Frenzel, B., M. Pecsi, A.A. Velichko: *Atlas of paleoclimates and paleoenvironments of the northern hemisphere*, (G. Fischer, Stuttgart 1992)
23. Ruddiman, W.F., S.J. Vavrus, J.E. Kutzbach: A test of the overdue-glaciation hypothesis, *Quaternary Sci. Rev.*, **24**, (2005), pp 1–10
24. Grove, J.: *The Little Ice Age*, (Routledge, London 1990)
25. Hormes, A., B.U. Müller, Ch. Schlüchter: The Alps with little ice: evidence for eight Holocene phases of reduced glacier extent in the Central Swiss Alps, *Holocene*, **11**, (2001), pp 255–265
26. Broecker, W.S.: The end of the present interglacial: how and when. *Quaternary Sci. Rev.*, **17**, (1998), pp 689–694
27. Grainger, R.G., J.E. Highwood: Changes in stratospheric composition, chemistry, radiation and climate caused by volcanic eruptions, in: Oppenheimer, C., D.M. Pyle, J. Barclay (eds.): *Volcanic degassing (Geological Society of London – Special Publications)* **213**, (2003) pp 329–347
28. Sadler, J.P., J.P. Grattan: Volcanoes as agents of past environmental change, *Global Planet. Change*, **21**, (1999), pp 181-196
29. Courtillot, V.E.: Die Kreide-Tertiär-Wende: verheerender Vulkanismus? *Spektrum der Wissenschaft*, **12**, (1990), p 4
30. Courtillot, V.E.: Die Kreide-Tertiär-Wende: verheerender Vulkanismus ? *Spektrum der Wissenschaft*, **12**, (1990), pp 60–69
31. Courtillot, V.E.: *Evolutionary catastrophes – the science of mass extinction*, (Cambridge University Press, Cambridge 2002)

32. USGS (United States Geological Survey): Earthquake hazard program, (2004) (see online resource: http://earthquake.usgs.gov/bytopic/lists.html; 10 Dec. 2004)
33. McCalpin, J.P.: *Paleoseismology*, (Academic Press, San Diego, CA 1996)
34. Grant, L.B., W.R. Lettis (eds.): Paleoseismology of the San Andreas Fault System. *Seismol. Soc. Am. Bull.*, **92-7**, (2002), pp 2551–2877
35. Grady, M.M.: *Catalogue of meteorites*, (Cambridge University Press, Cambridge 2000)
36. Grieve, R.A.F.: Impact cratering on the earth. *Sci. Am.*, **262**, (1990), pp 66–73
37. Kerridge, J.F., M.S. Matthews (eds.): *Meteorites and the early solar system*, (University of Arizona Press, Tucson, AZ 1988)
38. Lewis, J.S.: *Comet and asteroid impact hazards on a populated earth*, (Academic, San Diego, CA 2000)
39. Whitehead, J., J. Spray: Earth Impact database, (2004) (online resource: http://www.unb.ca/passc/ImpactDatabase/index.html; 10 Dec. 2004)
40. Whitehead, J., J. Spray: *Earth impact database; Barringer crater images*, (2004) (http://www.unb.ca/passc/ImpactDatabase/images/barringer.htm; 5 Dec. 2004)
41. Alvarez, L.W., W. Alvarez, F. Asaro, H.V. Michel: Extraterrestrial cause for the Cretaceous/Tertiary extinction, *Science*, **208**, (1980), pp 1095–1008
42. Alvarez, W., F. Asaro: Die Kreide-Teritär-Wende – ein Meteoriteneinschlag? *Spektrum der Wissenschaft*, **12** (1990), pp 52–59 (English version: *Sci. Am.*, **10**, (1990), pp 78–84)
43. Sharpton, V.L., P.D. Ward (eds.): Global catastrophes in Earth history, *Geol. Soc. Am. (Special Paper)*, **637**, (1990), pp 1–631
44. Herschy, R. (ed.): World catalogue of maximum observed floods. *Int. Assoc. Hydrolog. Sci. (publ.)*, **284**, (2003), pp 1–285
45. Snorasson, A., H.P. Finnsdottir, M. Moss (eds.): The extremes of the extremes – extraordinary floods. *Int. Assoc. Hydrolog. Sci. (publ.)*, **271**, (2002), pp 1–393
46. Baker, V.R., R.C. Kochel, P.C. Patton (eds.): *Flood morphology*, (Wiley, New York 1988)
47. Gregory, K.J., G. Benito (eds.): *Palaeohydrology – understanding global change*, (Wiley, Chichester, UK 2003)
48. O'Connor, J.E., G.E. Grant, J.E. Costa: The geology and geography of floods, in: House, P.K., R.H. Webb, V.R. Baker, D.R. Levish (eds.): *Ancient floods, modern hazards – principles and applications of paleoflood hydrology*, Washington (American Geophysical Union, Water Science and Applications 5, Washington, DC 2002), pp 359–385
49. Herget, J.: Reconstruction of ice-dammed lake outburst floods in the Altai-Mountains, Siberia. *Geol. Soc. Am. (Special Paper)*, **386**, (2005) (in press)
50. Herget, J.: Eisstausee-Ausbrüche – Ursache für katastrophale Hochwässer. *Geogr. Rundschau*, **55**, (2003), pp 14–20
51. Tweed, F.S., A.J. Russell: Controls on the formation and sudden drainage of glacier-impounded lakes: implications for jökulhlaup characteristics, *Prog. Phys. Geogr.*, **23/1**, (1999), pp 79–110
52. Baker, V.R., R.C. Bunker: Cataclysmic late Pleistocene flooding from glacial lake Missoula – a review. *Quaternary Sci. Rev.*, **4**, (1985), pp 1–41

53. Benito, G., J.E. O'Connor: Number and size of last-glacial Missoula floods in the Columbia River valley between the Pasco Basin, Washington, and Portland, Oregon. *Geol. Soc. Am. Bull.*, **115**, (2003), pp 624–638
54. Teller, J.T., D.W. Leverington, J.D. Mann: Freshwater outbursts to the oceans from glacial Lake Agassiz and their role in climate change during the deglaciation, *Quaternary Sci. Rev.*, **21**, (2002), pp 879–887
55. Clarke, G.K.C., D.W. Leverington, J.T. Teller, A.S. Dyke: Paleohydraulics of the last outburst flood from glacial Lake Agassiz and the 8200 BP cold event. *Quaternary Sci. Rev.*, **23**, (2004), pp 389–407
56. Von Grafenstein, U., H. Erlenkeuser, J. Müller, J. Jouzel, S. Johnsen: The cold event 8200 years ago documented in oxygen isotope records of precipitation in Europe and Greenland, *Clim. Dynam.*, **14**, (1998), pp 73–81
57. Walder, J.S., J.E. Costa: Outburst floods from glacier-dammed lakes – the effect of mode of lake drainage on flood magnitude, *Earth Surf. Proc. Land.*, **21**, (1996), pp 701–723
58. Schumm, S.A.: *To interpret the Earth – ten ways to be wrong*, (Cambridge University Press, Cambridge 1991)
59. Baker, V.R.: Cataclysmic processes in geomorphological systems, in: Scheidegger, A.E., M.J. Haigh (eds.): Dynamic system approach to natural hazards, *Z. Geomorphol. Suppl.*, **67**, (1988), pp 25–32
60. Mann, C.J., R.L. Hunter: Probabilities of geologic events and processes in natural hazards, in: Scheidegger, A.E., M.J. Haigh (eds.): Dynamic system approach to natural hazards, *Z. Geomorphol. Suppl.*, **67**, (1988), pp 39–52
61. Baker, V.R.: A bright future for old flows: origin, status and future of paleoflood hydrology, in: V.R. Thorndycraft, G. Benito, M. Barriendos et al (Centro de Ciencias Medioambientals, Madrid 2003) (eds.): *Palaeofloods, historical floods and climatic variability: applications in flood risk assessment*, Proc. PHEFRA Workshop, Barcelona, 16–19 October 2002, (2003), pp 13–18
62. Graßl H.: *Süddeutsche Zeitung*, **61**, V2/11 (2000)
63. Caspary H.J., A. Bardossy: *Wasser Boden*, **47/3**, (1995), pp 18–24
64. Gosnold W.D., J.A. LeFever, P.E. Todhunter et al: *Geotimes*, **45/5**, (2000) pp 20–23
65. Kempe S., P. Krahe: *Erdkunde*, **59**/2, (2005), pp 216–250
66. Pfister Ch.: *Wetternachhersage – 500 Jahre Klimavariationen und Naturkatastrophen*, (Paul Haupt, Bern 1999)
67. Saint-Laurent D.: *Prog. Phys. Geogr.*, **28/4**, (2004), pp 531–543
68. Chow V.T.: *Open-channel hydraulics*, (McGraw-Hill, Tokyo 1959)
69. Bork H.-R., H. Bork, C. Dalchow et al.: *Landschaftsentwicklung in Mitteleuropa*, (Klett-Perthes, Gotha 1998)
70. G. Benito: Palaeoflood hydrology in Europe, in: V.R. Thorndycraft, G. Benito, M. Barriendos et al. (Centro de Ciencias Medioambientals, Madrid 2003) (eds.): *Palaeofloods, historical floods and climatic variability: applications in flood risk assessment*, Proc. PHEFRA Workshop, Barcelona, 16–19 October 2002, (2003), pp 19–24
71. Jackson R.G.: *J. Fluid Mech.*, **77**, (1976), pp 531–560
72. Herget J., H. Agatz: Modelling ice-dammed lake outburst floods in the Altai Mountains (Siberia) with HEC-RAS, in: V.R. Thorndycraft, G. Benito,

M. Barriendos et al (Centro de Ciencias Medioambientals, Madrid 2003) (eds.): *Palaeofloods, historical floods and climatic variability: applications in flood risk assessment*, Proc. PHEFRA Workshop, Barcelona, 16–19 October 2002, (2003), pp 177–181

73. Carling P., R. Kidson, Z. Cao, J. Herget: Palaeohydraulics of extreme flood events: reality and myth, in: K.J. Gregory, G. Benito (eds.): *Palaeohydrology: understanding global change*, (Wiley, Chichester, UK 2003) pp 325–336

8 Wind and Precipitation Extremes in the Earth's Atmosphere

Andreas Hense and Petra Friederichs

Summary. This chapter presents an overview of some typical meteorological extreme events. For reasons of conciseness we restrict ourselves to wind and precipitation extremes. The major goal is to emphasize the fact that very different types of wind or precipitation extremes may occur on different scales in space and time. The main phenomenological presentation is supported by short descriptions of conceptual models, in order to help the reader to grasp some of the underlying physics. We show that it is debatable as to whether the concept of universality holds for extremes, even for those involving atmospheric motion alone.

8.1 Introduction

Weather extremes are natural Xevents that can affect anybody at any place on Earth. The extreme summers suffered in central Europe in 2002 and 2003 revealed the significant socio-economic problems weather extremes may cause. Furthermore, extreme weather events enter into the public debate on the "increasing" number of extreme weather events resulting from climate change.

Very different types of extremes may occur on different scales in space and time. These types are related to different variables like wind, temperature and precipitation and different processes involving these variables. On the smallest scales (1–100 m), meteorological extremes are dominated by wind extremes like breaking gravity waves downstream of mountain ridges or tornadoes with the highest windspeeds recorded on Earth. Extreme precipitation events, extreme not only in strength but also in appearance (such as hailstones), are connected with deep convective structures on scales of kilometers. Even larger wind and precipitation extremes with longer time scales (hours to days) are found in tropical-subtropical synoptic disturbances like hurricanes or typhoons. They are intrinsically linked to deep convection that occurs in the atmosphere whenever the vertical density or entropy distribution develops into an unstable situation. In contrast, mid-latitude synoptic disturbances, together with their types of extreme rainfall or wind events, originate in the baroclinic shear zones over the mid-latitude oceans, which have strong meridional temperature gradients.

In the following, an overview of some typical meteorological Xevents will be given. The major goal is to emphasize the fact that different types of extremes can occur on different scales in space and time. The phenomenological presentation of these extremes is supported by short descriptions of conceptual models, in order to understand some of the physics behind atmospheric extremes in wind and precipitation. The discussion will show that it is still debatable as to whether the concept of universality holds for extremes, even just for atmospheric motion alone.

8.2 Atmospheric Scales

The climate system that contains the atmosphere can be defined as that part of the total Earth system that exchanges energy with outer space and the interior of the Earth by natural processes. Besides the atmosphere, the climate system can be further separated into subsystems, including the hydrosphere, the cryosphere, bare land surfaces and the biosphere. These subsystems are coupled by the exchange of energy, mass and momentum. The subsystems are defined in terms of their bulk aggregate state, namely gaseous for the atmosphere, liquid water for the hydrosphere, frozen water for the cryosphere, the bare soil matrix and the (largely) plant-based communities on the land and in the ocean. The climate system in general and the atmosphere specifically can be considered to be a high-dimensional, nonlinear physicochemical system [1]. If we assume a minimum spatial scale of 10^{-3} m (in order to resolve viscous dissipation), the entire atmosphere can be covered by $\mathcal{O}(10^{28})$ of these unit volumes. If the state within each volume is furthermore described by $\mathcal{O}(100)$ state variables, such as temperature, pressure, concentrations or densities of bulk or trace substances, and velocities, the atmosphere has at least 10^{30} degrees of freedom (DOF). The complete instantaneous state of such a system (named $\boldsymbol{m}$) can be mathematically described by an element (vector) of a high-dimensional phase space, where the dimensionality $\dim(\boldsymbol{m}) = a$ is the number of DOF. It is not possible to compute the temporal evolution of such an object, nor is it possible to measure the data required to initialize such a computation. A description and analysis of such a complex system can only be achieved through statistical physics, where the state vector $\boldsymbol{m}$ is described by statistical measures like the probability density function (PDF). The aim of modern climatalogy is to analyze the temporal evolution of such statistical measures through internal dynamics or external forcings [2].

The assumption of thermodynamic equilibrium, which is inherent to statistical physics, allows the PDF of the state vector to be computed from first principles. However, due to variations in solar insolation, the atmospheric system and the other climate subsystems are far from being in thermodynamic equilibrium. Such thermodynamic systems are subject to large random fluctuations [3] and are therefore prone to develop Xevents. Due to asymmetries in the land-ocean distribution and orography and/or combinations of the Earth's geometry and rotation with the land-sea distribution (among other factors) high energy fluctuations can be localized to certain regions, while the variability in other regions is small. Hence, specific atmospheric Xevents can be observed with varying probabilities at different locations on Earth. These structures are called "teleconnections" and present further complications to the analysis because they do not allow for inherent symmetries in the climate system, such as translation invariance in the longitudinal direction.

The PDF of the atmospheric state vector $\boldsymbol{m}$ contains the full description of the statistics of $\boldsymbol{m}$. Thus it is a mathematical construct or model that can be used to describe the climate. Since this PDF and its temporal evolution cannot be specified from first principles or symmetry considerations, they must be estimated from observations. This introduces uncertainties into the estimates that have to be specified for further applications.

The dynamics of the atmospheric state vector $\boldsymbol{m}$ are observed to be concentrated at particular space and time scales. Figure 8.1 gives an overview of typical processes in the atmosphere and their related space and time scales. All of these processes are turbulent, chaotic processes with the ability to generate extreme values. The turbulent character arises from the nonlinear interactions of the flow components of velocity, temperature and mass as well as the step processes of the condensation or evaporation of water.

Extremes at different scales are often connected to each other, for example:

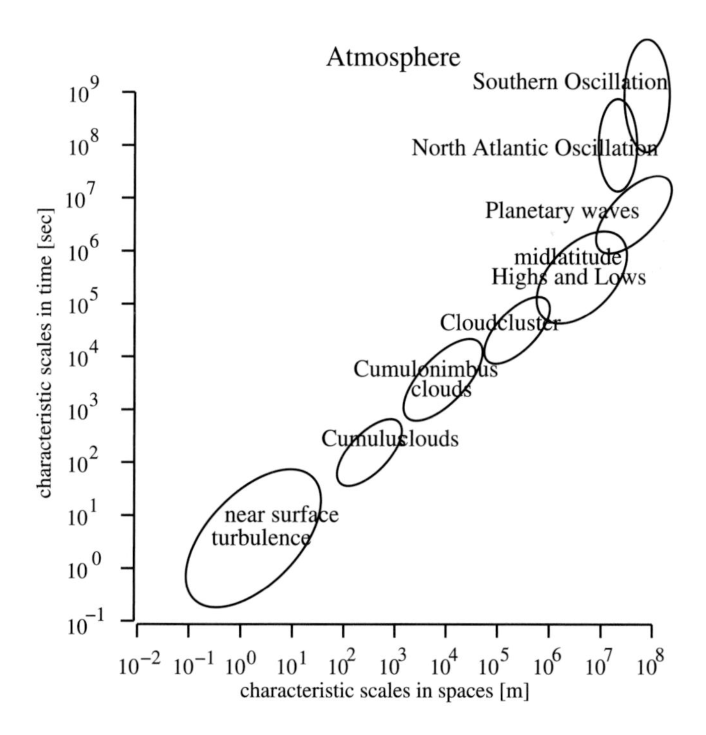

Fig. 8.1. Scale diagram for atmospheric motions

- Large (extreme) vertical gradients of entropy lead to convective instability on spatial scales of $\mathcal{O}(100 \text{ km})$. The resulting deep convection can lead to extreme vertical velocities on spatial scales of $\mathcal{O}(1 \text{ km})$. Heavy precipitation (including hailstones) can grow within these vertical circulation structures, which (depending on their size) can cause extreme damage.
- Large (horizontal) temperature gradients in mid-latitudes can lead to baroclinic instabilities at spatial scales of $\mathcal{O}(1000 \text{ km})$. Cyclones, which represent geophysical turbulence, can result, leading to extreme horizontal velocities along the frontal zones. Friction near the surface leads to strong vertical shear flow which causes dynamically induced turbulence with extreme wind bursts or gusts.

In the following we will present an overview of extreme weather events observed on different space and time scales, based on wind speed and precipitation. The aim is to show that different physical processes are responsible for generating wind extremes. This raises the question as to whether a universal behavior should be expected for extreme wind velocities. Furthermore, we will present a short summary of precipitation extremes over a wide range of accumulation periods ranging from minutes to years. In contrast to wind extremes, extremes in precipitation do show a scaling law, which is most probably related to the general space – time scaling of precipitation [4].

8.3 Wind Extremes

8.3.1 Small-Scale Extremes

The most prominent extremes at the smallest scales (a few tens of meters) are encountered during "clear air turbulence" (CAT). CAT is usually related to breaking atmospheric gravity waves initiated by airflows across large mountain ridges. CAT is also found in or near strong horizontal shear flows that develop in regions where there is strong curvature of the "jet stream". The jet stream meanders around the Earth at heights of about 10–12 km (see below). This atmospheric phenomenon provides an aviation hazard known as "air holes", which can lead (in extreme cases) to structural damage to aircraft. Figure 8.2 shows a photo of a DC-8 aircraft that landed in Stapleton Airport (Denver, CO, USA), after having experienced severe CAT in breaking mountain waves in the lee of the Rocky Mountains on 9th December 1992. The DC-8 lost its leftmost engine, including the wing tip. This indicates that these extreme turbulence elements have a spatial scale of roughly the size of an aircraft.

At slightly larger scales, atmospheric wind extremes are linked to deep convection or thunderstorms. If the vertical temperature and moisture structure is conditionally unstable and the moisture within the parcel condensates, rising air parcels will accelerate. The latent heat is passed to the parcel and

THE DENVER POST — Thursday, December 10, 1992

EMERGENCY LANDING: The damaged DC-8 cargo jet landed safely at Stapleton Airport yesterday morning.

Jet lands minus engine, wing tip

Fig. 8.2. The DC-8 aircraft that experienced severe CAT in the lee of the Rocky Mountains on 9th December 1992 (from http://www.etl.noaa.gov/about/review/as/ralph/10.html)

its buoyancy increases. The parcel also gains kinetic energy at the expense of the latent heat. This is called the convective updraft. Its spatial scale is typically $\mathcal{O}(1\ \text{km})$ and it may reach upward peak velocities of 10-30 m/s. Due to the strong drag exerted upon raindrops and hail or graupel, such large velocities will permit the formation of large quantities of water and ice in the upward-moving air parcel. Extreme updraft velocities allow extremely large hailstones to grow, which present hazards to farms, greenhouses and orchards (Fig. 8.3).

If the water and ice is moved outside the main updraft it will, due to its weight, fall and the ice and water will melt or evaporate. Thus, internal energy is extracted from the air and stored as latent heat which creates and increases the negative buoyancy of the parcel. A "downdraft" develops which can produce extreme wind gusts of up to 200 km/h when it reaches the surface. Furthermore, downdraft or downbursts are among the most dangerous meteorological hazards for aircraft during take-off or landing. During the 1970s and 1980s, more than 20 US aircraft accidents were related to downbursts, causing more than 400 fatalities [5]. Downbursts range in size from 1 km (microburst) to over 50 km (macroburst), and are often accompanied by specific cloud formations (Fig. 8.4).

Fig. 8.3. A grapefruit-sized hailstone (from http://www.photolib.noaa.gov/nssl)

Updrafts and downdrafts within a single thunderstorm lead to extreme horizontal shears in the vertical velocity. If this is combined with a strong vertical shear in horizontal wind velocity, vorticity is generated in the horizontal plane. This is one way in which a tornado can form, and these provide the most extreme wind events observed on the Earth. Tornado wind velocities of up to 600 km/h have been measured using Doppler radar. Tornadoes (Fig. 8.5) are always associated with strong convection and exhibit spatial

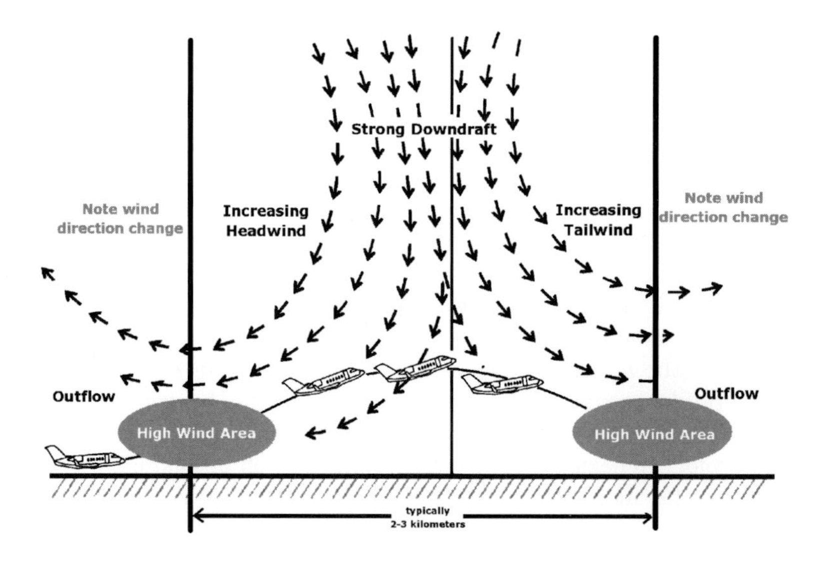

Fig. 8.4. Sketch of a convective downburst

Fig. 8.5. A tornado south of Dimmitt, TX on 2nd June 1995. Photographer: Harald Richter (from http://www.photolib.noaa.gov/nssl)

scales between $\mathcal{O}(100\text{ m})$ and $\mathcal{O}(1\text{ km})$. According to the Fujita scale, tornado strength is classified between F1 and F5 (F6) (see Table 8.1).

Tornadoes are frequent phenomena in the midwest of the US (in "Tornado Alley" in Texas, Oklahoma and Kansas). A careful analysis of observations made in Germany has shown that they are also present there, although the frequency of occurrence is about an order of magnitude smaller. More details can be found at http://www.tordach.org/de/.

8.3.2 Mesoscale Extremes

While the wind extremes described in the previous section are almost unaffected by the Earth's rotation, atmospheric motions and their extremes seen at larger scales, the meso- and macroscales, are influenced by the Coriolis force. This has important consequences. Assuming a stationary and horizontal (tangential to the sphere) flow field $\boldsymbol{v}_h$, the approximate momentum balance between advective, pressure gradient and Coriolis acceleration then reads:

$$(\boldsymbol{v}_h \boldsymbol{\nabla}_h)\boldsymbol{v}_h = -\frac{1}{\rho}\boldsymbol{\nabla}_h p - f\boldsymbol{e}_r \times \boldsymbol{v}_h \; . \tag{8.1}$$

($\boldsymbol{e}_r$ is the unit vector that is locally perpendicular to the sphere and points to the local zenith, while $f = 2\Omega \sin(\varphi)$ is the Coriolis parameter, where Ω is the Earth's rotation rate and φ is the geographical latitude). The left side can be modified using a well-known identity, and the momentum balance changes into

$$(f + \zeta)\boldsymbol{e}_r \times \boldsymbol{v} = -\boldsymbol{\nabla}_h \left(\frac{1}{\rho}p + \frac{1}{2}\boldsymbol{v}_h^2\right) \; , \tag{8.2}$$

Table 8.1. The Fujita tornado scale

F-Scale	Intensity phrase	Wind speed	Typical damage
F0	Gale tornado	40–72 mph	Some damage to chimneys Breaks branches off trees Pushes over shallow-rooted trees Damages signs
F1	Moderate tornado	73–112 mph	Lower limit on hurricanes Peels off roofs Mobile homes pushed off foundations Moving automobiles pushed off the roads Attached garages may be destroyed
F2	Significant tornado	113–157 mph	Considerable damage Roofs torn off frame houses Mobile homes demolished Boxcars pushed over Large trees snapped or uprooted Light object missiles generated
F3	Severe tornado	158–206 mph	Roof and some walls torn off well-constructed houses Trains overturned Most forest trees uprooted
F4	Devastating tornado	207–260 mph	Well-constructed houses leveled Structures with weak foundations blown some distance Cars thrown, large missiles generated
F5	Incredible tornado	261–318 mph	Strong frame houses lifted off foundations and carried considerable distances, disintegrating Automobile-sized missiles fly through the air in excess of 100 m Trees debarked Concrete structures badly damaged
F6	Inconceivable tornado	319–379 mph	Winds this strong are very unlikely Scale of damage they might produce would probably be difficult to envisage through the F4 and F5 winds surrounding the F6 winds

where ζ is the vertical component of the vorticity vector $\zeta = \boldsymbol{e}_r(\boldsymbol{\nabla}_h \times \boldsymbol{v}_h)$. The density ρ is assumed constant for simplicity. The total vorticity $f + \zeta$ on the left hand side, which is the sum of the planetary and flow vorticities, determines the existence of a stationary solution. If the local vorticity is of the same sign as the planetary vorticity, which results in cyclonic low pressure systems with counterclockwise circulation on the northern or clockwise on the southern hemisphere, any pressure gradient allows for a steady solution. This means that cyclonic pressure systems can be of any strength on those spatial scales where the planetary vorticity is important. However, if ζ is the opposite sign to the planetary vorticity, which means that the flow is anti-cyclonic, steady solutions are only possible for weak high pressure systems, because the total vorticity may become zero. Although processes other than those discussed here would never allow the full development of the steady solution, the momentum balance dictates that extreme circulations on the mesoscale in the atmosphere will be observed for cyclonic circulations only. Breakdown under anticyclonic conditions is strongly connected to internal gravity wave generation, which leads to the Xevents discussed in the preceeding section.

8.3.3 Tropical Cyclones

The most intense cyclonic circulations that occur on spatial scales larger than 100 km are found in tropical cyclones. They occur in all tropical oceanic regions and are called hurricanes in the Atlantic and eastern Pacific, typhoons on the Asian side of the Pacific, willi-willies around Australia and simply tropical cyclones in the Indian Ocean basin. They are classified into five classes depending on strength according to the Saffir-Simpson scale (Table 8.2).

To initiate cyclonic rotational flow in the lower part of a tropical cyclone (TC) $\zeta > 0$, a minimum value of planetary vorticity is needed. Therefore, TC's are always found at latitudes of about 10° above or below the equator. Figure 8.6 illustrates this effect. It shows the tracks of all of the TC's observed in the year 2002. Tropical cyclones start their life cycle near ±10°, initially move westward, and then gain in intensity as they start to

Table 8.2. The Saffir-Simpson scale for classifying hurricanes

Category	Wind speed (1 min av)	Barometric pressure in hPa	Storm surge in m	Damage potential
1 weak	120–155 km/h	> 980	1.2–1.5	Minimal damage to vegetation
2 moderate	155–180 km/h	965–980	1.8–2.4	Moderate damage to houses
3 strong	180–210 km/h	945–965	2.7–3.7	Extensive damage to small buildings
4 very strong	210–250 km/h	920–945	3.9–5.5	Extreme structural damage
5 devastating	> 250 km/h	< 920	> 5.5	Catastrophic building failures possible

move polewards. At the very end of its life cycle, a TC will often convert into a mid-latitude low pressure system moving eastward within the westerly winds zone.

The kinetic energy of a tropical cyclone is generated almost entirely through the release of latent heat of water vapor. This requires a sufficiently strong supply of water vapor by evaporation from the ocean surface, which can only be achieved when sea surface temperatures are higher than 26 °C. Once the TC makes land this evaporation is shut off and surface friction is enhanced, which leads to an almost immediate decrease in intensity.

Tropical cyclones are one of the few natural examples of a system that can be described rather successfully by a Carnot heat engine [6], [7]. If the standard Carnot assumptions are broadened by a simple dynamical model, an estimate for the intensity of a TC can be found using

$$\begin{aligned} \eta &= \frac{T_{sst} - T_{tpl}}{T_{sst}} \\ v_T^2 &= 2\eta E_{max} \\ E_{max} &= L_v(1-\mathcal{H})q_{sat}(T_{sst}) \end{aligned} \tag{8.3}$$

where T_{sst} is the sea surface temperature and T_{tpl} is the temperature at the tropopause level (approximately 15 km up at the tropics), η is the well known efficiency of a Carnot machine, while E_{max} is the maximum energy input into a TC through evaporation of the ocean surface. This is determined by the latent heat of evaporation L_v, the relative humidity of the atmosphere above the ocean $\mathcal{H}$, and the saturation-specific humidity at the sea surface temperature $q_{sat}(T_{sst})$, obtained through the Clausius–Clapeyron equation. Inserting this gives an estimate of the maximum wind speed v_T attainable in a TC

$$v_T = \sqrt{2\frac{T_{sst} - T_{tpl}}{T_{sst}} L_v(1-\mathcal{H})q_{sat}(T_{sst})} \tag{8.4}$$

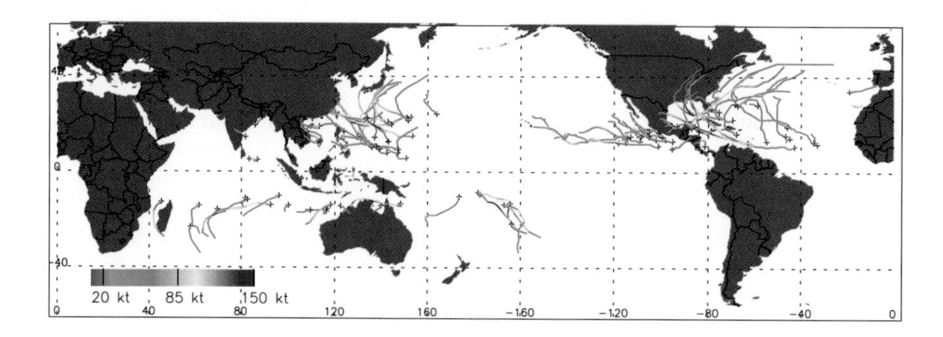

Fig. 8.6. All tropical cyclone tracks for the year 2005 (data from http://www.solar.ifa.hawaii.edu/Tropical/)

Using typical observed values, one arrives at $v_T \sim 210$–250 km/h, which is a Category 4 or 5 hurricane according to Table 8.2. The model can be refined to obtain a potential intensity index that is useful in forecasting.

8.3.4 Extratropical Cyclones

Extratropical low pressure systems (usually termed synoptic disturbances in meteorology) can be found at a slightly larger scale than tropical cyclones. Besides thunderstorms, these low pressure systems are responsible for extreme weather events in central Europe. The most prominent recent examples of this were the winter storms (known as Anatol, Lothar and Martin) of December 1999, or those of February 1990 (Vivian and Wiebke). In contrast to TC's, extratropical disturbances do not feed primarily on the latent heat of condensation but on the potential energy stored in the atmosphere, which occurs if a strong horizontal (mainly meridional) temperature gradient is present. This configuration is known as baroclinic, and it is the baroclinic instability mechanism that converts the potential energy into the kinetic energy of the flow. Baroclinic instabilities organize themselves so as to reduce the meridional temperature contrast by transporting heat from low to high latitudes. Those intense winter storms are characterized by extremely low central pressures and extreme wind velocities and wind gusts, which are related through the gradient wind balance of (8.2).

Typical examples of such exceptional events include the series of storms in the late winter of 1990 which led to the cancellation of major processions on Carnival Monday (26th February 1990) in western Germany, or the Anatol, Lothar and Martin storms in December 1999 that were associated with disasters in southwestern Germany and northern France. Table 8.3 (taken from the homepage of SwissRe) summarizes the extreme storm events that have occurred in central Europe over the past 30 years. It is evident that these storms typically occur in series, which is because the extreme baroclinicity of the atmosphere is not released by a single storm. Several storm events are necessary to transport a significant amount of heat northward to reduce the extreme temperature gradient between low and high latitudes. This slow variability is part of the North Atlantic Oscillation (NAO) over the North Atlantic–European region, which itself was in an extreme state during the early and late 1990s. The NAO was extremely positive (see Fig. 8.7), which was related to a deeper than normal Icelandic low and a higher than normal Azores high, with above-normal (below-normal) temperatures in central Europe (the East American coast).

8.3.5 Jet Streams

Extreme wind velocities on the largest scales (planetary scales) are found in the *jet streams* in the upper troposphere. These jet streams also owe their

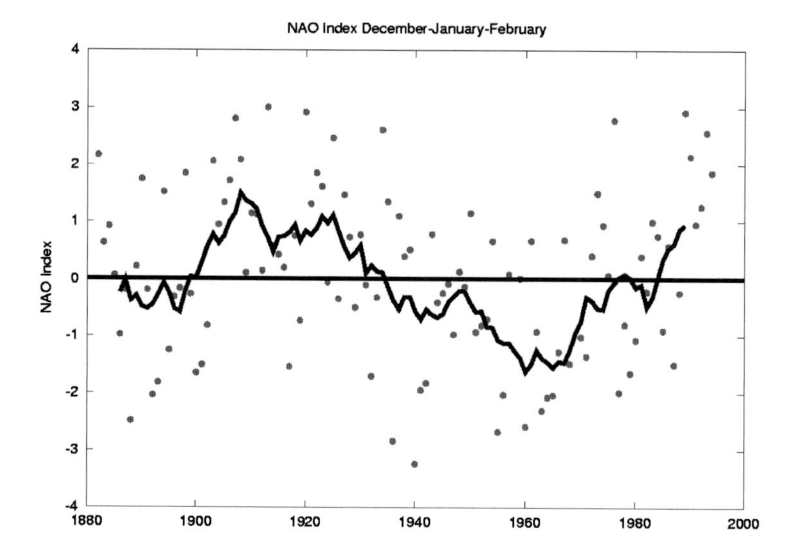

Fig. 8.7. Time series for the NAO index for winter (December to February) between 1880 and 2000; a positive NAO index is associated with above normal temperatures in Europe, individual winter values are represented by *dots*, the *line* shows the 11-year running mean, indicating the decadal variability [8]

existence to the meridional temperature gradient. The wind balance gradient on the planetary scale can be relaxed to the geostrophic balance with the pressure gradient force balancing the Coriolis force:

$$-\frac{1}{\rho}\boldsymbol{\nabla}_h p - f\boldsymbol{e}_r \times \boldsymbol{v}_h = 0 \,. \tag{8.5}$$

Table 8.3. Major historical events, insured losses and estimated return period for the winter storms that have occurred in central Europe over the past 30 years (according to http://www.swissre.com/)

Date	Event	Area affected	Insured loss (in 10^9 US $)	Return period (in years)
2./3.1.1976	Capella	UK, NL, B, D, UK	1.2	> 5
16.10.1987	87J	UK, F, NL	4.3	5
25.1.1990	Daria	Europe, F, D, UK	5.8	8–10
3./4.2.1990	Herta	F, D	1.1	< 5
26.2.1990	Vivian	UK, F, NL, B, D	3.4	< 5
28.2.1990	Wiebke	D, CH, A, S. Germany	1	< 5
21.1./2.2.1995	Div.storms	N. Europe, D, F, B, NL	1	< 5
3./4.12.1999	Anatol	DK, D, UK, SW, DK	1.5	< 5
26.12.1999	Lothar	F, D, CH	6.2	8–10 (F: 70)
27.12.1999	Martin	F, CH	2.6	< 5

For large-scale motion it is more convenient to use the pressure p as an alternative vertical coordinate instead of the height h. The appropriate transformation reads [9]

$$\nabla_h p = g\rho \nabla_p h \,. \tag{8.6}$$

Using the definition of the geopotential $\Phi = gh$ leads to the following formula for the geostrophic equilibrium:

$$\nabla_p \Phi - f \boldsymbol{e}_r \times \boldsymbol{v}_p = 0 \,. \tag{8.7}$$

Assuming that the atmosphere is an ideal gas at temperature T with a gas constant R_L, then, in pressure coordinates, the hydrostatic pressure balance

$$\frac{\partial}{\partial z} p = -g\rho \tag{8.8}$$

reads

$$\frac{\partial}{\partial p} \Phi = -\frac{R_L T}{p} \tag{8.9}$$

Solving 8.7 for $\boldsymbol{v}_p$ and combining it with 8.9 gives

$$\frac{\partial}{\partial p} \boldsymbol{v}_p = \frac{R_L}{fp} \boldsymbol{e}_r \times \nabla_p T \tag{8.10}$$

which shows that the vertical wind shear $\frac{\partial}{\partial p} \boldsymbol{v}_p$ (called the thermal wind) is related to the horizontal temperature gradient. If there is a strong poleward drop in temperature in the lower troposphere, with a less strong or even reversed temperature gradient above, the zonal (east-west) velocity forms a jet that is a local maximum of the wind velocity. In the northern hemisphere, the most prominent positions of these jets are found south of Japan and on the east coast of the US, near 35 °N at a height of about 12 km (pressure $p = 200$ hPa, see Fig. 8.8), with average maximum values of about 240 km/h. Note that Fig. 8.8 is a mean picture of the atmosphere and does not signify any extremes in the statistical sense. Extremes of jet strength are seen in the time series of observed wind velocities at positions near the average maximum (taken to be 35 °N and 130 °E) for two different years, 1983 and 1999 in Fig. 8.9. Extreme velocities in winter are seen to reach values of up to 350 km/h. Within these regions of strong horizontal as well as vertical wind shear, dynamical instabilities can develop that lead to the generation of extreme turbulence by breaking gravity waves (clear air turbulence CAT), which links back to the first Sect. 8.3.1 on small-scale extremes.

8.4 Precipitation Extremes

Precipitation in the atmosphere is the final result of a large cascade of scale interactions. The cascade begins with the interaction of single water molecules,

Table 8.4. World record point rainfall values. These were collected by the World Meteorological Organization and are published in [11]

Duration τ	Time units	Rainfall (mm)	Location	Date
1	min	38	Barot, Guadeloupe	26 Nov 1970
8		126	Füssen, Bavaria	25 May 1920
15		198	Plumb Point, Jamaica	12 May 1916
20		206	Curtea-de-Arges, Romania	7 Jul 1889
42		305	Holt, USA	22 Jun 1947
60		401	Shangdi, Nei Monggol, China	3 Jul 1975
2.17	hours	483	Rockport, USA	18 Jul 1889
2.75		559	D'Hanis, USA	31 May 1935
4.5		782	Smethport, USA	18 Jul 1942
6		840	Muduocaidang, China	1 Aug 1977
9		1087	Belouve, La Réunion	28 Feb 1964
10		1400	Muduocaidang, China	1 Aug 1977
18.5		1689	Belouve, La Réunion	28–29 Feb 1964
24		1825	Foc Foc, La Réunion	7–8 Jan 1966
2	days	2467	Aurere, La Réunion	7–9 Apr 1958
3		3130	Aurere, La Réunion	6–9 Apr 1958
4		3721	Cherrapunji, India	12–15 Sep 1974
5		4301	Commerson, La Réunion	23–27 Jan 1980
6		4653	Commerson, La Réunion	22–27 Jan 1980
7		5003	Commerson, La Réunion	21–27 Jan 1980
8		5286	Commerson, La Réunion	20–27 Jan 1980
9		5692	Commerson, La Réunion	19–27 Jan 1980
10		6028	Commerson, La Réunion	18–27 Jan 1980
11		6299	Commerson, La Réunion	17–27 Jan 1980
12		6401	Commerson, La Réunion	16–27 Jan 1980
13		6422	Commerson, La Réunion	15–27 Jan 1980
14		6432	Commerson, La Réunion	15–28 Jan 1980
15		6433	Commerson, La Réunion	14–28 Jan 1980
31		9300	Cherrapunji, India	1–31 Jul 1861
2	months	12767	Cherrapunji, India	Jun–Jul 1861
3		16369	Cherrapunji, India	May–Jul 1861
4		18738	Cherrapunji, India	Apr–Jul 1861
5		20412	Cherrapunji, India	Apr–Aug 1861
6		22454	Cherrapunji, India	Apr–Sep 1861
11		22990	Cherrapunji, India	Jan–Nov 1861
12		26461	Cherrapunji, India	Aug 1860–Jul 1861
2	years	40768	Cherrapunji, India	1860–1861

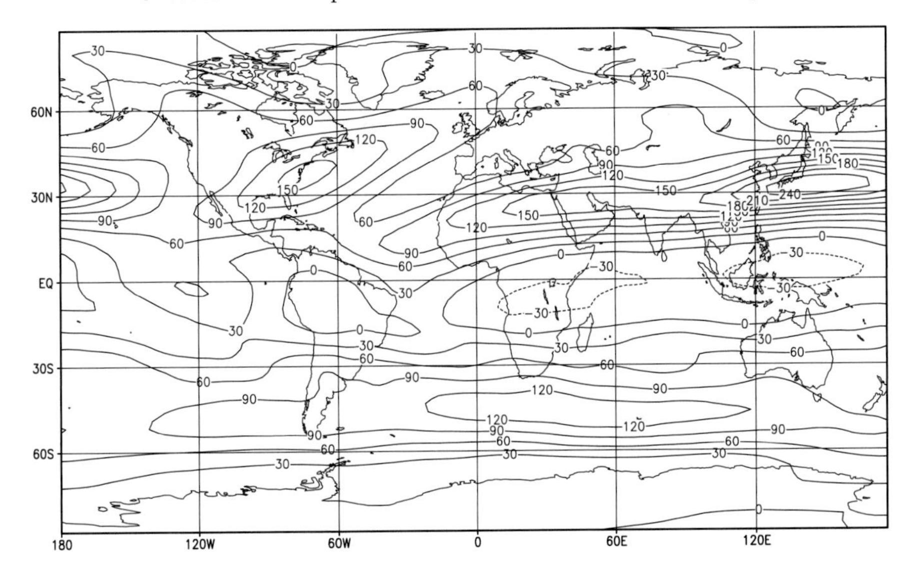

Fig. 8.8. Time-averaged distribution of zonal wind velocity in January between 1980 and 1999 in km/h at a pressure level of 200 hPa, and a height of about 12 km

continues on scales of cloud droplets and rain drops, and then passes through small-scale atmospheric turbulence, clouds and mesoscale flow variations embedded into synoptic and planetary circulations. All of these scales determine whether it rains or not at a specific point in space and time. Precipitation (and also clouds) therefore have a very complex and rich space–time structure [4]. Part of this structure derives from the fact that one cannot talk about a precipitation extreme per se. Rather, it is important to state the time interval T over which the precipitation is accumulated. From the physical point of view, precipitation is the vertical mass transfer rate of liquid or frozen water mass (assuming the discrete nature of the rain drops is ignored). Standard observation systems can only measure the aggregate mass over a specific time interval and a specific area. Both the time and the area have to be large enough compared to the droplet size in order to sample enough rain drops to measure the mass transfer rate accurately. Typical values can be as small as minutes, with areas of just a few hundred square centimeters. The World Meteorological Organization has collected reports on accumulated rainfall extremes over a wide range of aggregation times T. These are given in Table 8.4 together with the location and date of the measurement. A log-log plot (Fig. 8.10) of these data reveals that the extremes of accumulated precipitation scale like $T^{0.5}$ over several orders of magnitude [12]. This clearly shows that extreme precipitation events can only be defined with respect to a fixed accumulation time. The authors of [4] have shown that this remarkable scaling behavior is related to the general scaling of precipitation over time (similar scaling

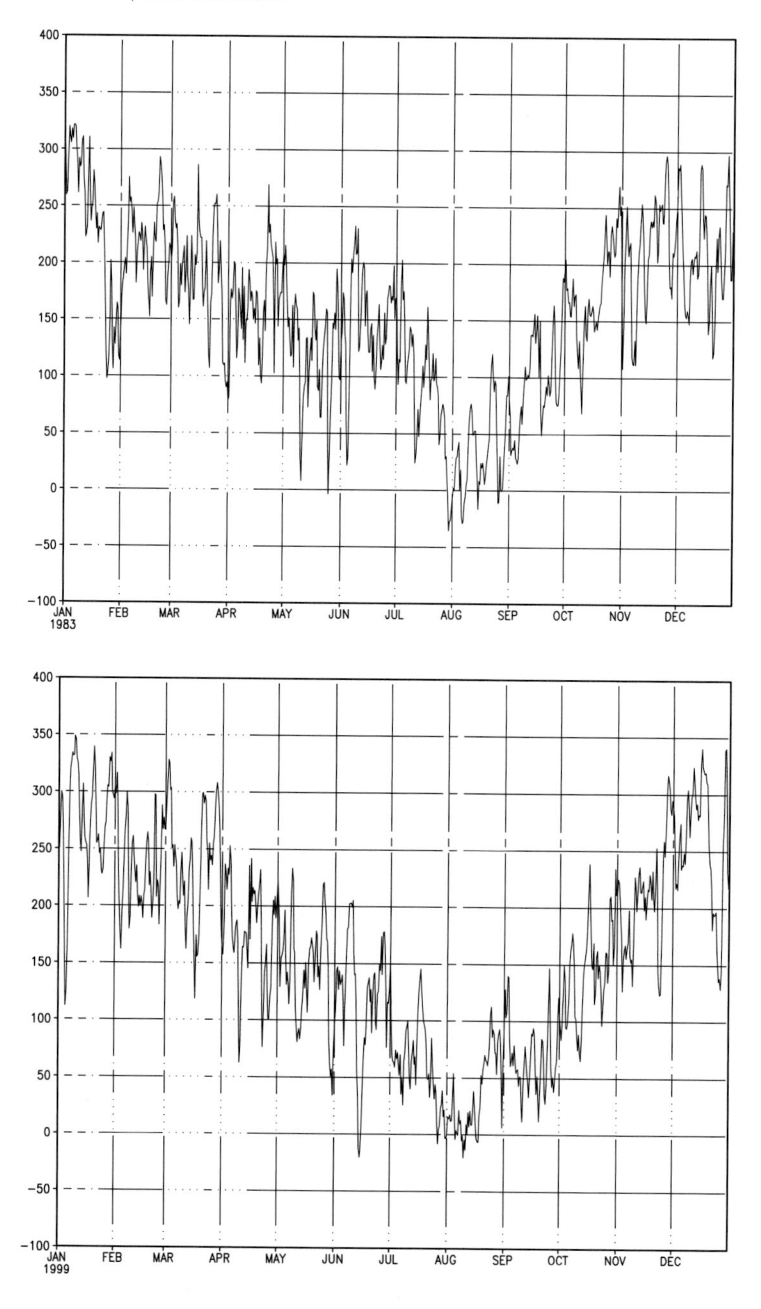

Fig. 8.9. Time series of half-daily observations of jet stream velocity near 35 °N and 130 °E at a height of about 12 km taken in 1983 (*top*) and 1999 (*bottom*) in km/h, from NCEP reanalysis [10]

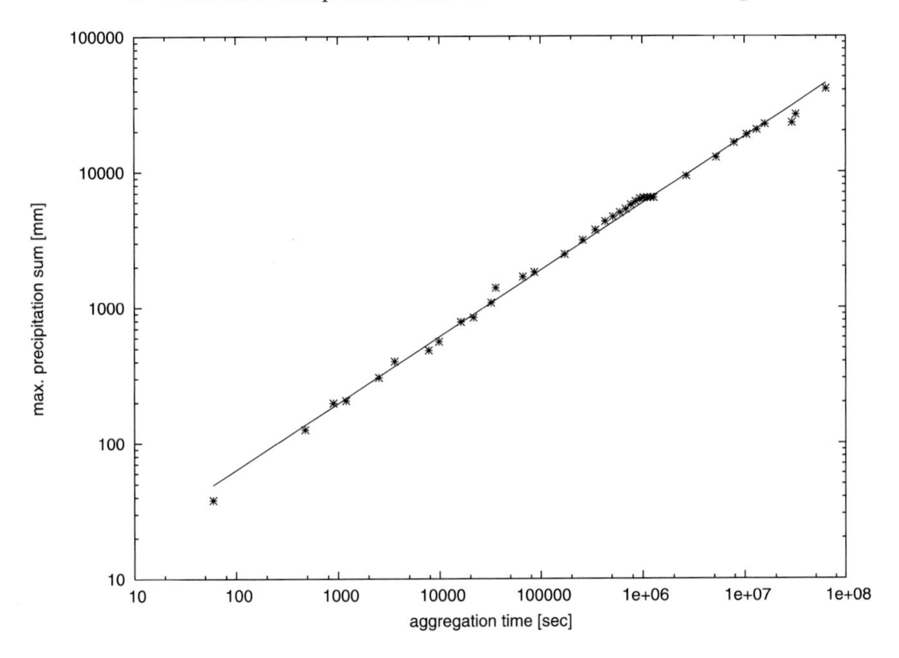

Fig. 8.10. A log-log representation of the extremes in rainfall accumulated over a time period T; the *dots* are the data from Table 8.4, and the *line* is the least squares fit $6.5\tau^{0.49}$

laws hold for space), which can be modeled using a random cascade model. Assume that the rainfall rate R_τ over a time τ scales like $q_\tau \tau^H$ where q_τ is a conserved random quantity (such as total water mass in the present case) and H is the Hurst exponent. If E defines the expectation, we have $E(q_\tau) \sim const$; in other words the expectation of q_τ is scale-independent. The random cascade can be understood as a process distributing an initial rain mass R over a large accumulation period $\tau = 1$ into one of its subperiods using a random scaling of the initial mass μR. In the simplest case, the subperiods are obtained by splitting τ into two identical subperiods. Then the redistribution is a Bernoulli process with $\text{Prob}(\mu R = \tau^{-\gamma_+}) = p_\tau = \tau^c$ when assigning the scaling of the rain mass to one of the subintervals and $\text{Prob}(\mu R = \tau^{-\gamma_-}) = 1 - p_\tau$ for the other.

Generalized scaling with more complex noise processes is developed on the basis of q_τ with [4]:

$$\text{Prob}(q_\tau \geq \tau^{-\gamma}) \sim \tau^{c(\gamma)} \tag{8.11}$$

where $c(\gamma)$ is the scaling exponent (co-dimension function) of the probability distribution characterizing the occurrence of rare (small probability) versus extreme (q_τ large) events. In [13] an expression is derived for $c(\gamma)$ for different noise processes in the random cascade model:

$$c(\gamma - H) = C_1 \left(\frac{\gamma}{\alpha' C_1} + \frac{1}{\alpha} \right)^{\alpha'}$$
$$\frac{1}{\alpha} + \frac{1}{\alpha'} = 1$$
$$\alpha \neq 1$$
$$C_1 = c(\gamma = C_1) \tag{8.12}$$

Here the parameter α characterizes the noise process for the random cascade model, where (for example) $\alpha = 2$ is the parameter for a Gaussian noise generator. The cases with $0 < \alpha < 1$ are more general noise processes: the so-called "Levy stable noises". The scaling for the data in Fig. 8.10 can now be derived from these results. The data given represent the accumulation $A_\tau = \tau R_\tau \sim \tau^{1-\gamma_{max}}$ over the time scale τ. Equation (8.4) can be used to estimate an upper bound for the scaling exponent γ_0 by calculating the maximum of the function c, which is also an upper bound for γ_{max}. This leads to

$$\gamma_{max} \leq \gamma_0 = \frac{C_1}{1-\alpha} + H \tag{8.13}$$

The constants C_1 and α can be evaluated from daily precipitation measurements (by box counting algorithms or related techniques [14]), giving $C_1 = 0.4 \pm 0.2$ and $\alpha = 0.5 \pm 0.1$ ([4] and citation therein). If H is assumed to be zero, the resulting scaling exponent for maximal accumulation is 0.2 ± 0.2. However, if the $C_1 = 0.2, \alpha = 0.5$ pair for observations made on the tropical island of La Réunion ([4], Table 3) are used, which constitute a major part of the data in Fig. 8.10, the scaling exponent for maximum accumulation with $H = 0$ is 0.6 ± 0.2, which is much closer to the results in Fig. 8.10 than the previous estimate.

8.5 Discussion

The aim of this contribution was to give an overview of typical atmospheric extremes and related processes. We took wind and precipitation as examples. For wind, it was shown that typical wind extremes are the result of very different processes. At the smaller spatial and temporal scales, wind extremes result from fully developed three-dimensional turbulent eddies resulting from dynamical (shear zones) and thermodynamical (deep convection) instabilities. At large scales, the effects of the Earth's rotation become important as well as the mean vertical density or entropy stratification (this was not discussed for reasons of brevity), meaning that wind extremes develop in a quasi two-dimensional flow at these scales. This means that it is debatable as to whether it will be possible to postulate a universal behavior for atmospheric wind extremes.

For precipitation, the observational results presented above seem to indicate a different message. Over a wide range of accumulation periods, and at different locations on Earth, the maximally accumulated rainfall follows a straightforward scaling law that can be interpreted as a universal behavior. However, based on the derivations and specific parameter values given in [4], a quantitative comparison leads to a significant (order of magnitude) difference between observation and theory (0.2 ± 0.2 vs. 0.5), except when the observational values for La Réunion are inserted (0.6 ± 0.2 vs. 0.5). Taking a close look at Table (8.4), it appears that the majority of the data points come from stations in areas and/or seasons where heavy precipitation is produced by deep convection. Therefore the results of Fig. 8.10 could indicate that a universal behavior for extreme precipitation does exist for convective processes, but not in general.

References

1. A.S. Monin: *An Introduction to the Theory of Climate*, (Reidel Publishing Company, Dordrecht 1986), pp 261
2. K. Hasselmann: Tellus, 28, 473–493 (1976)
3. I. Prigogine: *From Being to Becoming, Time and Complexity in Physical Systems*,(W.H. Freeman and Company, 1980), pp 272
4. S. Lovejoy, D. Schertzer: Multifractals and Rain. In: *New Uncertainty Concepts in Hydrology and Hydrological Applications*, ed. by A.W. Kundzewicz, (Cambridge University Press Cambridge 1995), pp 63–103
5. T.T. Fujita: *The Downburst – Microburst and Macroburst. Report of Projects NIMROD and JAWS* (University of Chicago Press, Chicago, IL, 1985) pp 122
6. K.A. Emanuel: Nature, 401, 665–669 (1999)
7. A. Fink, P. Speth: Naturwissenschaften, 85, 482–493 (1998)
8. R. Glowienka: Contrib. Atmos. Phys., 58, 160–170 (1985)
9. A. Gill: *Atmosphere–Ocean Dynamics*, (Academic, New York, 1982)
10. E. Kalnay, M. Kanamitsu, R. Kistler, W. Collins, D. Deavens, L. Gandin, M. Iredell, S. Saha, G. White, J. Woollen, Y. Zhu, M. Chelliah, W. Ebisuzaki, W. Higgins, J. Janowiak, K.C. Mo, C. Ropelewski, J. Wang, A. Leetma, R. Reynolds, R. Jenne, D. Joseph: Bull. Am. Meteorol. Soc., 77, 437–471 (1996)
11. WMO: *Guide to Hydrological Practises, 5th edn*, (No. 168, World Meteorological Organization, Geneva 1994), see http://www.bom.gov.au/hydro/has/notables.shtml#World_record_rainfall_table
12. J.L.H. Paulhaus: Mon. Weath. Rev., 93, 331–335 (1965)
13. D. Schertzer, S. Lovejoy: J. Geophys. Res., 92, 9693–9714 (1987)
14. H. Kantz, Th. Schreiber: *Nonlinear Time Series Analysis, 2nd edn*, (Cambridge University Press, Cambridge, 2003)

[illegible]

[illegible]
12. [illegible] Mon. Weath. Rev. [illegible]
13. [illegible] J. Geophys. Res. [illegible]
14. [illegible] University Press, Cambridge, [illegible]

9 Freak Ocean Waves and Refraction of Gaussian Seas

Eric J. Heller

Summary. Rogue or freak waves sink ships at an alarming rate – estimated at one large ship every few weeks worldwide. It is thought that vulnerable ships (light cargo ships) simply break in two when they plough into a 60 foot wave preceded by a 40 foot hole in the sea, as some sailors that have survived such experiences have called it. Wave refraction due to current eddies (which are ubiquitous in the oceans) has long been suspected to play a role in concentrating wave energy into rogue waves. Existing theories have been based on refraction of plane waves, not the stochastic Gaussian seas one finds in practice. Gaussian seas ruin the dramatic focal caustic concentration of energy, and this fact has discouraged further investigations. Although it was thought that chaos, or the extreme sensitivity to initial conditions displayed by individual ray trajectories would quickly wipe out all significant fluctuations, we show that this is incorrect, and the fluctuations are "structually stable" entities. Significant "lumps" of energy survive the averaging over wave directions and wavelengths. We furthermore demonstrate that the probability of freak waves increases dramatically in the presence of these lumps, even though most parameters, such as the significant wave height, are unchanged. We show here that a single dimensionless parameter determines the potential for freak waves; this is the "freak index" of the current eddies – a typical angular deflection in one focal distance, divided by the initial angular uncertainty of the incoming waveset. If the freak index is greater than 2 or so, truly spectacular enhancements of freak index waves can result, even though the caustics are washed out by the Gaussian nature of the impinging sea.

9.1 Introduction

It is tragically clear from recent events that mankind has not taken oceanic Xevents seriously enough. In the case of the recent tsunami, the loss of life could have been greatly diminished by simple instructions given as a part of one's schooling: if the water suddenly recedes from an open beach, run to high ground, immediately. People often believe that Xevents always happen to somebody else, or worse, that forecasts of Xevents may be rumors without substantiation. Until very recently, this was the case with extreme or "freak" waves in the sea. Old salts who (amazingly) survived them had difficulty convincing people on shore that a wave at least twice as high as normal, and very steep, came out of "nowhere" and smashed a ship to pieces.

Giant waves in deep water have now been photographed (see Fig. 9.1), imaged by satellite, and sensed via ocean buoys. There is no doubt about their existence.

How could ten large ships be sunk this way every year and most of us not hear about it? There are several reasons. First, it has been happening for a very long time: ships disappear, and this is chalked up to bad weather. Second, the ship usually sinks immediately, or the electronics are wiped out immediately, and no distress signal is sent before the ship sinks. The ship is not missed for some time; the idea of a tragic loss arises slowly and does not arouse the press. Third, there is no visible wreckage for everyone to see or photograph.

Cruise ships and military ships are more resilient than container vessels. They do not break in two so easily, as a huge wave will crash on the deck just as the hull passes over a deep trough. However, they are not immune from problems. There were two dangerous events in February, 2005, one in the Pacific, another in the Mediterranean, involving cruise ships with around 700 people aboard. In both cases the electronics on the bridge were wiped out by giant waves that smashed the windows. This lead to engine shut down, which is dangerous, because it results in the ship drifting parallel to the waves, which soon makes it roll dangerously.

What are your chances of getting hit "out of the blue" by a giant wave on the high seas? Freak wave lore has it that they can strike even on relatively calm days with a seemingly benign swell from a nearby storm. Or they can occur on a stormy day, when a 70 foot wave will suddenly appear when the higher ones have previously been a much more negotiable 30 feet.

Fig. 9.1. (*Left*) Photograph of an oncoming freak wave from the bridge of a ship. (*Right*) This poster from the movie "Perfect Storm" is probably not much of an exaggeration

There are several kinds of freak waves, according to witnesses. Sometimes they take the form of the "three sisters": three large waves traveling in a group. They can also be the feared "wall of water preceded by a hole in the sea". Other freak waves are said to arrive at a large angle with respect to the mean direction of the waves, something like 30° or 40°. Tsunamis are rare and are much more devastating than freak waves, but they arise from earthquakes, landslides, meteors, and other nonoceanic causes. Freak waves on the other hand are wind-generated and recurrent (there are probably several hundred of them occurring around the world as you read this!). They arise "naturally"; they are Xevents that arise from fairly routine circumstances. One would like to know their cause. Models for wave generation and propagation can be established. Statistical reasoning will have to play a role, but a good theory will give an idea of how high the waves can get, and how often they reach dangerous heights.

Since the physical laws that govern them are, in principle, well known, and the conditions leading to an Xevent are not hidden from us as they are when trying to predict an earthquake, there is hope of predicting their likelihood. these days we routinely forecast the weather; perhaps one day soon we will hear the marine forecaster say "...the chance of encountering a 80 foot or higher breaking freak wave today is 1% in the area 100 nautical miles southwest of ...".

Today though there is no widely accepted theory of how freak waves form in the open ocean. Three categories of models predominate discussions: (1) Gaussian statistical ("unlucky" constructive addition), (2) refraction leading to focal caustics, and (3) nonlinear growth and steepening. By combining aspects of statistical models and refraction we arrive in this chapter at an attractive hybrid theory that has many superior traits compared to either parent. It is expected that nonlinear wave evolution, not treated here, will also play a role, no doubt worsening the large waves that the linear theory we put forth here creates.

Each of the three models, Gaussian, refraction, and nonlinear, has difficulties: (1) Xevents appear to be too rare compared to observations in the Gaussian statistical model [1, 2]; (2) although White and Fornberg [3] showed that a plane ocean wave incident on random current eddies could yield focal cusps at realistic distance scales, the refraction model [3, 4] has the problem that the caustics, which are the loci of freak wave events, are smoothed away by realistic averaging over wave directions [5]; (3) nonlinear models [6, 7] can explain the growth of large waves but presumably need a "seed" event for parasitic nonlinear effects [8] to take hold. Aspects of different models have been combined before; for example, the interaction of nonlinear effects and focusing has been the subject of much investigation [9, 10]. It seems very likely that a "final theory" of freak waves will be a synthesis of statistical, refractive, and nonlinear effects. The present work is not so ambitious, concentrating instead on a combination of statistical and focusing effects.

9.2 Gaussian Seas

We begin with a review of the statistical theory of waves described, to a first approximation, by Gaussian wave height distributions. The first development in this field came from Longuet-Higgins in connection with water waves [1]. Later, quantum theory encountered very similar problems. Length scales in quantum mechanics are typically one angstrom to one micron, while in ocean waves, length scales of meters to a hundred meters or so are appropriate, giving a ratio of quantum scales to ocean wave scales of as much as 10^{12}! But quantum waves obey superposition, interference and diffraction just as classical (for example water) waves do. Their superposition involves the same mathematics, independent of scale. In quantum theory, the field of quantum chaos raised questions about the random superposition of different plane waves. This was taken to be the analog of classical chaotic motion, which involves classical trajectories or "rays" heading every which way. It was M. Berry [11] who first postulated that the eigenstates (stationary states or "standing waves") of classically chaotic systems would be like random superpositions of plane waves, in the limit of short wavelength. Since the kinetic energy is locally given by the fixed energy and the local potential, the plane waves would locally all have the same wavelength, differing in propagation direction, in amplitude, and in phase. It is clear that such a random linear combination of plane waves must give Gaussian statistics, by the Central Limit Theorem (CLT). In the case of a billiard, where the potential is flat across it, the wave is then given by

$$\psi(x,y) = \sum_{n=1}^{\infty} a_n \sin(x\cos\theta_n + y\sin\theta_n + \delta_n) \tag{9.1}$$

with independently random distributions a_n, θ_n, and δ_n, all independent of position (x, y). Within limits, it does not matter what the distribution a_n is drawn from, so long as θ_n, δ_n are random and independent. However normalization requires that, for a billiard enclosure of area A,

$$\int_A |\psi(\boldsymbol{x})|^2 d^2\boldsymbol{x} = 1 \tag{9.2}$$

which imposes that $\sigma = 1/\sqrt{A}$ is the central limit Gaussian distribution of wave heights,

$$P(\psi) = \frac{1}{\sqrt{2\pi\sigma^2}} e^{-\psi^2/2\sigma^2} . \tag{9.3}$$

No direction is preferred in such a sum. It is straightforward to show, as Berry first did, that the autocorrelation function of such a random wave is given by

$$\langle\psi(\boldsymbol{x})\psi(\boldsymbol{x}+\boldsymbol{\delta})\rangle = J_0(k\delta)/A; \quad \delta = |\boldsymbol{\delta}| \tag{9.4}$$

where k is the wavevector of the plane waves. In quantum chaos theory, the closed billiard is used so often that the assumption of no preferred direction is sometimes taken for granted, but it is not obvious and in fact would be misapplied if used to describe mixed systems, such as the lemon billiard, which have a phase space consisting of integrable and nonintegrable regions. The same can be said for open systems, which is the way we treat the ocean problem below. Such mixtures or open systems are far more commonplace than either totally integrable or totally chaotic systems, which have to be carefully chosen to adhere to one extreme or the other. Eigenfunctions of mixed systems tend to live in either the integrable domain with small tails in the chaotic one, or vice versa, with few exceptions. In describing the chaotic subset of eigenfunctions, the correspondence principle requires that we exclude the integrable domains and therefore a uniform distribution of directions (momentum) no longer exists in position space. A moment's thought reveals that this does not harm the central limit theorem locally, and the statistics of the chaotic subdomain eigenstate will still be given by (9.3), with a normalization (dispersion) chosen *locally* to reflect the classical density, which varies from place to place. (In a chaotic two-dimensional system, the classical density is independent of position.) This was incorporated into the theory of the statistics of such eigenstates by Bäcker and Schubert [12]. These eigenstates are locally Gaussian random by the central limit theorem, but globally they are not uniformly sampled in space. The dispersion changes from place to place to reflect the local classical density. We shall return to the consequences of this below, but such a correction to Gaussian statistics is the key idea from quantum chaos theory here, which we believe has important consequences for freak waves.

In oceanography, the standard reference for Gaussian seas is M.S. Longuett-Higgens [1]. Noting that a storm does not produce a plane wave, the idealization of a random superposition of plane waves with a mean direction of travel and some dispersion in angle, wavelength, and amplitude was introduced. Once again the CLT applies, and the distribution of amplitudes becomes Gaussian. It is well understood that seas "evolve", even with no wind, from shorter, steeper "young" waves to a longer wavelength, more rounded swell. The nonlinear process that causes this was identified by Benjamin and Feir [8]. Even so, the plane wave approximation may apply locally, with the wavelength and dispersion parameters varying slowly over tens or hundreds of kilometers. Assuming a stationary random process, the wave height distribution is again (9.3), with ψ the sea level displacement and σ^2 the variance. The Rayleigh distribution for wave heights can be derived in the limit of a narrow spectrum of frequencies:

$$P(h) = \frac{h}{\sigma^2} e^{-h^2/2\sigma^2} \tag{9.5}$$

To give physical significance to these distributions, some practical definitions are made. The significant wave height H_s is defined as the average height of the highest third of all of the waves. This is easily shown to be

$$H_s = (3\sqrt{2\pi} erfc\left(\sqrt{\log 3} + 2\sqrt{\log 9}\right) \sigma \approx 4.0043\sigma) , \tag{9.6}$$

in other words a trough-to-crest height of almost exactly 4σ, or a mean level-to-crest height of very close to 2σ. The commonly accepted definition of a freak or rogue wave is 2.2 times this height, or $H_{freak} > 8.8\sigma$ measured from trough to crest. The probability of a wave being this height or greater is 6.25×10^{-5}, according to Gaussian statistics.

The danger a wave poses is not merely a matter of its height. A 40 foot wave in a sea with $H_s = 40$ feet is likely to be far less dangerous than a 40 foot wave with $H_s = 15$ feet. The reason is the steepness. A $H_s = 40$ foot sea is likely to have a longer wavelength than a $H_s = 15$ foot sea, so that the sudden appearance of a 40 foot wave means a very steep wave. Steep waves tend to break, which makes them more dangerous to any ship or boat. For larger vessels, the steep wave (often described as a "wall of water" by lucky survivors, and indeed nonlinear processes may be at play to make it more fearsome) poses a problem, because the buoyant bow ploughs into it (especially if preceded by a large trough, or "hole in the sea"). The support under the vessel is lessened by the trough, and the final blow is delivered when the wave breaks on deck, with the result that the ship breaks in two.

9.3 Refraction

The refraction of waves due to random current eddies is an example of scattering in a weakly refracting random medium [13,14]. Similar problems have been investigated in many contexts, including acoustics [15], light scattering [13], and mesoscopic electron physics [16] in addition to water waves [3]. However, it appears that less work has been done for the present kind of Gaussian incident wave, which corresponds to a diffuse source of "radiation" (waves), and even less has been done concerning the Xevent statistics in this situation.

When waves move through a medium that is moving uniformly, they are Doppler-shifted in an obvious way, given by a Newtonian frame transformation. However, if there are velocity gradients within the medium, then the resulting phase velocity gradients will lead to the refraction of rays representing the group velocity. The language just used is essentially semiclassical; the wavelength must be short compared to significant changes in the velocity.

The refractive effects of current gradients on water waves have been of interest since Perigrine's analysis [4,9]. The dispersion relation for deep water waves with no currents is

$$\omega(\boldsymbol{k}) = \sqrt{g|\boldsymbol{k}|} \tag{9.7}$$

where g is the acceleration due to gravity. This is very different to quantum waves, where $\omega \propto k^2$, or the wave equation, where $\omega \propto k$. With currents, one has for deep water waves

$$\omega(\boldsymbol{k}) = \sqrt{g|\boldsymbol{k}|} + \boldsymbol{k} \cdot \boldsymbol{U}(\boldsymbol{x}) \tag{9.8}$$

where $U(\boldsymbol{x})$ is the velocity field. This equation has the obvious meaning for $\boldsymbol{U} = const.$, but defines the situation for non-constant $\boldsymbol{U}(\boldsymbol{x})$, in the eikonal sense:

$$\frac{d\boldsymbol{k}}{dt} = -\frac{\partial \omega}{\partial \boldsymbol{x}}\,; \quad \frac{d\boldsymbol{x}}{dt} = \frac{\partial \omega}{\partial \boldsymbol{k}}\,. \tag{9.9}$$

When the wavelength becomes of the order of the depth, the waves begin to feel the bottom so to speak, and the dispersion relation changes, becoming for shallow water the usual wave equation with $\omega = \sqrt{gz}\ k$ for small amplitude shallow water waves, where z is the depth. Therefore, changing depth contours also refract water waves, and lead to spectacular effects well known to surfers. However these are coastal effects, and here we are concerned with freak waves at sea. (Although it is true that continental shelves extend hundreds of kilometers in places, and heavy seas cannot be assumed to be totally free of the effect of the bottom.) We make one last comment before dropping the subject of bottom contour refraction: certain coastal or inland sea regions with known bottom contours might make good laboratories for studying the effects of refraction on the statistics of the sea state.

The oceans are filled with eddies and currents on various scales [17]. A typical eddy might have relative velocities of a knot or two and be 20–100 km across. Two of the most famous currents, the Gulf stream and the Aguhlas current off the coast of South Africa, are also the site of many freak wave events and many well known ship losses [18, 19], and many well documented strong eddies. Specific formations exist, like annular currents, spun off from current streams. These have been investigated for their refractive effects [20].

White and Fornberg [3] showed definitively that for a pure plane wave, a random eddy field could lead to focal caustics. Their work included a discussion of the parameters expected in the oceans. It turns out that random eddies can typically cause focal cusps 100–300 km downstream of the start of the eddies. In the present work, the deterministic certainty of the focal cusps is replaced again by probabilities, as in the Gaussian seas model. We show below that the high formerly caustic regions for rogue waves are smoothed and raised by typically a factor of 2 or so, to 50 or 150 km. By the "deterministic certainty" of the plane waves we mean that as long as a plane wave impinges on the refractive zone, large waves appear with the perfect regularity of the wave period, and always in the same place. This in itself is a strong argument against refraction of such a pure sea as a model for freak waves, because nobody suspects that rogue waves appear with such regularity in one spot. (One of the most interesting questions this work raises, however, is the issue of movement of the "lenses" causing the cusps (the current eddies), which would cause the cusps to move at an as yet unknown rate.)

Suppose waves with a mean velocity $\boldsymbol{U}$ are incident on random velocity fluctuations $\boldsymbol{u}(\boldsymbol{x})$ arising from eddies. The important parameters are 1) the fluctuation amplitude δu_0 of $\boldsymbol{u}(\boldsymbol{x})$, which we take as Gaussian, $P(u) = \delta u_0^2 \exp(-u^2/2)/\sqrt{(2\pi)}$, 2) the correlation length d of the isotropic velocity fluctuations, so $c(y) = \langle \delta\boldsymbol{u}(\boldsymbol{x}) \cdot \delta\boldsymbol{u}(\boldsymbol{x}+\boldsymbol{y})\rangle \equiv \delta u_0^2 \exp(-y^2/2d^2)$, and 3) the mean distance L "downstream" of the resulting first focal cusps measured from the beginning of the refraction. The mean distance between the focal cusps normal to the average flow is on the order of the correlation length d. The dimensionless ratio d/L, which is essentially the deflection angle in one focal distance, can be shown to scale as $d/L = (\delta u_0/U)^{2/3}$, where U is the mean wave velocity.

Figure 9.2 shows the geometrical situation:

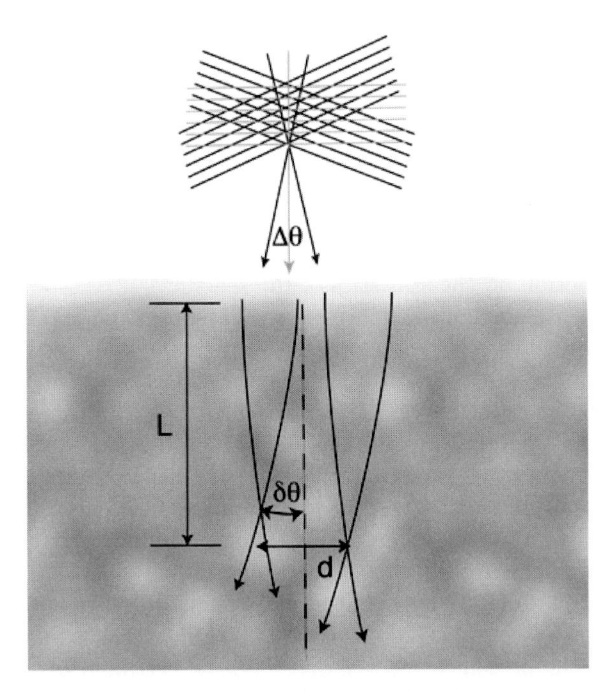

Fig. 9.2. Geometry of incidence and refraction in the model. The *gray area* is the random refracting medium

The essence of White and Fornberg's work is shown in Fig. 9.3, left. A random velocity field is shown on the right, with higher velocity (in the top-down direction of propagation) indicated by darker shades. The focusing caused by the random velocity field is evident; the structures seen in the center are cusp catastrophes that have developed. A dense set of parallel rays was launched from the top, all heading straight down; the random potential caused the cusps to develop. The middle panel reveals the phase space or "surface of

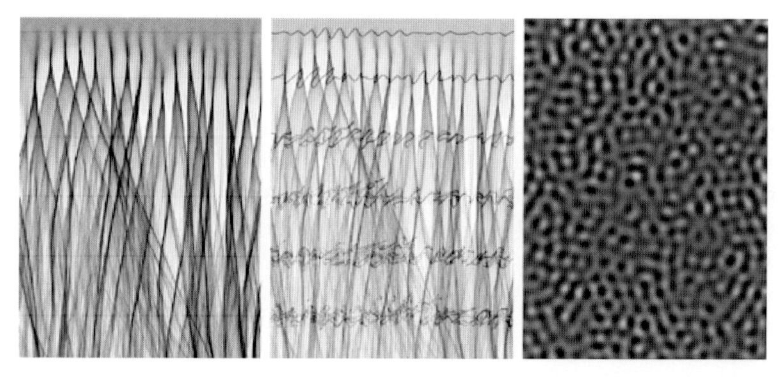

Fig. 9.3. Ray limit study of wave energy density in a random refracting field for an initial set of parallel rays corresponding to a single plane wave. Here *darker* means a higher density of rays, and the flow is from top to bottom. The *right hand panel* is the result of a dense set of trajectories launched uniformly from the upper edge in the direction straight down. The *middle panel* repeats this, with the surface of section phase space slices added

section" structure, obtained by slicing the flow along the transverse lines shown and plotting x *vs.* p_x on the $y = const.$ slices. The coordinate space density is the phase space density summed over all momenta p_x, so that it is projected onto the x-axis. Cusps are the result of the development of a kink in the phase space manifold which folds over on itself. As the central part becomes vertical (a tangent to the line $x = const.$), a cusp singularity develops, followed by two fold singularities which give the cusp its characteristic V-shape.

9.4 Refraction and Gaussian Seas

As we have seen, the refraction of an initially pure plane wave leads to focal cusps and other caustic structures where large waves appear with perfect regularity. (Ray theory suffers infinities at these cusps, but they are smoothed out by finite wavelengths [21].) The hybrid theory proposed here supposes a Gaussian sea (consisting of a random superposition of plane waves) impinging on refracting current gradients. When one averages over many plane waves at once, with different propagation directions, the cusps are smoothed over even in ray theory, so that a new regime is reached where the wavelength no longer limits the wave heights. The focus shifts, literally and figuratively, from a deterministic generator of freak waves at fixed regions to a stochastic picture, where regions simply have a higher or lower *probability* of freak wave formation.

Granted that the caustics are washed out, what *does* happen when a Gaussian sea is incident on a refracting region? Simple arguments suggest that random refraction of a Gaussian sea would have a negligible effect on its wave statistics. The random nature of the refraction introduces no bias, so the sea is Gaussian again after refraction with exactly the same wave height distribution as before. A second argument, made in [5], seems to reinforce this view. Due to the sensitive dependence of the ray motion on initial conditions, the caustics will move around rapidly as a function of incident ray angle, washing them out after averaging over even one or two degrees of angle of incidence.

Surprisingly, both arguments are misleading. While it is true that a perfectly Gaussian sea will re-emerge as Gaussian after random refraction, it is not quite Gaussian *during* and just after the refracting events, a fact that is well known in the theory of scattering by random media [13]. Furthermore, although caustics are eliminated by averaging over angle of incidence, the average stubbornly refuses to yield a uniform energy density in $\boldsymbol{x}$, the plane of motion of the waves. Our numerical results show that this holds true to a surprising extent even after averaging over a significant range of wavelengths.

We turn now to investigate what actually happens when a Gaussian sea is substituted for a plane wave. Figure 9.4 compares ray limit and wave studies of wave energy density in a random refracting zone, with diffuse source Gaussian waves incident (the left panel shows the limit of no wavelength or directional dispersion, for comparison). The grayscale gives the local energy density averaged over a long time. A lumpy wave energy variation has replaced the caustics. Figure 9.4 also shows that the time-averaged wave energy $\propto \int |\psi(x,y,t)|^2 dt$ in the right hand panel; the agreement with the ray density is remarkable, lending confidence to ray calculations of the energy density.

Increasingly smaller scale density fluctuations arise deeper in the refraction zone, including high-angle "runners" which form after two or so focal lengths L; these are streaks in the density traveling at around 30° to the mean flow direction. V-shaped "roostertails" also form, which are overlaps of fold catastrophes coming from adjacent cusps. (These superficially resemble focal cusps, but they are not the same.) It had been argued [5] that even a one or two degree average would wipe out the energy variations, but this is clearly not true; here at 5° there is still a roughly 5:1 ratio between the highest and lowest energy densities.

For the wave propagation we used a *linear* Schrödinger equation, ensemble-averaged over five or ten wavesets, each containing 700 randomly chosen plane waves, and propagated over a region 240 wavelengths in length and 80 in width for a total distance of 4800 wavelengths (the wave, 6000 wavelengths long, is incident from a non-refracting zone). A statistical analysis of the wave is performed at every time step. The Schrödinger equation has the advantage of being fast and simple to propagate, and our goal is simply to populate the region with waves in a statistically correct way. We have represented the

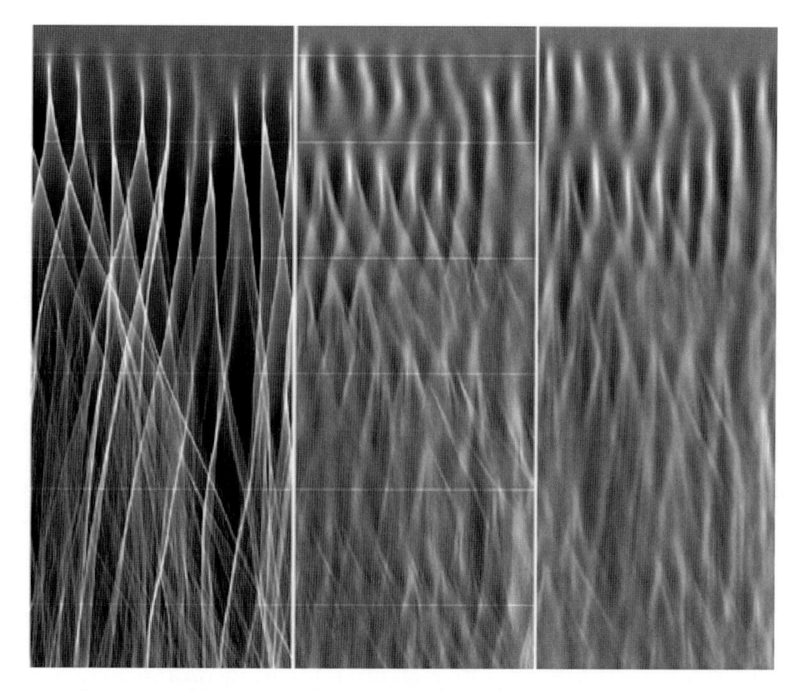

Fig. 9.4. Ray limit (*left, middle*) and wave (*right*) studies of wave energy density in a random refracting field. *Left*: an initial set of parallel rays corresponding to a single plane wave is shown. Here *darker* means a lower density of rays, and the flow is from top to bottom. The *middle panel* was given a 20% variation in velocity and a 5° range of initial directions (a Gaussian distribution of wavevectors with an uncertainty of 5°). The *right panel* displays the Schrödinger propagation of an ensemble of random wavesets each with 700 plane waves, averaged over a long time. The wavesets are chosen to correspond to the same average velocity, dispersion of velocity and angle as in the ray simulation. The *grayscale* gives the local energy density averaged over a long time. Lumpy wave energy variation has replaced the caustics

eddies with a potential field; this is a good approximation so long as the direction of propagation is maintained with relatively small dispersion. This is reinforced by the fact that only the statistical nature of the velocity gradients matters. There are small differences due to the different dispersion relations of water waves and matter waves, but these vanish in the limit of a narrow frequency band incident wave.

One critical dimensionless parameter that affects the energy density fluctuations is the "freak index" $\gamma = \delta\theta/\Delta\theta$ of the angular separation $\delta\theta \sim d/L$ of adjacent first focal cusps to the angular spread $\Delta\theta$ of the incoming waveset as measured from the start of the refraction. When $\gamma \sim 1$, the first focal cusps have been smeared over about one correlation length, and we may expect the angle averaging to strongly attenuate the energy fluctuations. Using

White and Fornberg's [3] parameters, this critical angle $\Delta\theta$ would typically be 10–20°.

9.5 Structure of the Density Fluctuations

9.5.1 Phase Space and Real Space

Why do the density variations survive the angle averaging? Dysthe's argument had been that trajectories are extremely sensitive to initial conditions, which is certainly true. He showed a simple calculation demonstrating the sensitivity of the focal cusp positions to initial angle. However, this does not guarantee the spatial uniformity of the resulting average. Figure 9.5 demonstrates the mechanism for the persistence of energy lumps, and shows why small-scale fluctuations are generated. Defining k_x as the transverse wavevector, we have $\Delta\theta \sim \Delta k_x/k_y$. The combined vertical and horizontal shears in the phase plane (k_x, x), which are the result of passing through weakly refracting regions, preferentially pile up density at certain positions ("lumps") in x. The fold and cusp singularities are smoothed away. The thinning out of the initially thick distribution into tendrils due to mixing in the phase plane generates fine structure in the phase plane and in its projection onto position. The energy lumps are structurally stable (universal) features of a random refractive region. As the wave progresses the energy contrast decreases slowly, since the *local* $\Delta\theta$ increases diffusively, decreasing *local* γ and with it the contrast. See Sect. 9.5.3 below.

The mean and variance of incoming wave directions are shown by the thick line at the top of section plot 9.5A; a line with a Gaussian profile and a variance Δk is used to represent a Gaussian distribution of initial angles. As the first perturbations that speed up or slow down the flow are encountered, the transverse velocities are affected and the k_x wavevector magnitudes are distorted randomly up and down by an area preserving vertical nonlinear shear by an amount we call δk_x. This is seen in the second wavy line in Fig. 9.5A. Later, the changes to k_x cause a drift right or left, depending on whether k_x was increased or decreased. This is a linear shear in the horizontal direction. At all times the position space energy density (ray density) is found by projecting the distribution onto the position. At the bottom of Fig. 9.5A the projection of the double sheared line reveals lumps that are the residual energy concentrations. To see what happens past the region of the first cusp, which is the domain of Fig. 9.5A, we examine Fig. 9.5B. The top panel shows a larger slice of coordinate space at about the same stage of development as the bottom of Fig. 9.5A. Subsequent evolution again shears vertically and horizontally, building an intricate structure out of the initially featureless horizontal diffuse line. The finer features show up in coordinate space as more rapid variations in density; these eventually become much more rapid than the correlation length d of the random perturbation.

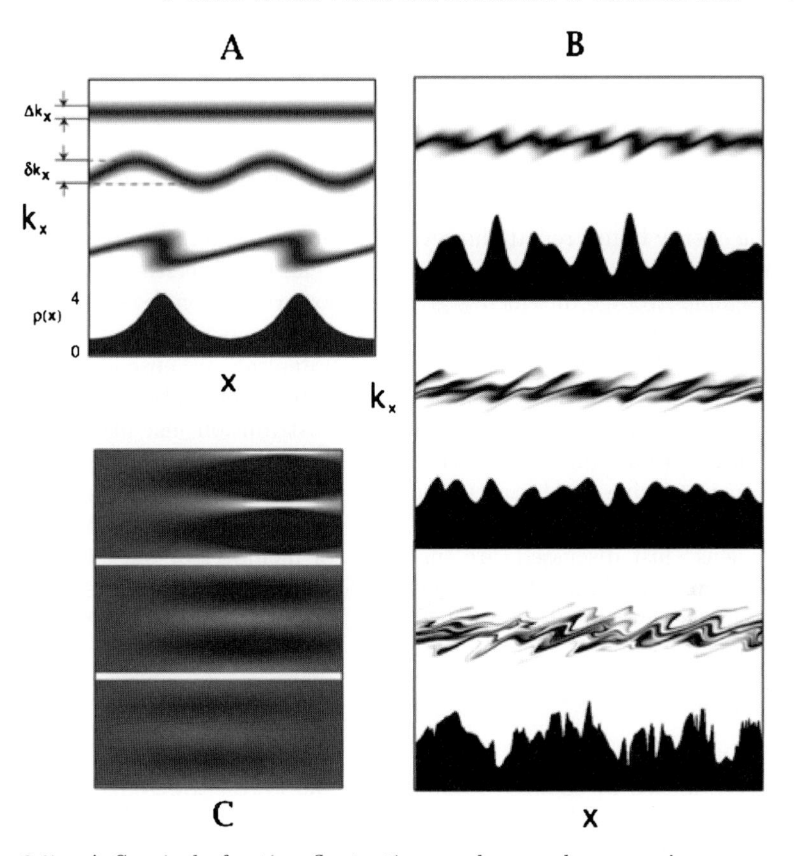

Fig. 9.5. **A** Survival of action fluctuations under angular averaging, as seen in phase space and coordinate space projection. The *top line* shows that the initial transverse spread is Δk_x; the *next line* shows that the first focal refraction introduces a variation δk_x; the *third line* shows the state after free drift shearing, and the projection onto coordinate x is shown by the *bottom line*. **B** Three sequential snapshots of typical evolution in the refractive zone seen in phase space and projected into in real space. This shows how sharp features develop despite the "averaging" implied by the initial spread Δk. **C** Real (x, y) space Gaussian beam simulations of first focal region averaging for $\gamma = \Delta k_x/\delta k_x = 20, 4$, and 2. Note that the maximum moves left as averaging increases

In Fig. 9.5C we see a coordinate space density plot of a focal cusp, for $\gamma = \Delta k_x/\delta k_x = 20, 4$, and 2. Note how the remnant of the cusp moves left; the cusp region has been thoroughly averaged by $\gamma = 2$, but significant density variation remains further toward the source, here on the left.

The variance of the energy density depends on $\gamma = \delta k_x/\Delta k_x = \delta\theta/\Delta\theta$ (the ratio of the deflection angle in one focal distance L to the incident angular uncertainty). The contrast ratio R in local wave action near the first smoothed focal maxima is found numerically to scale as $R \sim 1.6\gamma$ and the maximum density at distance F moves closer to the beginning of the gradient field approximately as $F \sim F_0(1 - 2\gamma)$ for $\gamma > 1$.

The fold and cusp singularities are smoothed away by angle averaging. The thinning out of the initially thick distribution due to mixing in the phase plane generates fine structure (tendrils) in the phase plane and in its projection onto position.

The energy lumps remain however, and are structurally stable (universal) features of a random refractive region. This can be understood by imagining an ensemble average in which a given random refraction region of size $\sim L$ in the longitudinal direction and $\sim d$ in the transverse direction is preserved in the ensemble, while all the "upstream" potentials are ensemble-averaged. The fixed region will then be simply sampled by an ensemble-averaged phase space density, such as a smooth Gaussian distribution just like the one we have initiated our calculations with. But we have seen in those calculations that lumps survive the destruction of the singular cusps; what we see deeper into the refracting region in the figures, which are not ensemble-averaged in the sense just discussed, are snapshots of members of the ensemble. On average, there are lumps behind the accidentally focusing regions, decorated by the projections of all the tendrils and "runners" that have developed.

9.5.2 Runners and Rooster Tails

The runners seen in real space plots are the result of phase space tendrils that have been kicked by chance to the highest and lowest transverse momentum by the random perturbation. Therefore, they travel at the highest angle with respect to the mean flow. Sometimes they are not thin and they carry substantial energy (area), so it is tempting to associate them with high angle rogue waves. However as yet we have no strong evidence that the runners might generate or somehow correspond to high angle rogue waves reported at sea.

For larger γ, some of the V-shaped fold caustics remain, not as singularities, but as regions of higher density, and when these collide we get the "rooster tails" mentioned earlier.

Figure 9.6 shows the specific correlation between the two-dimensional ray or energy density plots, A, and slices through those plots, B, showing the density fluctuations quantitatively. The base of each subplot is the location of the slice. C shows the phase space density corresponding to an alternate set of slices. Runners can be seen in the density plots and in the phase space plots , for example at the bottom, as tendrils of high transverse velocity.

Figure 9.7 gives the phase picture for runners and rooster tails. Rooster tails are most prominent at large γ, as in the case seen here. The vertical lines indicate, in each case, two phase space zones with reasonably high density overlapping in space (with a third weaker density in the middle), causing a temporary spike in the coordinate space density as the upper one moves

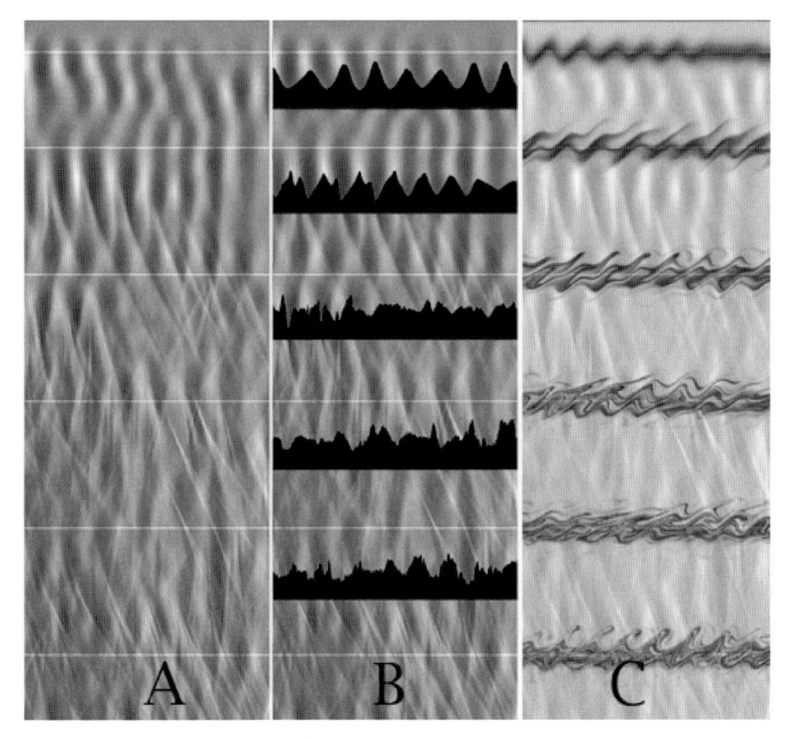

Fig. 9.6. The specific correlation between the two-dimensional ray or energy density plots (**A**) and slices through those plots (**B**) showing the density fluctuations quantitatively. The base of each subplot is the location of the slice. **C** shows the phase space density corresponding to an alternate set of slices

right and the lower one left. To the right of the thick black line we see a typical runner at high wavevector, here also with significant amplitude.

9.5.3 Diffusion and the Freak Index

As the wave progresses, the energy contrast decreases slowly, since the *local* $\Delta\theta$ increases diffusively, decreasing *local* γ and with it the contrast. Under angle averaging sufficient to smooth the caustics over more than a wavelength, the significant wave height scales as the square root of the ray density, independent of the wavelength. With little or no averaging, the ray caustics are still present, and the wave height is instead limited by wavelength [21]. With a some initial smoothing, it is straightforward to show that the variance $\Delta\theta(y)$ develops as

$$\Delta\theta(y) \sim \Delta\theta \sqrt{1 + \bar{y}\,\gamma^2} \tag{9.10}$$

where $\bar{y}$ is measured in units of the focal distance L along the mean flow direction. Assuming the nature of the perturbation remains the same as y

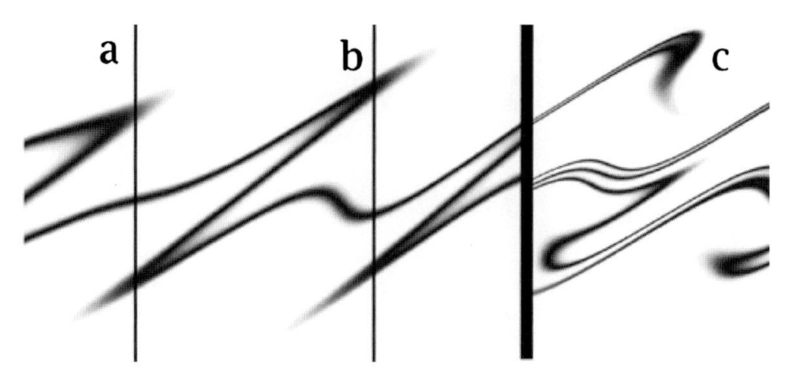

Fig. 9.7. Rooster tails are most prominent at large γ, as in the case seen here, *left*. The *vertical lines* indicate, in each case, a and b, two phase space zones with reasonably high density overlapping in space (with a third weaker density in the middle), causing a temporary spike in the coordinate space density as the upper one moves right and the lower one left. To the right of the thick black line we see a typical runner (c) at high wavevector, here also with significant amplitude

increases, the contrast decreases and with it the freak wave excess, as the dimensionless parameter that controls the contrast decreases:

$$\gamma(\bar{y}) \sim \frac{\gamma}{1 + \bar{y}\,\gamma^2} \to \frac{1}{\sqrt{\bar{y}}} \tag{9.11}$$

where $\gamma(y)$ is the evolving freak index and γ is the initial freak index upon entering the refractive region. The freak index declines like the inverse square root of the distance traveled in the eddies. Remarkably, the implication is that seas with large angular variance tend to be safer (at least from freak waves), because such seas are more resistant to subsequent refraction (the freak index will tend to be lower).

9.6 Implications for Wave Statistics

9.6.1 Nonuniform Sampling

We are ready to discuss and compute the combined attributes of initially Gaussian seas and refraction. We have learned that the energy density fluctuates on various scales. The variance $\sigma^2(x, y)$ of the sampled waves is proportional to the energy density at (x, y). Our model will generalize the Longuet-Higgins picture to include these fluctuations, which correspond to nonuniform sampling over space. In this model, strictly Gaussian statistics will apply over very small patches of smaller extent than the energy density lumps. For a narrow band (a small spread in angle and energy) Gaussian random wave with dispersion σ, the Rayleigh distribution [1]

$p(h) = h/\sigma^2 \exp[-h^2/2\sigma^2]$ describes the statistics of wave heights. The energy or action lumps correspond to the spatial variation of $\sigma^2 = \sigma^2(x,y)$, which is proportional to the grayscale density in the time average; for example in Fig. 9.4, right. Suppose over some region A we measure the probability density of $\mathcal{P}(\sigma^2) = \int_A dxdy\ \delta(\sigma^2 - \sigma^2(x,y))$. Then, assuming local Gaussian statistics adjusted to the local σ^2, the averaged wave height distribution is

$$P(h) = h \int \frac{\mathcal{P}(\sigma^2)}{\sigma^2} e^{-h^2/2\sigma^2}\ d\sigma^2 \tag{9.12}$$

The distribution $\mathcal{P}(\sigma^2)$ is log-normal, deep enough in the refractive zone [13–15], for a single incident angle, as in Fig. 9.4, left, but is found to approach a Gaussian in σ^2 after two or so focal distances L for angle-averaged cases. Near the first focus distance a remnant of the log-normal tail is seen in $\mathcal{P}(\sigma^2)$. In the numerical studies, $\mathcal{P}(\sigma^2)$ was determined from the data and used as in (9.12). For variations in $\sigma^2(x,y)$ that are not too severe, the result of the integral in (9.12) will appear Gaussian to the eye. Indeed its second moment is the energy density and by energy conservation (and in calculations) it does not change. Higher moments will be affected by the averaging over different energy density regions. However, the fourth moment (or the Kurtosis) is only very weakly affected for the cases we have studied so far, γ =1-6. The statistic for water level is

$$P(\psi) = \int \frac{\mathcal{P}(\sigma^2)}{\sqrt{2\pi\sigma^2}} e^{-\psi^2/2\sigma^2}\ d\sigma^2 \tag{9.13}$$

Since we are interested in 4.4σ to 6σ events, it is interesting to note that $\psi^{20} \exp[-\psi^2/2\sigma^2]$ peaks around $\psi = 4.5\sigma$, and $\psi^{36} \exp[-\psi^2/2\sigma^2]$ peaks at 6σ. This gives us some idea about which moments of the distribution might be most useful to collect. Normally such high moments would be statistically unreliable, but we generate an enormous amount of data with our simulations (see the discussion near the end of Sect. 9.4).

From a mathematical point of view, the nonuniform sampling is perfectly well defined. We can average over domains larger than the lumps if we wish, and in doing so the overall statistics cannot be strictly Gaussian. However, we must ask which of the following is *physically* correct approach to use for the oceans: to gauge the statistics (SWH) only locally, in which case the statistics are always Gaussian according to our model, or to use the *average* SWH to judge which events are Xevents? Are the statistics properly measured locally or globally? One extreme is clear: In the White and Fornberg refraction model [3], using a single plane wave, very high waves occur at the classical caustics. One would not want to renormalize away the caustic regions by noting the SWH is larger there! What is dangerous about the caustics is the *sudden* accumulation of wave action, leading to steep waves which the sea has insufficient time or distance to accommodate through slower nonlinear evolution, increasing the wavelength and lowering the slopes [8]. The energy lumps we have shown to exist gather their energy on a scale no larger than the order of $\sim L/3$ in the propagation direction near the first focal region, but much more sudden gatherings occur later, for example in the onset of a rooster tail. For 50 km eddies, a sudden doubling of the energy density over 5 km or less is possible. Moreover, it is likely that the lumps are moving (because the eddies are not stationary), so that even a long sample at one place would constitute an average over high and low energy regions. Our point of view, then, is that it is often essential to average as in (9.12).

9.6.2 Freak Wave Events

We have already shown the time-averaged energy density derived from wave propagation in Fig. 9.4, and now we delve more deeply into the statistics and occurrence of freak waves in such simulations. We emphasize that the Schrödinger propagator used does not give realistic space–time "movies", but it does populate the energy density (ray tube density) with waves in a statistically satisfactory way. (It is interesting that the oceanography community uses a nonlinear Schrödinger equation (NLSE) to describe real-time nonlinear water wave evolution, but the equation is somewhat removed from what is called the NLSE in quantum mechanics of Bose condensates for example.)

Figure 9.8 is useful in that it probes events and statistics before, during, and after entering a refractive region, shown in Fig. 9.8A. In panel B, we notice the smooth average obtained by sampling several wave sets with the same dispersions in wavelength and angle over a long time. In the middle refraction zone we see the smoothed lumpiness that is the remnant of the caustics. The return to a uniform density is nearly complete at the bottom of B, after it has propagated some distance beyond the last refractions. (This corresponds to horizontal shearing only in the phase space plots, eventually wiping out significant fluctuations.) When the density is again uniform, the sea returns again to the same energy density (the same σ^2) as it started with, but a wider angular variance σ_θ^2 than it started with). Note the rapid drop in 6σ events past the refractive region, in C. In D, the 4.4σ events are much more frequent in the refractive zone (in fact, much more than can be conveyed by this simple grayscale plot). There are some 4.4σ events prior to the refractive zone, as expected from the Gaussian distribution. After the refractive zone, we see an enhanced probability of 4.4σ freak waves compared to before the zone. In this example, $\gamma \sim 2$.

9.6.3 Statistical Evidence

Let $\mathcal{R}_a$ be the ratio of the local probability of an $a\sigma_0$ event to the probability of the same event using a Rayleigh distribution with dispersion σ_0. Then $\mathcal{R}_a$ is given by

$$\mathcal{R}_a(x,y) = \exp\left[\frac{-a^2}{2}\left(\frac{\sigma_0^2}{\sigma^2(x,y)} - 1\right)\right] \tag{9.14}$$

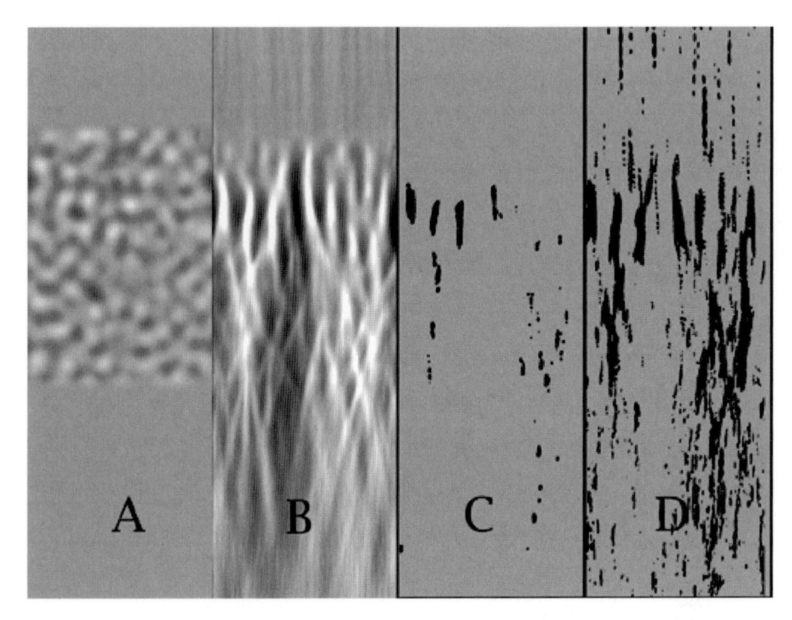

Fig. 9.8. Freak events before, during, and after interaction with a refracting region. **A** Random refraction zone. **B** Energy density average over a long run. **C** Loci of 6σ waves during the simulation. **D** Loci of 4.4σ waves during the simulation. Clearly the high-energy areas are the danger zones, with a high probability of freak wave formation. In this example, $\gamma \sim 2$

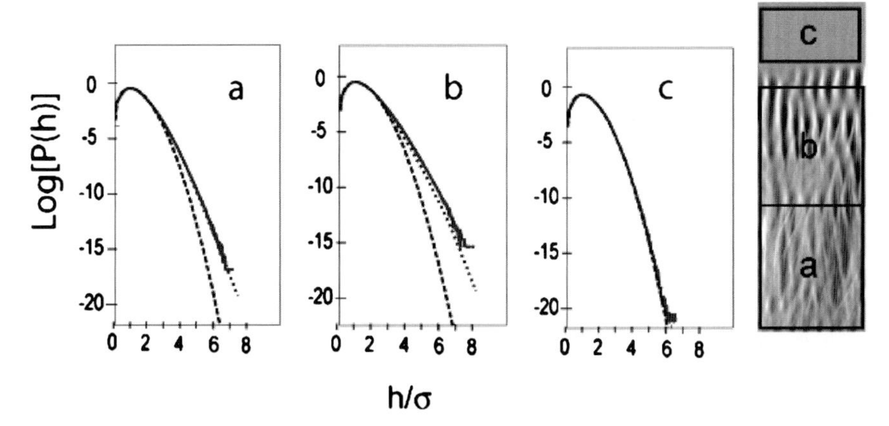

Fig. 9.9. Log of wave height data by region (compared with Rayleigh and theory). *Dashed line*: Rayleigh distribution, based on the average SWH, *solid*: data from wave propagation, *dotted*: theory based on (9.12) and measurement of the energy variation. Each of the three regions (as indicated) were analyzed similarly. The SWH differs in the three cases by less than 3%. $\gamma = 3.4$ (see Table 9.1)

Table 9.1. Average extreme wave enhancement factors. Three cases are shown; the parameter γ controlling the angular smearing and $\delta k/k$ controlling the wavelength averaging are given; the two regions a and b have different statistics; b is the region of the smoothed first and second cusps. The $6\sigma, 5\sigma$, and 4.4σ enhancement factors giving the ratio of the predicted extreme wave events (based on (9.12)) higher than (for example 6σ to the expected events based on a Rayleigh distribution fitted over the region are shown in **bold**). For example, waves of height 5σ or above are 54 times more likely on average over region b in trial 1 than the prediction based on the Rayleigh distribution fitted to all of the wave heights in zone b

Case	γ	$\delta k/k$	μ_a	$6/5/4.4\sigma$	μ_b	$6/5/4.4\sigma$	Remarks
1	3.4	0.16	0.22	**101/ 12/05**	0.32	**892.54.15**	Figs. 1,3
2	2.2	0.16	0.17	**16/ 4/2.5**	0.22	**69. 9.4.5**	
3	1.2	0.18	0.12	**6/2.5/1.5**	0.15	**11. 4.2**	$\gamma \approx 1$

where $\sigma^2(x, y)$ is the local action density including the effects of refraction. Equation (9.14) shows that spot enhancements and suppressions at particular places are significant, reaching a factor of 25 for a threshold 4.4σ wave where the local energy is 50% above the mean, and 400 for a 6σ wave. At the same time, the low energy zones are remarkably quiescent: 4.4σ events are 20 times less likely in a patch only 25% lower in energy density from the mean. The unlucky ship that finds herself in one of the bright lumps in Fig. 9.8 will have approximately 1500 times the probability of encountering a 4.4σ rogue wave there than in a zone with the average energy density. A fearsome 5σ event is 12,000 times more likely there than in a zone with average energy density.

9.7 Conclusions

The key point in this paper is that remnant refractive effects cause a lumpy spatial energy or wave action distribution, skewing the formerly Gaussian height distribution. Gaussian seas have no lumps. Low-order moments such as the Kurtosis show little effect, while Xevent tails are sometimes enhanced by many orders of magnitude. Dangerous seas can result when the angular deflections of the eddies exceeds the initial angular dispersion (the freak index $\gamma > 1$). The freak index is a new concept that can be used to characterize the danger level of the seas. A freak index of 2 or 3 on heavy seas is a dangerous situation. Seas with large angular variance tend to be *safer* from freak waves. We are not speaking here of "cross seas" with two or three well defined directions from which the waves are arriving; these and other non-Gaussian incident seas will need their own studies, and will certainly enhance freak waves compared to Gaussian seas with the same γ. Thus, the present results should be viewed as the least that can happen when mixed seas meet a refractive zone.

The question of the relation of the lumps to transient wave groups [2, 22] arises. The high energy lumps discussed here are typically much larger than the extent of a wave group, and are therefore regions where high-amplitude wave groups are more common.

A number of other issues suggest themselves but are not addressed here, such as the experimental detection of energy lumpiness, the rate of movement of lumps (due to changing eddy positions and velocities), and the nonlinear evolution of waves through a region with lumpy energy density. It will be especially important to test the behavior of the nonlinear models [5–7] under the conditions of Gaussian seas meeting refractive zones.

It is worth looking at the developments here from a broad perspective. As mentioned, the Gaussian seas hypothesis, which was proposed in the 1950s, provides a statistical theory for the occurrence of freak waves. If this was the whole story we would already have weather forecast-like estimates of the probabilities of freak waves, but it gradually became clear that refraction played a role in many events, since regions such as the Gulf stream and Aguhlas current are notorious freak wave areas. This does not contradict the statistical effects, but rather adds a new mechanism. It is crucial to note that it is not an *independent* mechanism; refraction can still apply in connection with statistics, as we have noted here. For many years now, nonlinear processes have held the attention of researchers. It is clear they are important: without them one can't develop the fearsome breaking shapes that some freak waves have. An example of a nonlinear wave equation involves terms like $|\psi|^2\psi$, which is cubic in the wave amplitude. Nonlinear processes surely happen and do not contradict either statistics or refraction, and they therefore represent another tool for reaching a complete understanding of freak waves. However, models involving the refraction of "ideal" wave fields like plane waves and models relying on nonlinear processes have not produced good estimates for the frequency of freak waves as a function of the state of the sea, and have tended to lead thinking away from statistical predictions. If energy lumping and the freak index are important, then a statistical theory is again crucial, one that might permit forecasts. Whether or not we have identified a typical scenario for freak wave production (enhancement due to energy lumping), it seems clear that such effects should be considered more carefully.

References

1. Longuet-Higgins, M.S., "The Statistical Analysis of a Random, Moving Surface", Philos. Trans. Roy. Soc. Lond. A, 249, 966 (1957), pp. 321–387
2. Dankert, H., J. Horstmann, S. Lehner and W. Rosenthal, "Detection of Wave Groups in SAR Images and Radar-Image Sequences", IEEE Trans. Geosci. Remote Sens., 41, 6 (2003)
3. White, B.S. and B. Fornberg, "On the Chance of Freak Waves at Sea", J. Fluid. Mech., 355, (1998), pp. 113–138

4. Peregrine, H., "Interaction of Water Waves and Currents", Adv. Appl. Mech., 16 (1976), pp. 9–117
5. Dythse, K.B., in "Rogue Waves 2000 (Proceedings of a Workshop held in Brest, France, 29–30 November 2000)", Olagnon, M. and G.A. Athanassoulis, eds. (Editions Ifremer, Plouzane, France, 2001)
6. Onorato, M., A.R. Osborne, M. Serio, and S. Bertoni, "Freak Waves in Random Oceanic Sea States", Phys. Rev. Lett., 86 (2001), pp. 5831–5834
7. Trulsen, K. and K.B. Dysthe, "A Modified Nonlinear Schrödinger Equation for Broader Bandwidth Gravity Waves on Deep Water", Wave Motion, 24 (1996), pp. 281–289
8. Benjamin, T.B. and J.E. Feir, "The Disintegration of Wave Trains on Deep Water", J. Fluid Mech., 27 (1967), pp. 417–430
9. Peregrine, D. H., "Approximate Descriptions of the Focusing of Water Waves", Proc. 20th Intl. Conf. Coastal Eng., vol. 1 (1986), Ch. 51, pp. 675–685
10. Gerber, M., "The Benjamin–Feir Instability of a Deep-water Stokes Wavepacket in the Presence of a Non-uniform Medium", J. Fluid Mech., 176, (1987), pp. 311–332
11. Berry, M.V., in "Chaotic Behaviour of Deterministic Systems", Iooss G., R.H.G. Helleman, and R. Stora, eds. (North-Holland, Amsterdam, 1983), p. 172
12. Bäcker, A. and R. Schubert, "Amplitude Distribution of Eigenfunctions in Mixed Systems", J. Phys. A, 35 (2002), pp. 527–538
13. Uscinski, B.J., "The Elements of Wave Propagation in Random Media" (McGraw-Hill, New York, 1977)
14. de Wolf, D.A., "Waves in Random Media: Weak Scattering Reconsidered", J. Optical. Soc. Am., 68 (1978), pp. 475–479
15. Wolfson, M.A. and S. Tomsovic, "On the Stability of Long-range Sound Propagation Through a Structured Ocean", J. Acoust. Soc. Am., 109 (2001), pp. 2693–2703
16. Topinka, M.A., B.J. LeRoy, R.M. Westervelt, S.E.J. Shaw, R. Fleischmann, E.J. Heller, K.D. Maranowski, A.C. Gossard, "Coherent Branched Flow In a Two-Dimensional Electron Gas", Nature, 410 (2001), pp. 183–186
17. Colling, A., "Ocean Circulation, 2nd edn" (Butterworth Heinemann, Oxford, 2001)
18. Gutshabash, Ye. Sh and I.V. Lavrenov, "Swell Transformation in the Aguhlas Current", Izv. Atmos. Ocean. Phys., 22 (1986), pp. 494–497
19. Lavrenov, I., "The Wave Energy Concentration at the Agulhas Current of South Africa", Nat. Hazards, 17 (1998), pp. 117–129
20. Gerber, M., "The Interaction of Deep Water Gravity Waves and an Annular Current: Linear Theory", J. Fluid Mech., 248 (1993), pp. 153–172
21. Berry, M.V., "Focusing and Twinkling: Critical Exponents from Catastrophes in non-Gaussian Random Short Waves", J. Phys. A, 10 (1977), pp. 2061–2081
22. Longuet-Higgins M.S., "Wave Group Statistics", in: "Oceanic Whitecaps", Monahan E.C. and G. MacNioceill, eds. (Reidel, Dordrecht, 1986), pp. 15–35

10 Predicting the Lifetime of Steel

Matz Haaks and Karl Maier

Summary. Even today, lifetime predictions of construction parts are still based on the Wöhler method, which is almost 150 years old. To construct a reliable Wöhler diagram, it is necessary to perform alternating load fatigue experiments on a huge number of equivalent samples for up to 10^8 or 10^9 load cycles. The lifetime under a specific applied load is then deduced from this diagram using statistical techniques.

Physically, the reason for fatigue and finally fracture is the accumulation of lattice defects like dislocations, vacancies and vacancy clusters, which are produced even when the load is significantly below the material's yield strength. The progress of fatigue can be observed from its earliest stages – after only a few load cycles – up to the final state of fracture by employing positrons as extremely sensitive lattice defect probes. In situ experiments can be performed to study test samples or real construction parts under realistic conditions. In steels a critical defect density is reached just before fatigue failure occurs. The point of failure can therefore be extrapolated from the early stages of fatigue by monitoring the defect density.

Spatially resolved experiments performed on a simple carbon steel and employing the Bonn Positron Microprobe indicate significant variations in defect densities over the region under stress even after just a few load cycles. These inhomogenieties grow from a typical starting size of less than a millimeter to encompass the entire volume after further fatigue. With more experimental experience and a better theoretical understanding of this process, this new prediction method should lead to much simpler and more reliable predictions of the lifetimes of metallic materials in the near future.

10.1 Introduction

Failure of construction materials due to fatigue is a phenomenon well known to the public, since it can lead to serious accidents involving airplanes, trains and cars, and so it can make the front pages of newspapers. From our own experiences we know that the lifetimes of metals, alloys and polymers under repeated loads are finite. Generally, material failure due to fatigue occurs if the load exceeds 50% of the yield strength, which is within the "reversible" elastic region (so Hooke's law applies). Mechanical parts, critical to machine stability or security, that are affected by fatigue are normally replaced during expensive maintenance long before the end of the useful lifetimes of the parts involved. To reduce the need for such a wasteful and expensive procedure, precise lifetime prediction is highly desirable. For almost 150 years the lifetimes of mechanical parts have been determined via the Wöhler test [1]. In a Wöhler diagram, the stress level is plotted against the logarithm of the number of load cycles that lead to failure by fatigue fracture. It is obvious that to obtain a sufficiently accurate Wöhler diagram, a huge number of identical samples must be tested in a very time-consuming process. To achieve realistic conditions of 10^8 to 10^9 load cycles, each fatigue tests must be run for a period of months. New ways to simplify this problem have come from computational physics. The huge progress in computer power and simulation algorithms has allowed us to obtain a realistic description of stress and defect production in a small region – for instance around the tip of a fatigue crack. But an ab initio lifetime prediction of a rivet in an airplane is still some way off.

Here we describe new ideas for realistic lifetime prediction. The physical reason for material fatigue is the production of defects in the lattice, even under conditions that are fully reversible macroscopically. A crack is caused when a critical density of defects has accumulated locally. Since the defect density can be observed, even at the very beginning of deformation, using positron annihilation spectroscopy (PAS), we can use it as a precursor for material failure. It is therefore possible to predict the lifetime of a sample after only a few load cycles in steels. At the moment PAS is a technique with both unique sensitivity to lattice defects and a large dynamic range. The method is nondestructive, does not need advanced sample preparation, and can be applied in situ during tensile or fatigue experiments. Due to its properties, PAS is well suited to lifetime predictions of mechanically stressed parts. In comparison, classical experimental methods like transmission electron microscopy, hardness testing, flow stress measurement or measurement of the internal stress with X-rays or neutrons are either not sensitive enough or are far from nondestructive. A fundamental knowledge of the defect spectrum is not necessary for lifetime prediction. To investigate the level of deformation it is sufficient to assign the positron signal to defects related to material fatigue. The determination of a complete reliable Wöhler diagram using only a single sample and 1% of the load cycles until failure may become possible in the near future.

10.2 The Search for Defects: Positrons in Solids

The positron's sensitivity to lattice defects has been well known to the scientific community since the 1960s [2, 3]. Vacancy-like defects in the metal's lattice constitute an open volume, which acts as an attractive potential for positively charged particles. This potential trapping can be described by a temperature-dependent trapping model [4, 5]. Point-like defects in metals like monovacancies can be detected at concentrations of 10^{-7} to 10^{-6} per atom. Hence, the positron acts as a probe on the atomic scale. Due to its diffusion (in the range of several hundred nanometers), a volume of about $0.5\,\mu m^3$ is scanned for defects by one single positron. There are some comprehensive articles about the potential applications of positron annihilation spectroscopy (PAS) to material research; for example see [6–8]. These days PAS is an established method in the field of defect analysis and nondestructive material testing [9–12]. The interaction of positrons with matter can be divided into four sections: thermalization, diffusion, trapping, and finally, annihilation.

Implanted in condensed matter, a positron loses all its kinetic energy within a few picoseconds, which is rather short compared to its lifetime in matter (from a few 100 ps in metals to several ns in polymers). Its energy loss is due to bremsstrahlung and scattering processes with electrons, plasmons and phonons. At the end the positron is in thermal equilibrium with the lattice ($E_{\mathrm{kin}} = 3/2\, k_B T \approx 0.04\,\mathrm{eV}$ at room temperature (RT)) [13–15]. This is possible despite the fact that the positron is a fermion because there is only one positron inside the sample at a time under the experimental conditions. The implantation profile is determined by the scattering processes during thermalization. For monoenergetic positrons from a slow positron beam, the implantation profile reaches its maximum below the surface. The profile and the position of the maximum can be calculated according to [16–18]. For transition metals (Fe, Cu, Ni) and a positron energy of 30 keV the maximum is located 1 μm below the surface.

Once thermalized, the positron diffuses through the lattice and behaves like a free particle. Repelled from the positively charged nuclei, its probability of occurrence is a maximum in the interstitial regions of the lattice [19], while its motion can be described as a three-dimensional random walk [20]. The positron is highly mobile, with a diffusion constant at RT of the order of $10^{-4}\ \mathrm{m^2/s}$. Hence a positron scans about 10^6 atomic positions within its lifetime, which explains the high sensitivity of positrons to lattice defects.

Every open volume in the lattice that causes a local increase in the distance between atoms acts as an attractive potential for the diffusing positron. Lattice defects created by plastic deformation (like dislocations, vacancies and vacancy clusters) form this kind of open volume. For instance, an atomic vacancy constitutes a deep positron trap with a binding energy of around 1 eV due to the missing repulsion by the nucleus. For a detailed discussion of positron trapping in open volume defects, see [4, 5, 21, 22]. Once trapped in a vacancy, a positron cannot escape since its kinetic energy at RT is too small to overcome the barrier. The lifetime of a positron in solid matter is reciprocal to the electron density at the site where it annihilates. In open volume defects the electron density is lower than in the interstitial region, which results in a higher defect specific lifetime.

Ideal edge dislocations are assumed to be shallow traps with a long range potential well and a binding energy of around 100 meV [24–26]. Due to the positron's kinetic energy of 40 meV at RT, escape from such a trap is probable. Experimentally measured positron lifetimes characteristic of dislocations differ significantly from lifetimes obtained for defect-free materials and are almost equal to the characteristic lifetimes measured for vacancies in materials with a high vacancy concentration (caused by irradiation for instance). Hence, trapping into dislocations obviously produces an intermediate state from which the positron is trapped by the deep potential of the associated vacancy (see Fig. 10.1). The transition rates between the states depend on the temperature and can be described by a three state trapping model [24,25,27,28] that is consistent with many experiments (for example [29]). Additionally, dislocations may act as fast diffusion paths for positrons, which enhances their sensitivity to vacancy-like defects [29].

When the positron annihilates with an electron, the rest masses of both particles are transformed into two γ-quanta of 511 keV emitted in antiparallel orientation[1], when considering the center-of-mass-system. Upon transformation into the laboratory system, the longitudinal component of the electron momentum causes a Doppler shift in the γ-energy while the transverse component produces a perturbation of 180° angular correlation. Here the momentum of the thermalized positron (~ 40 meV) can be neglected in comparison to the electron's momentum (1–10 eV) [30]. The contribution of electrons with

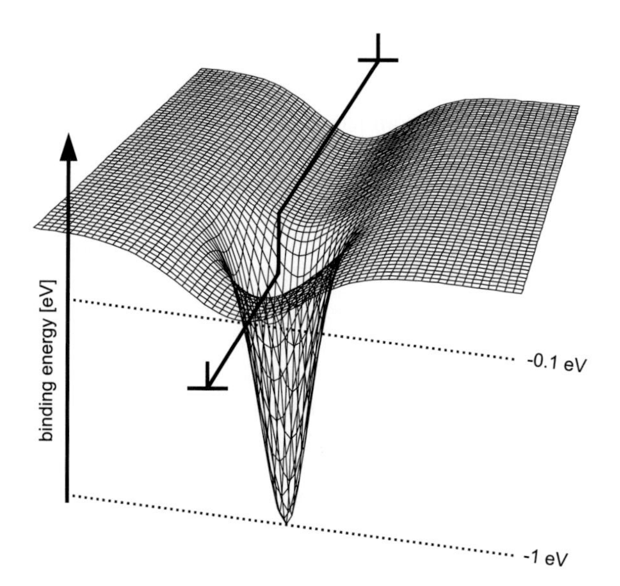

Fig. 10.1. Attractive potential formed by an edge-dislocation with an associated vacancy-like defect (jog). The shallow potential of the dislocation line constitutes a fast diffusion path for the positron. This is assumed to increase the trapping rate into the deep potential of the vacancy [23]

[1] Other possible decays into 1γ or 3γ are suppressed by $1/\alpha$ or $1/\alpha^3$, respectively ($\alpha = 1/137$).

Dopplershift due to the longitudinal component of the electron momentum p_L:

$$E_{1,2} = \frac{E'_{tot}}{2}\left(1 \pm \frac{p_L}{2m_0 c}\right)$$

$$E'_{tot} = E_{tot}\sqrt{1 - v^2/c^2}$$

v: realtive velocity between the center of mass of the annihilating electron-positron pair and the gamma detector.

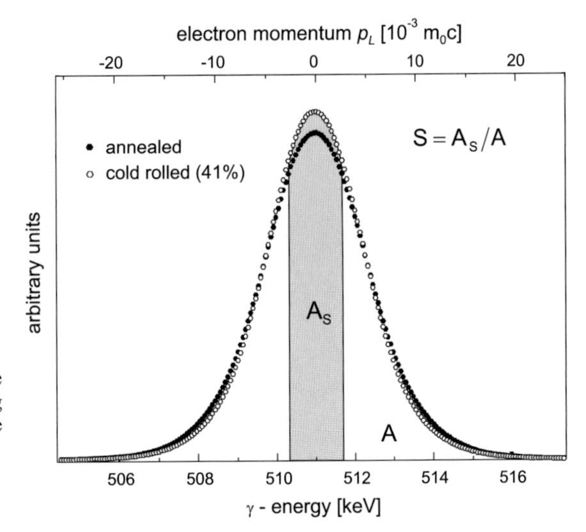

Fig. 10.2. Definition of the S-parameter. The longitudinal component of the electron momentum causes a Doppler broadening in the 511 keV photopeak. The S-parameter is defined as the ratio of the inner part of the photopeak, corresponding to low momentum electrons, to the integral over the whole peak

low momenta to the Doppler broadening is quantified by the shape parameter of the momentum distribution, the S-parameter, which is the quotient of the inner part of the photopeak and the integral over the whole peak (see Fig. 10.2).

The S-parameter also depends on the arbitrary choice of the borders used to determine the area A_S and on the energy resolution of the gamma spectrometer. To make measurements comparable, the S-parameter must be normalized to an appropriate reference value. For an investigation on plastically deformed metals, this would be the S-parameter of the well annealed state of the same material.

Plastic deformation is based on the movement and multiplication of dislocations. The creation of dislocation is always accompanied by the production of vacancies and interstitial atoms, where the most important processes for vacancy production due to plastic deformation are jog dragging and the annihilation of edge dislocations [31,32]. Since dislocations are shallow traps for positrons at room temperature, the signal for plastic deformation obviously stems from the associated vacancies. Due to the missing core electrons, the electron momentum density at the vacancy site is lower than in the undisturbed lattice, so a higher density of vacancies leads to a higher S-parameter, since an increasing fraction of positrons annihilates in vacancies.

In summary, employing the positron as a highly mobile probe which provides information about the electron momentum distribution in the atomic range, gives the defect density, a mesoscopic parameter. The shape of the

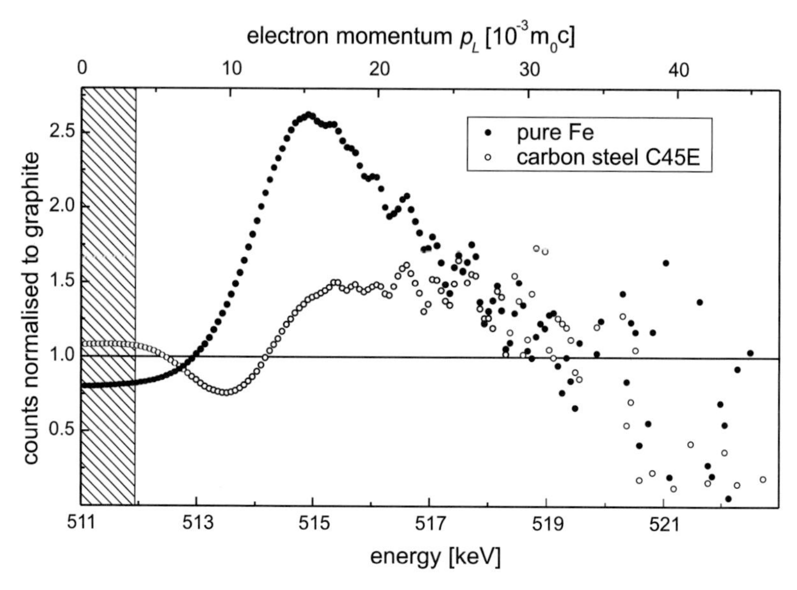

Fig. 10.3. High momentum distribution of pure Fe and carbon steel C45E normalized to graphite. The right half of the photopeak is given with the left side mirrored in. The *hatched region* shows the area used to evaluate the S-parameter. The spectra for pure Fe and C45E are divided by the spectrum of graphite. The peak in the iron spectrum seen at 515 keV is due to positrons annihilating with electrons from the 3d orbitals of Fe

wings of the photopeak is determined by annihilation with core electrons from atoms surrounding the annihilation site. As the momenta of these electrons differ for each element, the chemical environment of a trapping site can be studied by analyzing the high momentum part of the photopeak. Due to the low event rate in this momentum range, the spectrum is strongly disturbed by any background (caused mainly by gammas with an energy higher than 511 keV). Background reduction can be achieved by using two gamma detectors in coincidence [33] or via an accurate background calculation [34].

Figure 10.3 shows annihilation spectra for pure iron and carbon steel C45E, both normalized to the spectrum of pure graphite. In this diagram the spectra are mirrored at the 511 keV axis into the right half of the spectrum. The peaks in the data at 515 keV are caused by the annihilation of positrons and electrons from the 3d orbitals of Fe Despite the low concentration of carbon in C45E (0.45%), Fig. 10.3 shows strong evidence for the presence of carbon in the surroundings of the positron annihilation sites, since the contribution from 3d electrons is significantly reduced in C45E. This may be due to trapping into the interfaces between ferrite (α-iron) and cementite (Fe_3C) in the perlitic phase of the two-phase state of annealed C45E.

10.3 The Bonn Positron Microprobe

Most effects of plastic deformation and material fatigue show a strongly inhomogeneous defect distribution over the sample volume. Hence, to understand this processes through positron annihilation, it is crucial to achieve a spatial resolution in the micron range. Common positron sources with diameters of 0.5 to several millimeters cannot be employed for highly spatially resolving measurements.

The Bonn Positron Microprobe (BPM) provides a fine focused positron beam in the micron range with adjustable beam energy and a beam diameter that can be lowered to a few micrometers (see Fig. 10.4). The BPM [35] is a combination of a tiny positron sources with a small phase space and a conventional scanning electron microscope (SEM). The positrons are emitted from a ^{22}Na source and moderated using a tungsten moderator [36] employing the advantages of both transmission and reflection moderation. The moderated positrons are then accelerated in two steps up to their working energy, which is adjustable from 4.5 to 30 keV. The positron and the electron source are mounted on the opposite sites of a magnetic prism which bends both beams downward by 90° into the entrance plane of a SEM condenser zoom. Finally, the objective lens focuses the beams onto the sample, which is mounted on a motorized table movable laterally with an accuracy of 1 μm.

The positron beam diameter can be adjusted to between 5 and 200 μm. There is no need for any additional focusing using a strongly inhomogeneous magnetic field behind the sample position. This allows the study of ferromagnetic materials like iron, nickel and ferritic steels. The annihilation radiation is recorded by a high resolution Ge detector, mounted 10 mm below the sample position. During measurement, fluctuations of the experimental setup are minimized by stabilizing the electronics on the decay gamma of 7Be (477.8 keV) which is detected simultaneously.

10.4 Detection of Plastic Deformation

In plastic deformation the production of dislocations and vacancies are always interconnected. Jog-dragging of screw dislocations and annihilation of edge dislocations are very effective processes for producing vacancies and vacancy-like defects [31, 32]. During tensile testing or cyclic fatigue experiments, the dislocation density rises by several orders of magnitude from the well annealed state until fracture occurs. The binding energy of positrons to dislocations is below 100 meV for most metallic materials at room temperature and hence too low for effective binding [24–26]. The high sensitivity of positron annihilation to changes in the dislocation density can be ascribed to the sensitivity to the associated vacancies. In a tensile test the sample is elongated with a well defined axial stress.

The total axial elongation ε and the axial stress σ, which can be calculated from the applied force and the cross-sectional area in the waist of the sample, are recorded in a stress-strain diagram. We performed tensile tests on ferritic steel C45E (equivalent to AISI 1045), which is a pure carbon steel (0.42–0.50 weight% C) with a low contamination of phosphorus and sulfur.

Even though C45E is a very common tool steel with widespread applications, its mechanical properties are fortunately based only on iron and carbon, which makes it an ideal simple system to use to understand the underlying mechanisms of deformation. Before testing, all samples undergo temperature treatment for three hours at 860 °C under high vacuum conditions. As shown by a series of isochronal annealing tests, this treatment anneals out all of the vacancy-like defects in C45E that can be seen by positrons [11]. When slowly cooled down to room temperature ($\dot{T} = 1\,\mathrm{K/min}$) C45E appears to consist of a two-phase mixture composed of 60% ferrite (α-iron) and 40% perlite. Despite the fact that most of the plastic deformation takes place in the ferritic phase, the interfaces between α-iron and cementite (Fe_3C) in the

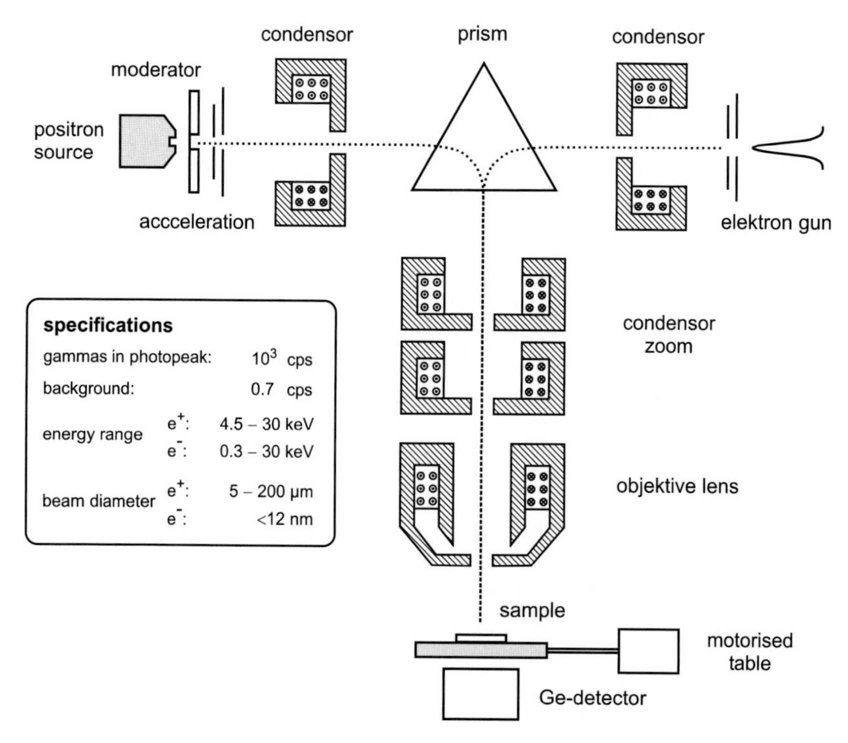

Fig. 10.4. Design of the Bonn Positron Microprobe. The particle electron and positron beams, produced by the electron gun and the positron source, respectively, are projected by condenser lenses into the entrance plane of a magnetic prism. Both beams are then focused by the condenser zoom and the objective lens onto the sample, which is mounted on a motorized table

perlitic phase may be responsible for competitive trapping of positrons. This leads to a higher S-parameter in the well-annealed state of C45E than for well-annealed pure iron (see Fig. 10.3).

To demonstrate the correlation of the S-parameter to plastic deformation, during tensile testing the sample is dismounted at several deformation stages and an annihilation spectrum is taken (see Fig. 10.5).

Below the yield strength around 350 MPa, the deformation is reversible and follows Hooke's law of elasticity $\sigma = E\,\varepsilon$, where the proportionality constant is given by the Young's modulus E. Previous results have shown that elastic strain has no influence on the S-parameter [23]. In most metallic materials there is a gradual transition between elastic and plastic behavior, but in the case of a mild steels such as C45E, there is a discontinuity in the stress–strain curve (Lüders strain [37]), which is due to the rupture of dislocations from pinning centers formed by carbon atoms, which are aggregated at the dislocation site by diffusion (Cottrell clouds) [38, 39] (see the inset in Fig. 10.5). Deformation in this region leads to only a small increase in the dislocation density and, hence, only a slight increase in the S-parameter. Above this region the deformation proceeds with the multiplication of dislocations and hence the generation of vacancies. This is reflected in an increase in the

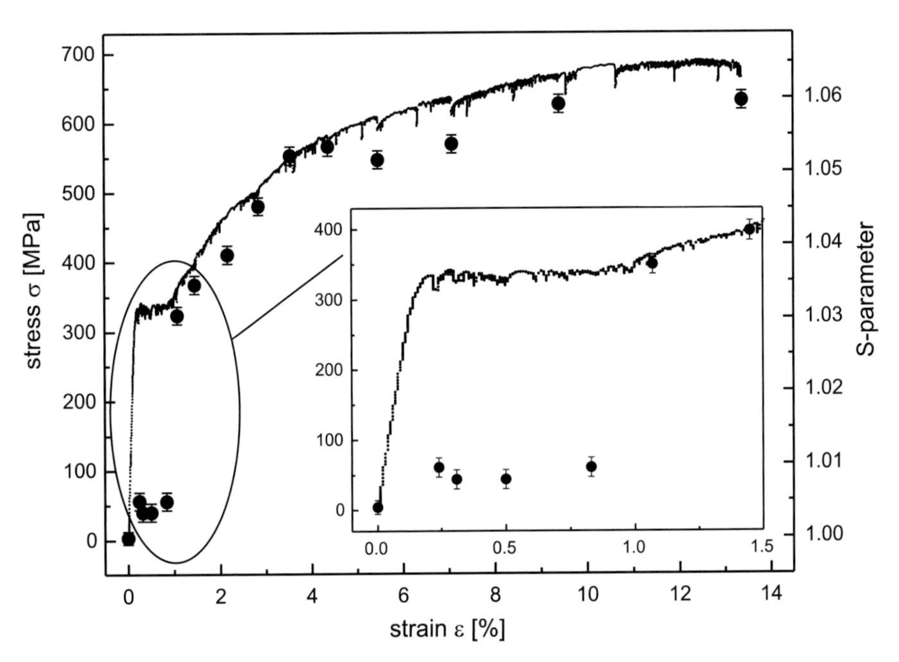

Fig. 10.5. Stress-strain diagram of the ferritic steel C45E. The stress is given on the left axis (*black line*) while the S-parameter relative to the well annealed state (*full circles*) is given on the right axis, both in relation to the strain. The *inset* zooms in at the region of low deformation [11]

S-parameter with an applied stress of more than 350 MPa. The S-parameter reaches saturation at the upper sensitivity limit, corresponding to a dislocation density of $\varrho \approx 1 \times 10^{11}\,\text{cm}^{-2}$ [40] until rupture occurs at 13.5% strain. Similar experiments have been performed on the technically eminent stainless steel AISI 316L [10] and a variety of pure metals [23].

A more distinct insight into the relevance of the S-parameter can be obtained by performing an in situ measurement of the gamma spectrum during a tensile test. Here two tubular samples containing a positron source inside the bore hole are investigated. The spectra are taken when the straining is interrupted and the stress is released, while the sample stays mounted in the deformation machine. This avoids incidental sample damage, which is important especially in the early states of deformation. Figure 10.6 shows the S-parameter plotted versus the stress amplitude for two different sample waist diameters. The plot provides evidence of a sensitivity threshold at $280 \pm 20\,\text{MPa}$. Above this, the dependence of the S-parameter on the stress amplitude is described well by a linear relation.

Like many other material properties, the surface hardness of a metal depends strongly on the deformation state and hence the dislocation density. The hardness can be measured by pressing an indenter into the surface when

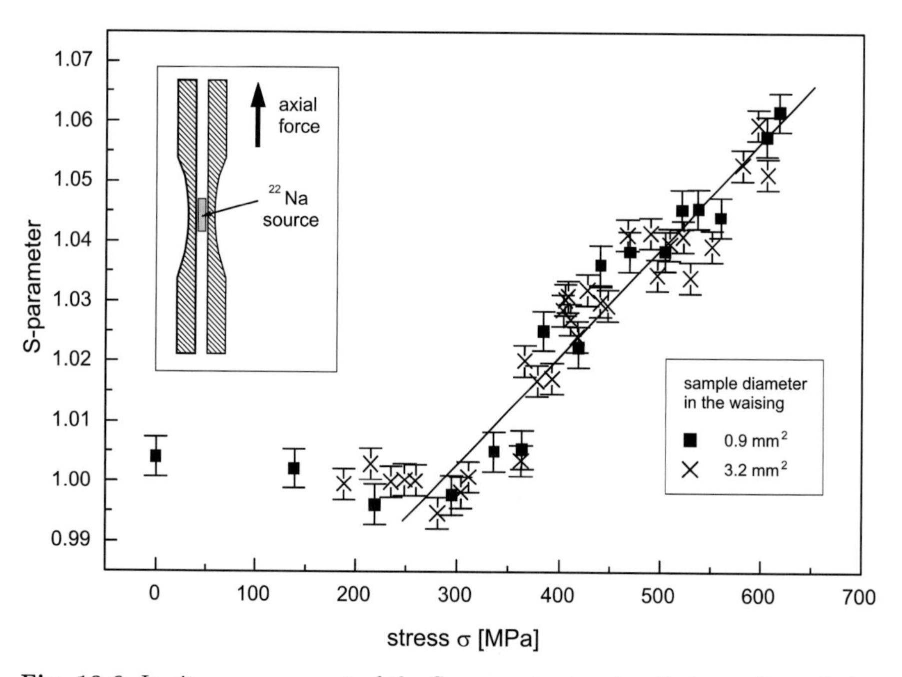

Fig. 10.6. In situ measurement of the S-parameter in a tensile test using tubular shaped samples of C45E. The positron source is mounted inside the borehole of the sample. After the sensitivity threshold at 280 MPa is reached, the S-parameter increases almost linearly with the applied stress

applying a defined static load. Common indenters are hardened steel spheres (Brinell) or pyramidal diamonds (Vickers). The hardness is given by the ratio of the applied load to the remaining impression area, which is evaluated using a calibrated microscope or an image recognition system. The variety of different hardness tests are somewhat equivalent and the hardness values can be interconverted or converted into other material properties such as the tensile strength [41]. Figure 10.7 shows a comparison between the S-parameter and the Vickers hardness, both measured in C45E chips produced by high-speed cutting [42]. As an example, Fig. 10.7a shows a line scan from a heavily deformed area (cutting position = 0 μm) into the undeformed bulk material, evaluating the S-parameter and the Vickers hardness at the same positions.

The damage penetration is detectable up to 550 μm below the cutting position with the S-parameter, but a significant increase in hardness is only evident down to 400 μm. The distributions of both values are similar, but the S-parameter shows a higher sensitivity to small deformations that cause no

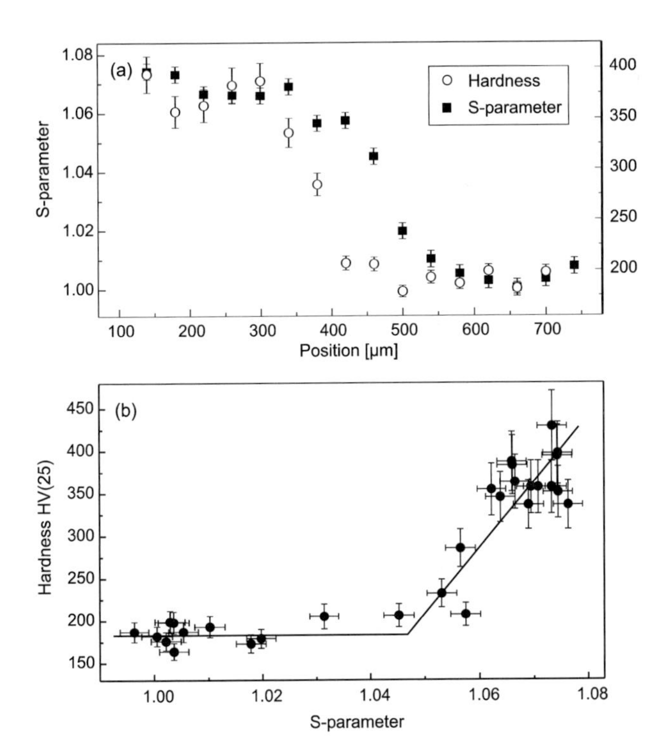

Fig. 10.7. Comparison between Vickers hardness and S-parameter in C45E. (**a**) shows a spatially resolved line scan taken with the BPM on a plastically deformed area in a chip produced by high-speed cutting. S-parameter (full squares) and hardness (open circles) were measured at the same positions (*open circles*). (**b**) Hardness versus S-parameter from several similar scans. The *straight line* is just a guide for the eye

significant change in the surface hardness. This becomes clearer upon plotting the hardness versus the S-parameter, as shown in Fig. 10.7b, which plots data accumulated from several line scans. Areas that underwent minor to medium deformations exhibit no significant increase in hardness, but they do show a distinct rise in the S-parameter up to 1.04.

10.5 Damage Prediction

If a material is subjected to repeated or cyclic stress it may fail by fatigue fracture even though the maximum stress in each single cycle is considerable less than the yield strength of the material. This is due to microscopically nonreversible movements of dislocations [43]. The dislocation structure created in the phase of elongation does not return to its initial state in the compressive phase. Thus, a certain amount of the deformation work is not dissipated into heat but is stored as mechanic energy in the material through the production of defects like dislocations and vacancies. This energy is released when the sense of deformation is reversed (Bauschinger effect [44,45]), resulting in a hysteresis in the stress–strain diagram (see Fig. 10.8b). Since the deposition of energy and, hence, the production of defects is cumulative, after a certain number of deformation cycles a macroscopical hardening of the sample occurs.

Many components are subjected to alternate loading cycles during service. A technique to estimate the useful life is therefore highly desirable. Almost 150 years ago August Wöhler conducted the first destructive tests

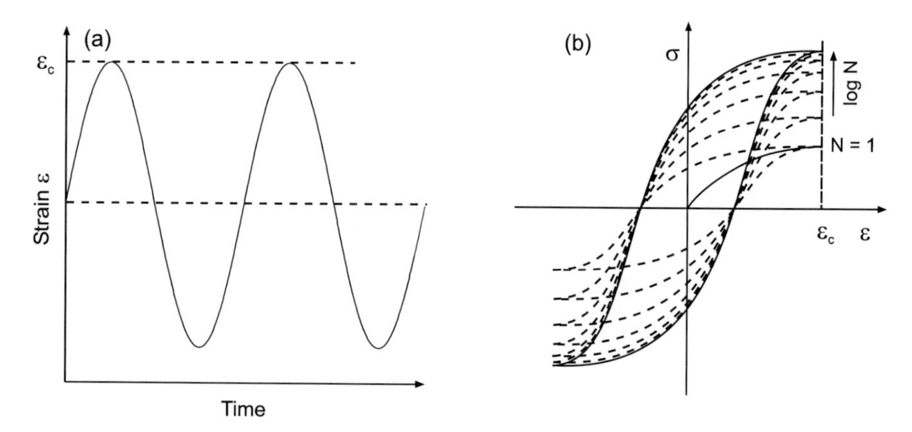

Fig. 10.8. (**a**) A sinusoidal command wave in an alternating load fatigue test with a controlled strain amplitude typically adjusted to 60–80% of the yield strength in the sample. (**b**) Schematic of the mechanical hysteresis curve under strain control. The area inside the hysteresis loop equals the amount of deformation work w_d deposited per cycle. Due to cyclic hardening, w_d increases as the test progresses

to determine material fatigue and the remaining useful lifetimes of rail vehicle axles [1]. The component was tested under alternating load by applying a controlled stress or strain amplitude at a frequency of 0.1 to several 100 Hz. The results are shown in a Wöhler diagram (see Fig. 10.9), where the stress amplitude σ is plotted against the number N of cycles before failure. The probability of failure (lower limit: 10%, upper limit 90%) is estimated statistically from the diagram.

Up to now all of the methods used for lifetime prediction are based on a similar principle, and a Wöhler diagram must be determined in an extraordinarily time-consuming test series for any industrially manufactured part important for stability or safety. This effort can be reduced significantly by assessing the useful life of an individual part via positron annihilation. Since the defect density rises during fatigue it provides a precursor for failure, which is accessible in a nondestructive way by measuring the S-parameter.

We performed rotating bending tests on several alloys of industrial importance. In rotating bending fatigue, a cylinder of the material to be examined is fixed on one side in a rotating holder and on the other side in a floating bearing, which is charged with an adjustable load. We used cylindrical samples with a diameter of 10 mm, which were cut spheroidically (radius = 30 mm) reducing the diameter at the waist to 6 mm. The tests were performed under total stress control, applying several levels of load all of which were

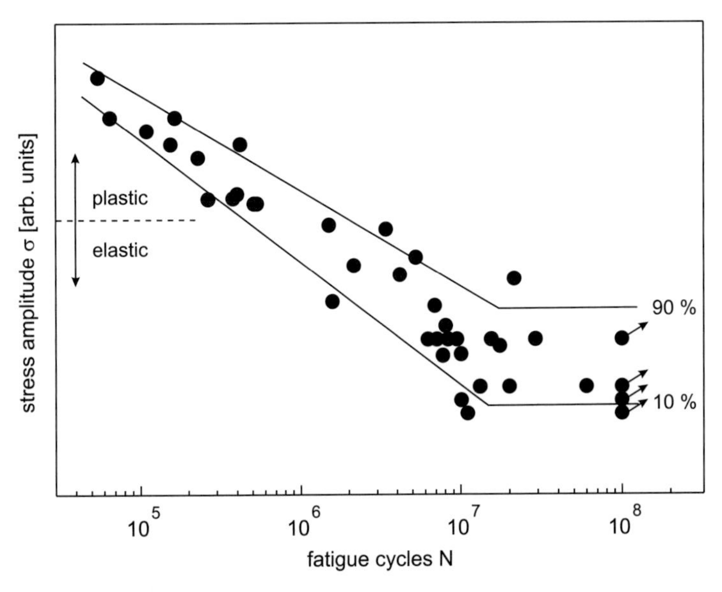

Fig. 10.9. Wöhler diagram (schematic). The points of failure at a given stress amplitude are denoted in a semilogarithmic plot. Samples that had not failed by the end of the test are marked with an *arrow*. The *upper (lower) line* gives the statistical estimate of a 90% (10%) probability of failure at a given stress amplitude

significantly below the yield strengths of the analyzed materials. After reaching a particular number of load cycles, the tests were paused and a Doppler spectrum was taken using a tiny positron source ($\varnothing = 2\,\mathrm{mm}$). The tests were finally stopped when a fatigue crack appeared on the surface and its visible length had reached 1/3 of the sample circumference at the waist [11]. Below 100 cycles the machine was rotated by hand, and above a command frequency of 100 Hz was used. All PAS data were obtained using a ^{22}Na source. Hence the signal originated from a 30 µm thick subsurface layer. The results for four technically relevant materials are shown in Figs. 10.10 and 10.11. All of the S-parameters displayed are given relative to the S-parameter of the same material in the well annealed state.

Figure 10.10a shows the S-parameter versus the cycle number in a semilogarithmic plot for several loads applied to a sample of the austenitic steel X6CrNiTi 18-10 (AISI 321). The applied load is given as the maximum stress at the surface at the waist. For all loads the S-parameter shows an almost linear increase with the logarithm of the cycle number, while the slope depends on the load. Finally, before fatigue failure the S-parameter saturates at a similar value (around 1.08) independent of the applied load. This linear dependence is also observed for a sample passing the test at 210 MPa up to a cycle number of 5×10^8 (not displayed in Fig. 10.10a). Figure 10.10b shows a similar test series for the ferritic steel C45E. The almost linear dependency is even evident here, despite the first few cycles where the effects of Lüders strain show up. But compared to the austenitic steel, there is no saturation of the S-parameter before failure at a well-defined value. Instead, the saturation level reached before failure depends on the load.

Figure 10.11 shows the results for non-alloyed titanium grade II (Ti2) (a) and the technically important titanium alloy TiAl4V6 (b). For titanium, the S-parameter shows a sensitivity threshold where its logarithm is almost linearly dependent on the applied load. Beyond that threshold, the S-parameter shows a linear relation to the fatigue cycle with the same slope independent of the applied load. The sample loaded with 212 MPa did not break until the test was stopped at 10^8 cycles. All of the samples that failed during the test showed a similar S-parameter of around 1.047, but saturation was not reached.

For the titanium alloy TiAl4V6 (Fig. 10.11b), no significant relation between the S-parameter and the number of load cycles was found. Only one sample showed a slight increase before failure (upward triangles, 730 MPa). After cracking begins, measurement of the S-parameter on the tip of the crack provides an inconsistent value of between 1.007 and 1.015.

In the three cases of the iron-based materials X6CrNiTi 18-10, C45E and Ti2, the S-parameter is a measure of the deformation state and it can be understood as a precursor for imminent failure. Combined with knowledge of the structure of these metals and an estimate for the emerging load amplitudes, the point of failure can be predicted from a short time fatigue experiment employing only a small number of fatigue cycles. For the single-phase materials X6CrNiTi 18-10 and Ti2, a simple linear relationship between the S-parameter and the logarithm of the cycle number appears. In the two-phase alloy C45E, the relation is more subtle due to the influence of either

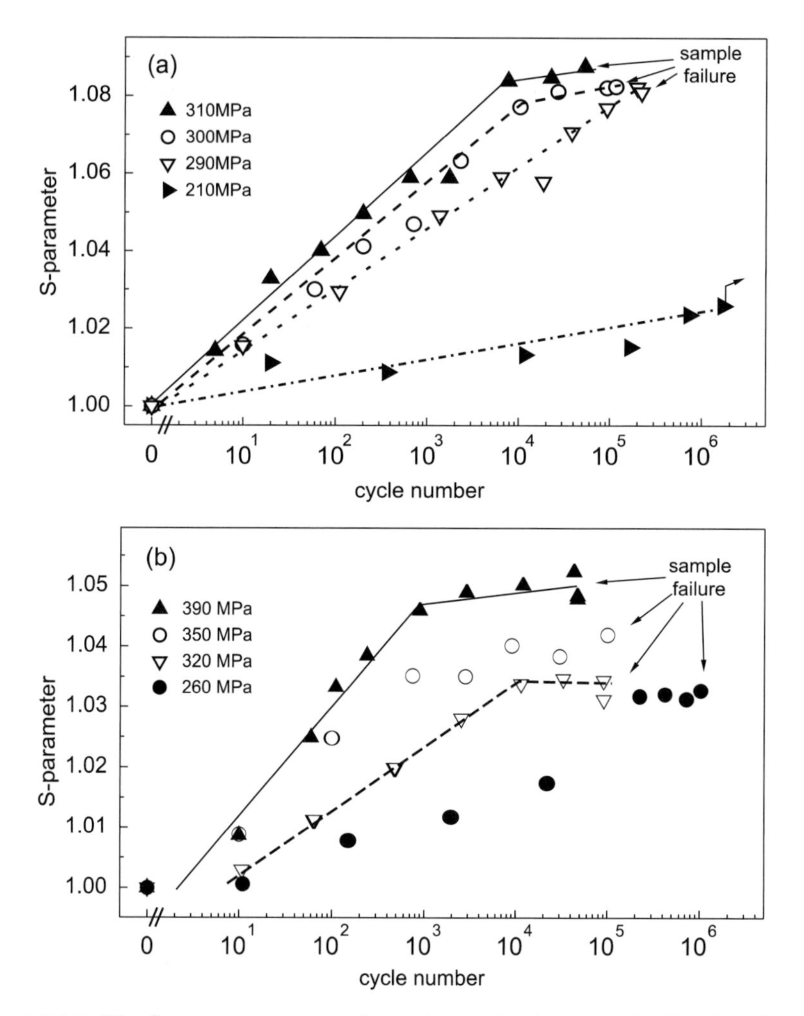

Fig. 10.10. The S-parameter versus the cycle number in a rotating bending fatigue test at various loads for the austenitic steel X6CrNiTi 18-10 (**a**) and the ferritic steel C45E (**b**). (**a**): There clearly is a linear relation between the S-parameter and the logarithm of the number of fatigue cycles, while the slope depends on the applied load. Thus, it should be possible from only a small number of fatigue cycles to estimate the remaining useful life of the sample. (**b**): There is also a linear relation for C45E, but due to the failure of the sample at load-dependent levels of saturation in the S-parameter, lifetime prediction is more difficult

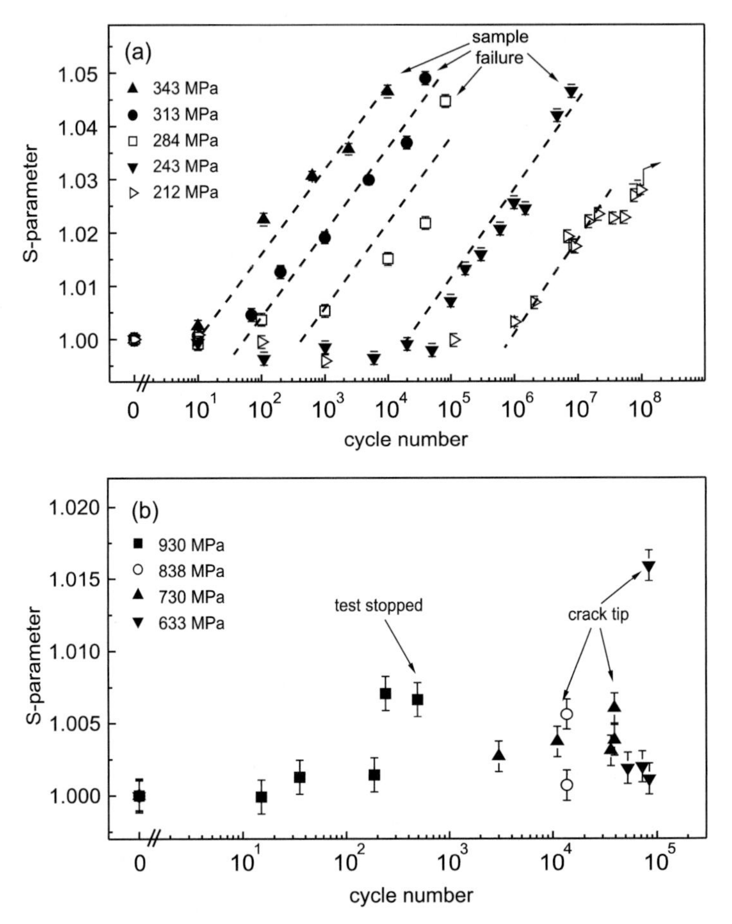

Fig. 10.11. The S-parameter versus the cycle number in a rotating bending fatigue test at various loads for non-alloyed titanium grade II (**a**) and the titanium alloy TiAl4V6 (**b**). (**a**): Beyond a sensitivity threshold the S-parameter depends linearly on the logarithm of the fatigue cycles, showing almost the same slope independent of the applied load. The sensitivity threshold is strongly dependent on the load, which hinders a straightforward lifetime prediction. (**b**): In the titanium alloy TiAl4V6 no dependence of the S-parameter on the number of load cycles is observed. An increase in the S-parameter can only be observed after failure at the tip of the crack

the Lüders strain or the competitive positron signal from interfaces in the perlitic phase.

This behavior can be explained as follows. Consider amplitudes of stress globally below the yield strength. The local stress will only be above the yield strength only in some favorably orientated grains, where the available slip-systems can be activated. The production of dislocations is only possible in those grains. During any additional cycle, those grains respond elastically (work hardening) up to an increased stress level, meaning that the elastic fraction of the total strain increases while the plastic fraction decreases. Hence, the dislocation production per cycle decreases with an increasing number of cycles. Additionally, adjacent grains are plastically deformed due to work hardening in the previously activated grains. Both the defect density in the grains involved and the affected volume of the sample increase.

A lifetime prediction is not possible for the titanium alloy TiAl4V6 using a positron source with a diameter in the range of millimeters. In this case the plastic deformation is intensely localized in a tiny region of just a few 100 μm^2 (see Fig. 10.13). In this type of alloy, the lifetime can only be predicted using a positron microbeam.

Figures 10.12 and 10.13 show scanning positron images of the plastically deformed region in front of a fatigue crack (plastic zone) generated in the compact tension (CT) geometry [46] in X6CrNiTi 18-10 (Fig. 10.12) [47] and TiAl4V6 (Fig. 10.13). The crack nucleation occurs in the final phase of fatigue, when the stress has locally exceeded the tensile strength at one particular point in a sample (for a detailed discussion, see any textbook, for example [46]). Both fatigue cracks are produced from the well-annealed state under equivalent conditions in a symmetric fatigue experiment, by applying a load below the yield strength of the materials. The images are taken after the

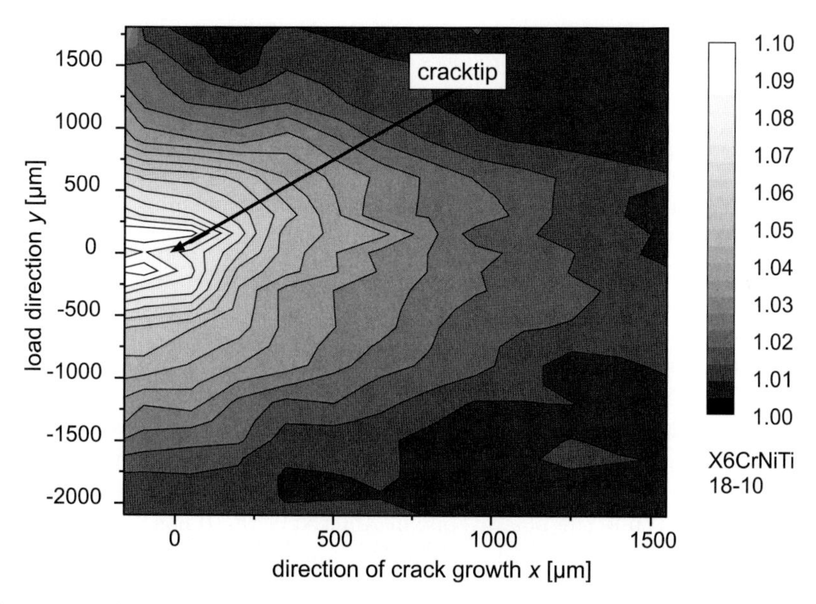

Fig. 10.12. Scanning positron image of the plastically deformed region in front of a fatigue crack in the austenitic steel X6CrNiTi 18-10 produced in CT geometry. The coordinate origin is located at the position of the tip of the crack. The S-parameter is coded in greyscale. The effects of plasticity are evident in a region extending from the cracktip up to 1.5 mm

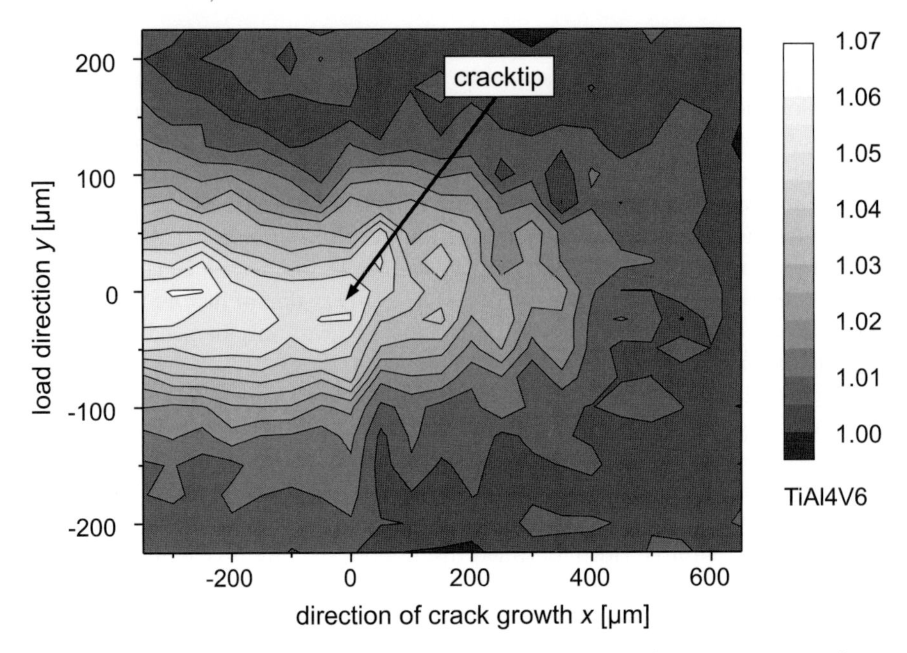

Fig. 10.13. Scanning positron image of the plastically deformed region in front of a fatigue crack in the titanium alloy TiAl4V6 produced in CT geometry. The coordinate origin is located at the position of the tip of the crack. The S-parameter is coded in greyscale. TiAl4V6 produces a very tiny plastic zone with an extension of only $300 \times 500\,\mu\text{m}^2$

crack has propagated several millimeters into the sample. The S-parameter in Figs. 10.12 and 10.13 is coded into greyscale, white meaning the maximum S-parameter corresponding to a dislocation density above $\varrho \approx 1 \times 10^{11}\,\text{cm}^{-2}$ and dark grey meaning the reference S-parameter of the well-annealed state ($\varrho \leq 2 \times 10^{8}\,\text{cm}^{-2}$). In the plastic zone the dislocation density decreases with the distance from the tip of the crack. Its border can be defined as the area where the S-parameters become significantly higher than 1.0 (S = 1.007 for X6CrNiTi 18-10, and S = 1.005 for TiAl4V6). The extension of the plastic zone reveals the ductility of the material and hence the response of an undeformed region in the sample to a work-hardened region in its vicinity.

A comparison of both images clarifies the difference in ductility between the two alloys. While the plastic zone in X6CrNiTi 18-10 has an extension of $1.5 \times 3.2\,\text{mm}^2$, the plastic zone in TiAl4V6 appears in a comparatively tiny region of $0.3 \times 0.5\,\text{mm}^2$. This has to be taken into account when interpreting the results of the fatigue tests on TiAl4V6 shown in Fig. 10.11b. Before crack nucleation occurs, a plastically deformed area (weak spot) will appear on the sample's surface. If the extension of this weak spot is too small compared to the positron source used, the increased dislocation density may not be observable.

In the rotating bending fatigue and the CT samples, the location of the maximum stress concentration was determined by the sample's geometry. Upon investigating a geometry where the affected area is loaded by a uniform stress field, the location of appearance of a weak spot is seen to be totally random. Figure 10.14 shows the results from a cyclic fatigue test performed on a flat sample made from C45E with a tapered central part that has a homogeneous cross-sectional area extending over the scanned 6 mm. The layout of the geometry is displayed in the top left corner of Fig. 10.14. The test was performed using a piezo-translator fatigue machine [29] under total strain control employing a strain ratio of $\Delta\varepsilon/\varepsilon = 1.52 \times 10^{-3}$, which corresponds to a stress amplitude of 330 MPa at the beginning of the test. The test was paused several times and a positron image of the central part

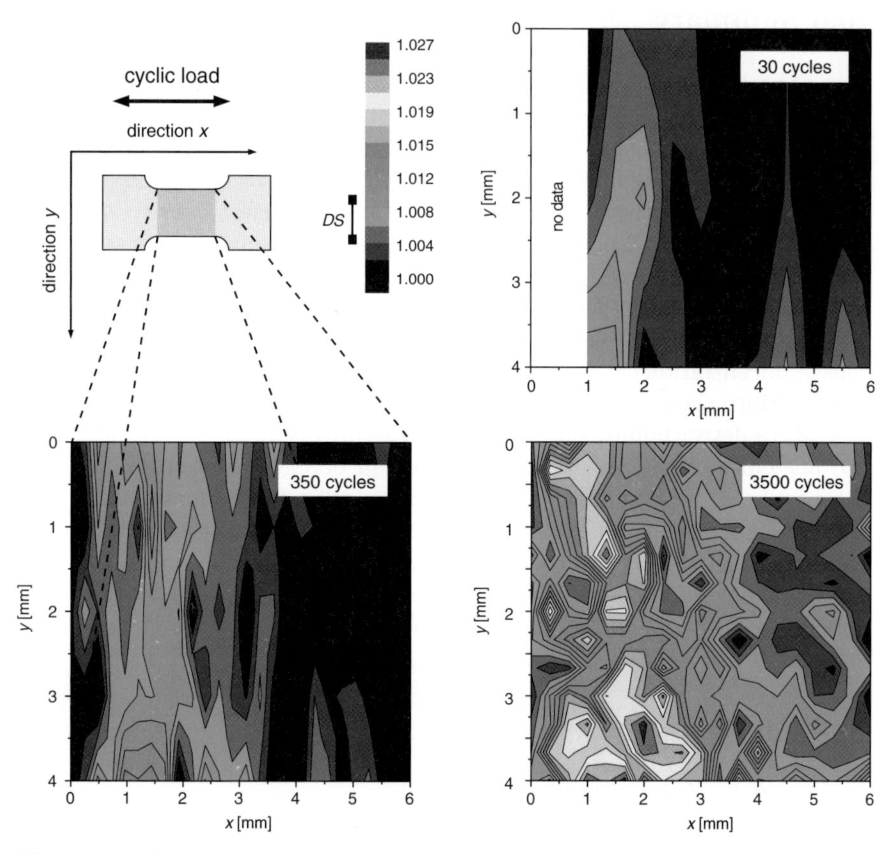

Fig. 10.14. Scanning positron images of the plastic deformation in a cyclic fatigue test performed on C45E. The S-parameter is coded is coded in colors, where. In the central part of the sample, where the images were taken, the stress amplitude has an uniform distribution. The formation of a weak spot is evident in a significant increase of the S-parameter even after the first 30 cycles

was taken. As an example, Fig. 10.14 shows the results for 30, 350, and 3500 fatigue cycles. Already after 30 deformation cycles the appearance of a weak spot is evident in a significant increase of the S-parameter to 1.011. As fatigue progresses, increased dislocation density arises in the weak spot in the similar manner to that expected in the rotating bending fatigue test performed on C45E (see Fig. 10.10b). The defect-rich region increases laterally at the same time and grows into as-yet unaffected area. As deformation progresses still further (3500 cycles), this region spreads over the whole sample area, becoming more structured. This is shown by a more inhomogeneous distribution of the S-parameter, denoting several spots of enhanced dislocation density.

10.6 Summary

Material fatigue is accompanied by the creation of defects like dislocations, point defects and small vacancy clusters. These irregularities in the crystal lattice act as trapping centers for positrons. The annihilation radiation of a positron changes significantly when trapped into a defect, which means that it is easy to detect the presence of defects, and the defect density is easy to determine using the S-parameter calculated from the annihilation spectrum. In fact, the well-known stress–strain diagram can be reproduced if the defect concentration – detected by positron annihilation – is plotted against the applied stress. Due to the extreme sensitivity of positrons to defects, a rise in the defect concentration can be observed even in the very early stages of material fatigue. Since these defects act as precursors to the final state – failure – it is possible to detect failure long before any fracture occurs. In industrial ferritic and austenitic steels, the point of failure can be extrapolated from a small database using positron annihilation measurements. Hence, material lifetimes can be reliably predicted by analyzing the defect density at the earliest stages of fatigue. In the near future, we can expect that it will become possible to determine a complete and reliable Wöhler diagram with just one sample and using only about 1% of the number of load cycles until failure.

References

1. A. Wöhler, Z. f. Bauwesen **8**, 642 (1858)
2. I. Dekthyar, D. Levina, V. Mikhalenkov, Soviet Phys. Dokl. **9**, 492 (1964)
3. I. MacKenzie, T. Khoo, A. McDonald, B. McKee, Phys. Rev. Lett. **19**, 946 (1967)
4. B. Bergensen, M.J. Stott, Solid State Commun. **7**, 1203 (1969)
5. W. Frank, A. Seeger, Appl. Phys. **3**, 61 (1974)
6. V.I. Goldanski, Positron Annihilation, Atomic Energy Review **6**, 183 (1968)
7. R. West, Adv. Phys. **22**, 263 (1973)

8. A. Seeger, J. Phys. F **3**, 248 (1973)
9. P. Hautojärvi, *Positrons in Solids, Topics in Current Physics, Vol. 12*, Springer, Berlin Heidelberg New York (1979)
10. U. Holzwarth, P. Schaaff, Phys. Rev. B **69**, 094110 (2004)
11. K. Bennewitz, M. Haaks, T. Staab, S. Eisenberg, T. Lampe, K. Maier, Z. Metallkd. **93**, 778 (2002)
12. Ch. Zamponi, St. Sonneberger, M. Haaks, I. Müller, T. Staab, G. Tempus, K. Maier, J. Mat. Sci. **39**, 6951 (2004)
13. R. Ritchi, Phys. Rev. **114**, 644 (1959)
14. A. Perkins, J. Carbotte, Phys. Rev. B **1**, 101 (1970)
15. R. Nieminen, J. Oliva, Phys. Rev. B **22**, 2226 (1980)
16. A. Makhov, Sov. Phys. Sol. Stat. **2**, 1934 (1961)
17. V. Ghosh, Appl. Surf. Sci. **85**, 187 (1995)
18. A. Vehanen, K. Saarinen, P. Hautojärvi, H. Huomo, Phys. Rev. B **35**, 4606 (1987)
19. W. Brandt, R. Paulin, Phys. Rev. B **5**, 2430 (1972)
20. M. Puska, R. Nieminen, Rev. Mod. Phys. **66**, 841 (1994)
21. D.C. Connors, R.N. West, Phys. Lett. **30A**, 24 (1969)
22. A. Seeger, Appl. Phys. **4**, 183 (1974)
23. T. Wider, K. Maier, U. Holzwarth, Phys. Rev. B **60**, 179 (1999)
24. C. Hidalgo, S. Linderoth, J. Phys. Metal Phys. **18**, L263 (1988)
25. L.C. Smedskjaer, M. Manninen, M.J. Fluss, J. Phys. F **10**, 2237 (1980)
26. K. Petersen, I.A. Repin, G. Trumpy, Cond. Mat. **8**, 2815 (1996)
27. B. Pagh, H.E. Hansen, B. Nielsen, G. Trumpy, K. Petersen, Appl. Phys. A **33**, 255 (1984)
28. E. Hashimoto, J. Phys. Soc. Japan **60**, 552 (1993)
29. T. Wider, S. Hansen, U. Holzwarth, K. Maier, Phys. Rev. B **57**, 5126 (1989)
30. S. DeBenedetti, C. Cowan, W. Konneker, Phys. Rev. **76**, 440 (1949)
31. G. Saada, Acta Metal. **9**, 965 (1961)
32. P. Hirsch, D. Warrington, Philos. Mag. **6**, 735 (1961)
33. K.G. Lynn, J.R. McDonald, R.A. Boie, L.C. Feldman, J.D. Gabbe, M.F. Robbins, E. Bonderup, J. Golovchenko, Phys. Rev. Lett. **38**, 241 (1977)
34. M. Haaks, T.E.M. Staab, K. Saarinen, K. Maier, Phys. Stat. Sol. A **202**, R38 (2005)
35. H. Greif, M. Haaks, U. Holzwarth, U. Männig, M. Tongbhoyai, T. Wider, K. Maier, J. Bihr, B. Huber, Appl. Phys. Lett. **71**, 2115 (1997)
36. L. Madansky, F. Rasetti, Phys. Rev. **79**, 397 (1950)
37. E.W. Hart, Acta. Met. **1**, 146 (1955)
38. A.H. Cottrell, *Dislocations and Plastic Flow in Crystals*, Clarendon, Oxford (1953)
39. J.P. Hirth, J. Lothe, *Theory of Dislocations*, McGraw-Hill, New York (1968)
40. T.E.M. Staab, R. Krause-Rehberg, B. Kieback, J. Mat. Sci. **34**, 3833 (1999)
41. DIN Norm 50150, *Umwertung von Härtewerten*, Beuth Verlag, Berlin (2000)
42. M. Haaks, J. Plöger, *Abbildung der Schädigung in der Randzone mit Positronen als Sondenteilchen*, in: H.K. Tönshoff, C. Hollmann (eds.), *Hochgeschwindigkeitsspanen metallischer Werkstoffe*, Wiley-VCH, Weinheim (2004)
43. A. Seeger, *Handb. d. Physik VII, Vol. 2*, Springer, Berlin Heidelberg New York (1958)

44. J. Bauschinger, Mittheilungen aus dem Mechanisch-Technischen Laboratorium der königlichen Technischen Hochschule in München **13**, 1 (1886)
45. A. Abel, H. Muir, Philos. Mag. **26**, 489 (1972)
46. D. Broek, *Elementary Engineering Fracture Mechanics*, Sijthoff & Noordhoff, Alphen a/d Rijn (1978)
47. M. Haaks, K. Bennewitz, H. Bihr, U. Männig, Ch. Zamponi, K. Maier, Appl. Surf. Sci. **149**, 207 (1999)

11 Computer Simulations of Opinions and their Reactions to Extreme Events

Santo Fortunato and Dietrich Stauffer

Summary. We review the opinion dynamics in the computer models of Deffuant et al. (D), of Krause and Hegselmann (KH), and of Sznajd (S). All of these models allow for consensus (one final opinion), polarization (two final opinions), and fragmentation (more than two final opinions), depending on how tolerant people are to different opinions. We then simulate the reactions of people to Xevents, in that we modify the opinion of an individual and investigate how the dynamics of a consensus model diffuses this perturbation among the other members of a community. It often happens that the original shock induced by the Xevent influences the opinion of a large part of society.

11.1 Introduction

Predicting Xevents is very important when we want to avoid losses due to earthquakes, floods, stock market crashes, and so on. But it is not easy, as we can see when we read a newspaper. It is much easier to claim that one has an explanation for this event after it has occurred. One important area of investigation in this field focuses on the opinions people have after an Xevent. Do they now take objective risks more seriously than before? Do people tend to exaggerate the risks and prefer to drive long distances by car instead of airplane, shortly after a plane crash happened? How do these opinions change as the time since the event or the geographical distance increases? It is plausible that people take the risk less seriously as the time since the last catastrophe increases. Less clear is the influence of geographical distance; for example whether the probability of dying in a terror attack in a distant country is comparable with the risk of dying from a traffic accident in this country.

Geipel, Härta and Pohl [1] looked at the geography question in a region of Germany where a volcano erupted 10^4 years ago and left the Laach lake. The closer the residents were to that lake, the more seriously they took the risk. However, their general political orientation was also correlated with their risk judgment. On the other hand, scientific announcements have led to some newspaper reactions within Germany that are independent of the distance, but these died down after a few months. Other examples are the reactions to nuclear power plants and accidents associated with them. Volker

Jentsch (private communication) suggested that such reactions to Xevents could be simulated on computers; such simulations are only possible if we have a reasonable model of opinion dynamics.

A very recent application of this would be to study the influence of a deadly tsunami after an earthquake on opinions. Those who live on the affected coasts after the tsunami Xevent of December 2004 will remember it as a clear danger. Those who live further inwards, away from the coast, know that tsunamis do not reach them, but they will still have learned about the thousands of people killed from the news. Will they judge the danger as being higher than or lower than those on the affected coast line? And what about those who live on the coast of a different ocean, where such events are also possible but last happened long ago? This example shows how the influence of an Xevent on general opinion can depend on distances in time and space. This is the question we want to simulate here in generic models.

It would not be desirable to invent a new dynamic model for opinions just for the purpose of studying reactions to Xevents. Instead, it would be nice if we could have one generally accepted and well tested model, which then could be applied to Xevents. No such consensus is evident from the literature. We thus concentrate here on three models, D, KH and S (Deffuant et al. [2–5], Krause and Hegselmann [6,7] and Sznajd [8]) which are currently extensively used to simulate opinion dynamics; we ignore the older voter models [9, 10] or those of Axelrod [11], of Galam [12, 13] and of Wu and Huberman [14], to mention some other examples. We will not claim that we can use these simulations one may predict public reaction; we merely claim that simulations like these may be a useful starting point in this research field.

Of course, one may, in general, question whether human beings can be simulated on computers where only a few numbers are used to describe the whole person. More than two millenia ago, the Greek philosopher Empedokles paved the way to this type of computer simulation by stating (according to J. Mimkes), that some people are like wine and water, mixing easily, while others are like oil and water, refusing to mix. Thus he reduced the complexity of human opinions to two choices, like hydrophilic or hydrophobic in chemistry, spin-up or spin-down in physics, and 0 or 1 in computer science. And in today's developed countries, we take regular polls on whether people like the government, allowing only a few choices like: very much, yes, neutral, no, or not at all. Simplifying Mother Nature like this is also common in sociology, and has been quite successful in physics.

11.2 General Opinion Dynamics

In this section we review the dynamics of the models D (of Deffuant et al.), KH (of Krause and Hegselmann), and S (of Sznajd) [2–8]. Their results are quite similar but they differ in the rules used to change the opinions. An earlier review of these models was given in [15], with emphasis on the Sznajd model. In that model two people who agree in their opinions convince suitable neighbours to adopt this opinion. In model D, each person selects a suitable partner and the two opinions converge. In KH, each person looks at all suitable partners and takes their average opinion. "Suitable" means that the original opinions are not too far from each other.

11.2.1 The D Model

In model D [2–5], all N agents have an opinion O that can vary continuously between zero and one. Each agent selects one of the other agents randomly and checks whether an exchange of opinions makes sense. If the two opinions differ by more than ϵ ($0 < \epsilon < 1$), the two refuse to discuss and no opinion is changed; otherwise each opinion moves partly in the direction of the other, by an amount $\mu \Delta O$, where ΔO is the opinion difference and μ the convergence parameter ($0 < \mu < 1/2$). The parameter ϵ is called the confidence bound or the confidence interval. For $\epsilon > 1/2$ all opinions converge towards a centrist one, while for $\epsilon < 1/2$ separate opinions survive; the number of surviving opinions in the latter case varies as $1/\epsilon$. Besides simulations, analytical approximations are also made [16] which agree well with the simulations.

Figure 11.1 shows a consensus formation with the number of simulated people close to that in the European Union, 450 million, and $\epsilon = 0.4$, $\mu = 0.3$. To plot the results, the opinions were binned into 20 intervals. We show intervals 1, 2, 10 and 11 only. Initially the number of opinions were the same in all intervals; soon two centrist opinions began to dominate until finally one of them swallows the other. Independent of this power struggle, some extremist opinions survive in the intervals close to zero and close to one. These extremist wings [17] are a general property for $\epsilon < 1/2$ but are not the theme of this "extreme" book.

Variants on this standard version have been published. It is numerically easier to look at integer opinions $O = 1, 2, 3, \ldots, Q$ instead of continuously varying O; a precursor of such work was given by Galam and Moscovici [18], where the discrete opinions (0, 1) and opinions in-between were allowed. If the opinions $O = 1, 2, \ldots, Q$ are integers, one can unambiguously determine whether two opinions agree or differ. The above expression for $\mu \Delta O$ then needs to be rounded to an integer. If two opinions differ by only one unit, one randomly selected opinion is replaced by the other one, whereas this other opinion remains unchanged.

The idea of everybody talking to everybody with the same probability is perhaps realistic for scientific exchanges via the internet, but, in politics, discussions on city affairs are usually restricted to the residents of that city and do not extend over the whole world. Putting agents onto a square lattice [2–5] where interactions are only available between lattice neighbours [19] is one possibility. In recent years, small-world networks and scale-free networks [20] have been simulated intensively as models for social networks. In the standard version of the Barabási–Albert model, the most popular model of scale-free

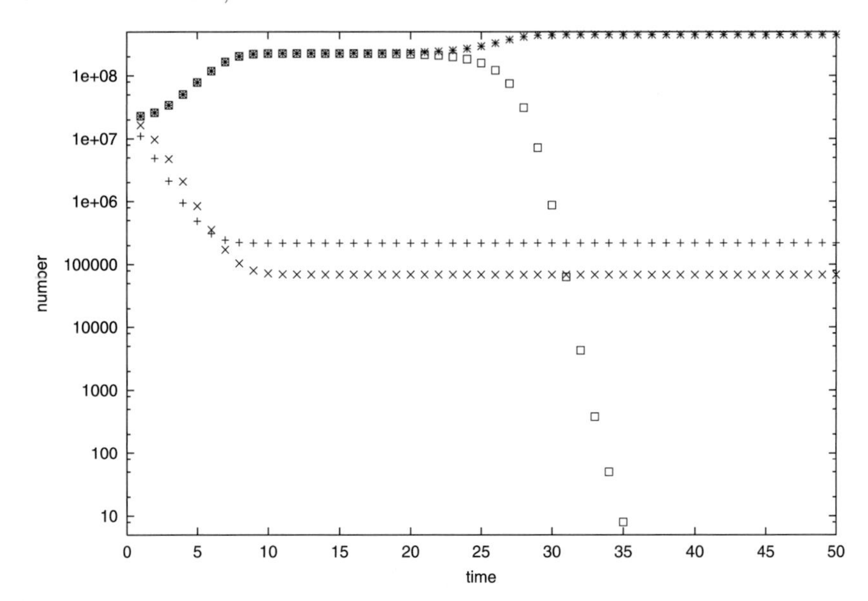

Fig. 11.1. Standard D model, 450 million agents, $\epsilon = 0.4$, $\mu = 0.3$, opinions divided int o 20 intervals. Intervals 1 (*crosses*), 2 (*x's*), 10 (*stars*), and 11 (*squares*) are shown

networks, one starts with a small number m of agents all connected to each other. Then, one by one, more members are added to the population. Each new member randomly selects m previous members as neighbours such that the probability of selecting one specific agent is proportional to the number of neighbours this agent had before. In this way, the well connected people get even more connections, and the probability of one agent being selected as a neighbour by k later members is proportional to $1/k^3$. (In contrast, on the square lattice and on the Bethe lattice, each agent has the same number of neighbours, and for random graphs the number of neighbours fluctuates slightly but its distribution has a narrow peak.) In opinion dynamics, only network neighbours can influence each other.

Putting Deffuant agents [21] onto this Barabási–Albert network, then, with continuous opinions, complete consensus is again found for large confidence intervals ϵ, whereas for small ϵ the number of different surviving opinions varies roughly as $1/\epsilon$. An opinion cluster is a set of agents sharing the same opinion in final equilibrium, independent of whether these agents are connected as neighbours or separated. Varying the total number N of agents one finds that the number of small opinion clusters with 1, 2, 3, ... agents is proportional to N, while the number of large opinion clusters comprising an appreciable fraction of the whole network is of order unity and is independent of N. This result reminds us of the cluster size distribution for percolation [22] above the threshold: there is one infinite cluster covering a finite fraction of the whole lattice, coexisting with many finite clusters whose number is proportional to the lattice size. One may compare this distribution of opinions with a dictatorship: the imposed official opinion coexists with a clandestine opposition fragmented into many groups.

This scale-free network can be studied in either a complicated or a simple way. In the complicated way, if a new agent Alice selects a previous agent Bob as a neighbour of Alice, then Alice is also a neighbour of Bob, as in mutual friendships. This is the undirected case. The directed case is the simpler way: Bob is a neighbour of Alice but Alice is not a neighbour of Bob; this situation corresponds more to political leadership: the party head does not even know all party members, but all party members know the head. Aside from simplifying the programming, the directed case seems to have the same properties as the undirected one [21].

Also, changing from continuous to discrete opinions $O = 1, 2, \ldots, Q$ does not change the results much but it does simplify the simulation [23], particularly when only people differing by one opinion unit discuss with each other (corresponding to $\epsilon \sim 1/Q$). Again the number of opinion clusters is proportional to N for $N \to \infty$ at fixed N/Q. A consensus is reached for $Q = 2$, but not for $Q > 2$. A scaling law gives the total number of final opinions as being equal to N multiplied by a scaling function of N/Q. This law has two simple limits. For $Q \gg N$ there are so many opinions per person that each agent has its own opinion, separate from the opinions of other agents by more than one unit; with no discussion, nobody changes opinion, giving N clusters of size unity. In the opposite case $Q \ll N$, all opinions have lots of followers and thus most of them survive up to the end. These simple limits also remain valid if people differing by up to ℓ opinion units (instead of $\ell = 1$ only) influence each other; a consensus is then formed if ℓ/Q (which now plays the role of the above ϵ) is larger than 1/2. (The more general scaling law for arbitrary N/Q now becomes invalid.) This threshold of $\epsilon = 1/2$, which has emerged so often in the previous examples, is supposed to be a universal feature of Deffuant dynamics, so long as the symmetry of the opinion spectrum with respect to the inversion right $\leftrightarrow$ left is not violated [24]. The symmetry means that the opinions O and $1 - O$ ($Q - O$ for integer opinions) are equivalent and can be exchanged at any stage of the dynamics without changing the corresponding configuration. In this way, at any time the histogram of the opinions is symmetric with respect to the central opinion 1/2 ($Q/2$ for integer opinions). If we instead let O and $1 - O$ ($Q - O$) play different roles, the threshold will in general be different. As a matter of fact, in [25] one introduced such an asymmetry in that the "convincing power", expressed by the parameter μ, is no longer the same for all agents but depends on the opinion of the agent. More precisely, μ increases with the opinion of the individual, and this implies that those agents with low values of O are less convincing than those with high values of O. In this case the opinion distribution is no longer symmetric with respect to $O = 1/2$ ($Q/2$) and the consensus threshold is larger than 1/2.

In all of this work, first the scale-free network was constructed, and then the opinion dynamics were studied on the fixed network. Not much is changed if opinion dynamics takes place simultaneously with network growth [26], in agreement with the models of Ising and Sznajd [27].

11.2.2 The KH Model

The KH model [6,7] has been simulated to a lesser degree than the D model since, until recently, it was only possible to study small systems with KH. However, for discrete opinions, an efficient algorithm was recently found that could be used to study millions of agents [28], compared with at most 300,000 for continuous opinions [29]. Again we have a continuous number of opinions O between zero and one, or discrete $O = 1, 2, \ldots, Q$. At each iteration, each agent looks at all other agents, and averages over the opinions of those that differ by not more than ϵ (continuous opinions) or ℓ (discrete opinions) from its own opinion. Then it adopts that average opinion as its own. As in the D model, the KH model shows a complete consensus above a particular threshold and many different opinions in the final configuration if ϵ is very small. However, in this case, there are two possible values for the threshold [30], depending on how many neighbours an agent has on average: if this number of neighbours, or average degree, grows with the number of agents in the community, there is consensus for $\epsilon > \epsilon_0$, where $\epsilon_0 \sim 0.2$; if instead the average degree remains finite when the population diverges, the consensus threshold is $1/2$ as in the D model. Various ways of averaging opinion have been investigated [31]. Hegselmann and Krause [6,7] have also simulated asymmetric ϵ choices, which may depend on the currently held opinion.

Figure 11.2 shows that the same scaling law as that used for the discrete D model also holds for the discrete KH model [28] on a scale-free Barabási-Albert network. For the usual version of the model, in which all individuals talk to each other, but with discrete opinions and discussions only allowed between agents differing by one opinion unit, a consensus is reached up to $Q = 7$, while several opinions remain for $Q > 7$. (The role of well-connected leaders in a similar opinion model on a Barabási-Albert network was studied in [32].)

As we mentioned above, by using discrete opinions it is possible to speed up the algorithm compared to the continuous case. The implementation of an algorithm for KH with discrete opinions must be probabilistic, because the value of the average opinion of compatible neighbours of an agent must necessarily be rounded to an integer and this would make the dynamics trivial, as in most cases the agent would keep its own opinion. We start with a community where everybody talks to everybody else, with opinions $O = 1, 2, \ldots, Q$ and a confidence bound of ℓ. After assigning opinions to the agents at random in the initial configuration, we calculate the histogram n_O of the opinion distribution, by counting how many agents have opinion O for any $O = 1, 2, \ldots, Q$. Suppose we want to update the status of agent i,

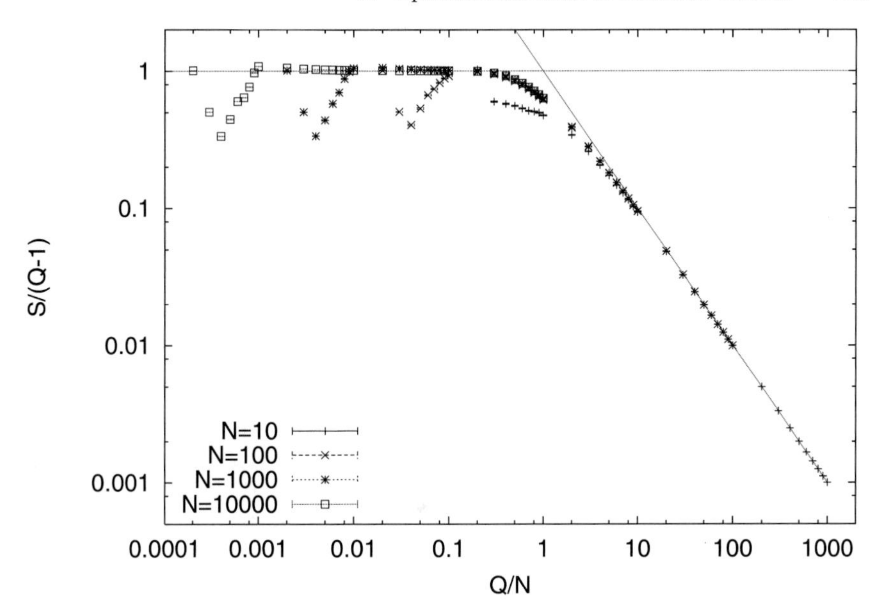

Fig. 11.2. Scaling law for the number S of surviving opinions in the discrete KH model, from [28]. The figure looks similar for the D model [23] except that the downward deviations at the left side of the data sets are weaker

which has opinion k. The agents that are compatible with i are all agents with opinion $\overline{k} = k - \ell, k - \ell + 1, \ldots, k, k + 1, \ldots, k + \ell - 1, k + \ell$. Let $n_{k\ell} = n_{k-\ell} + n_{k-\ell+1} + \ldots + n_{k+\ell-1} + n_{k+\ell}$ be the total number of compatible agents. Then we say that agent i takes opinion $\overline{k}$ with probability $p_{\overline{k}} = n_{\overline{k}}/n_{k\ell}$, which just amounts to randomly choosing one of the agents compatible with i and taking its opinion. Let k_f be the new opinion of agent i. We simply need to withdraw one agent from the original channel k and add it to the channel k_f to get the new opinion histogram of the system, and then we can move to the next update. Notice that, in this way, the time required for a sweep over the whole population goes like $(2\ell + 1)N$, where N is as usual the total number of agents and $2\ell + 1$ the number of compatible opinions. In the original algorithm with continuous opinions, however, the time to complete an iteration goes as N^2, because to update the state of any agent one needs to make a sweep over the whole population to look for compatible individuals and calculate the average of their opinions. The gain in speed of the algorithm with discrete opinions is then remarkable, especially when $\ell \ll N$.

We have seen that the presence of the second factor N in the expression of the iteration time for the continuous model is exclusively due to the fact that we consider a community where every agent communicates with all others. If one instead considers social topologies where each agent interacts on average with just a few individuals, like a lattice, the iteration time will onlygrow linearly with N, and the algorithm will compete in speed with that of D. As a matter of fact, in many such cases the KH algorithm is much faster than the D algorithm.

11.2.3 The S Model

The S model [8] is the most commonly studied model, and the literature on it (up to mid-2003) has been reviewed in [15]. Thus we concentrate here on more recent literature.

The most widespread version uses a square lattice with two opinions, $O = \pm 1$. If the two opinions in a randomly selected neighbour pair agree, then these two agents convince their six lattice neighbours of this opinion; otherwise none of the eight opinions changes. If less than half of the opinions initially have the value 1, at the end a consensus is reached with no agent having opinion 1; if initially the 1's have the majority, then at the end everybody follows their opinion. Thus a phase transition is observed, which grows sharper as the the lattice increases in size. The growth of nearly homogeneous domains of -1's and 1's is very similar to the spinodal decomposition of spin 1/2 Ising magnets.

With $Q > 2$ possible opinions ($O = 1, 2, \ldots, Q$), a consensus is always found except when only people with a neighboring opinion $O \pm 1$ can be convinced by the central pair of opinion O; then a consensus is usually possible for $Q \leq 3$ but not for $Q \geq 4$ in a variety of lattice types and dimensions, see Fig. 11.3 (from [15]).

The greatest success of the S model is the simulation of political election results. The number of candidates receiving v votes each varies roughly as $1/v$ with systematic downward deviations for large and small v. This was obtained on both a Barabási-Albert [33] and a pseudo-fractal model [34]. Of course, such simulations only give averages, not the winner in one specific election, just like physics gives the air pressure as a function of density and temperature, but not the position of one specific air atom one minute from now.

Schulze [35] simulated a multilayer S model, where the layer number corresponds to the biological age of the people in it; the results were similar to those for the single-layer S model. More interesting was his combination of global and local interactions on the square lattice: two people of arbitrary distance who agree in their opinions convince their nearest neighbours of this opinion. Similar to the mean field theory of Slanina and Lavicka [36], the times needed to reach consensus are distributed exponentially and are quite small. Therefore, up to 10^9 agents could be simulated. The width of the phase transition (for $Q = 2$, as a function of initial concentration) vanishes reciprocally with the linear lattice dimension [35].

If the neighbours do not always follow the opinion of the central pair, and instead do so with some probability [8], one may describe this probability through some social temperature T: the higher the temperature, the higher

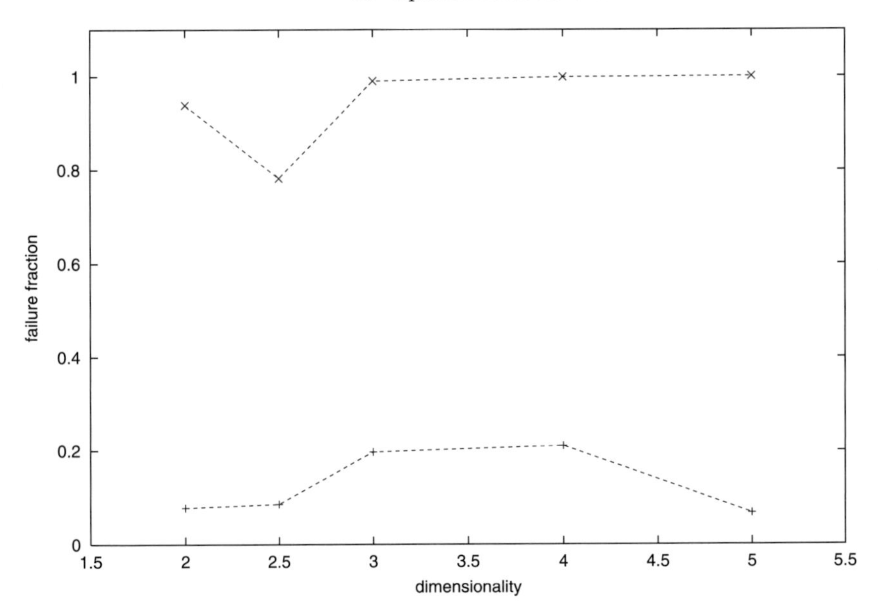

Fig. 11.3. Variation with dimensionality of the probability that a complete consensus will not be reached. $d = 2.5$ represents the triangular lattice. The *upper* data refer to four opinions, the *lower* ones to three opinions, in small lattices of size 19^2, 7^3, 5^4, 5^5. For larger lattices, the failures for three opinions vanish. Opinion O can only convince opinions $O \pm 1$

the probability that opinions will change [37]. In this case $T = 0$ means that nobody changes opinion, and $T = \infty$ means that everybody follows the S rule. Alternatively, one may also assume that some people permanently stick with their opinion [37,38]. In this way, a more democratic society is modelled, even for $Q = 2$, where not everybody ends up with the same opinion.

In an S model with continuous opinions and a confidence bound ϵ similar to the D and KH models, a consensus was always found independently of ϵ [39].

11.3 Damage Spreading

How is it possible to describe the reaction of people to Xevents in quantitative terms? From the previous discussion we have learnt that opinions can be treated as numbers: integer or real. A change of opinion of an arbitrary agent i is thus simply the difference between the new opinion and the old one. During the dynamical evolution, as we have seen above, opinions change, due to the influence of people on their acquaintances. This is, however, the "normal" dynamics within a community. What we would like to investigate instead is how much a sudden perturbation (an extreme event) would alter the opinion variables of the agents of the system. The concept of perturbation need not be defined exactly: for us it is whatever causes opinion changes in one or a few[1] agents of the system. We have localized events, like strikes, accidents, decisions involving small areas, and so on, in mind. We assume that people shape their own opinions only through the interactions with their acquaintances, without considering the influence of external opinion (sources like the mass media, which act at once on the whole population).

In order to evaluate the effect of a perturbation on the public opinion it is necessary to know the opinion distribution of the agents when nothing anomalous takes place (the "normal state"), and compare it with the distribution determined after the occurrence of an Xevent. From the comparison between these two replicas of the system we can evaluate, among other things, the so-called Hamming distance: how many agents have changed their mind, and how the influence of the perturbation spreads as a function of time and distance from the place where the Xevent occurred.

This kind of comparative analysis is by no means new in science, and it is commonly adopted to investigate a large class of phenomena, so-called *damage spreading* processes. Damage spreading (DS) was originally introduced in biology by Stuart Kauffman [40], who wanted to estimate the reactions of gene regulatory networks to external disturbances ("catastrophic mutations") quantitatively. In physics, the first investigations focused on the Ising model [41, 42]. Here one starts from some arbitrary configuration of spins and creates a replica by flipping one or more spins; after that one lets both configurations evolve towards equilibrium according to the chosen dynamics under the same thermal noise (identical sequences of random numbers). It turns out that there is a temperature T_d, near the Curie point, which separates a phase where the damage heals from another phase in which the perturbation extends to a finite fraction of the spins of the system.

The simplest thing one can do is just to follow the same procedure for opinion dynamics models. The perturbation consists of changing the opinion variable of an arbitrarily selected agent in the initial configuration. After that, the chosen opinion dynamics apply for the two replicas. Preliminary studies in this direction already exist, and they deal with the Sznajd model on the square lattice. In [43], one adopted a modified version of Sznajd where the four agents of a plaquette convince all of their neighbours if they happen to share the same opinion; here the perturbed configuration is obtained by changing the opinions of all agents that lie on a line of the lattice. In [44], the shock consists of the sudden change of opinions of some finite fraction g of the whole population and the time evolution of the number of perturbed agents is studied as a function of g. More importantly, the authors of the latter paper show that in several cases critical shocks in social sciences can be used as probes to test the cohesion of society. This recalls the strategy of natural

[1] Here "a few" means that the agents represent a negligible fraction of the total population, which vanishes in the limit of an infinite number of agents.

sciences: if we hit an iron bar with a hammer, we can derive its density from the velocity of the sound in the bar. In Sect. 11.3.2 we will present new results on damage spreading for Sznajd opinion dynamics [45]. Here we focus on the other two consensus models, D and KH. We shall first analyse the models for real-valued opinions, and then we will move on to integer opinions. In all our simulations we have defined the amount of damage as the number of agents differing in their opinions in an agent-to-agent comparison of the two replicas; we ignored the amount by which they differ.

An important issue is the choice of a suitable social topology. A bidimensional lattice lends itself to a geographical description of the damage spreading process: we can assume that the sites represent the spatial positions of the agents, and that the "acquaintances" of an agent are its spatial neighbours. In this way the lattice would map the distribution of people in some geographic area and the distances between pairs of agents on the lattice can be associated with physical distances between individuals. On the other hand, the regular structure of the lattice and the prescription of nearest-neighbour friendship endow the system with features that never occur in real communities. In fact, in the lattice, each agent has the same number of friends, and people who are geographically far from each other are never friends. These unrealistic features can be removed by adopting a different kind of graph to describe the social relationships between the agents. A Barabási–Albert (BA) network [46] could be a good candidate: it is a nonregular graph where the number of acquaintances of an agent varies within a wide spectrum of values, with a few individuals having many friends but most people have just a few. On the other hand, the BA network is a structure with a high degree of randomness and can hardly be embedded in an Euclidean bidimensional surface, so a geographical characterization of the damage propagation would be impossible. In our opinion the ideal solution would be a graph that includes both the regular structure of the lattice and the disorder of a random graph. A possibility could be a lattice topology where the connection probability between the agents decays with some negative power of the Euclidean distance, being unity for nearest neighbours. In what follows we shall however consider only the square lattice and the BA network.

11.3.1 Continuous Opinions

If opinions are real numbers, we need a criterion to state when the opinion of an agent is the same in both replicas or different due to the initial perturbation. Since we use 64-bit real numbers, we decided that two opinions are the same if they differ by less than 10^{-9}. In order to determine with some precision the fraction of agents that have changed their opinions, it is necessary to repeat the damage spreading analysis many times, by starting out from a new initial configuration each time without changing the set of parameters that constrain the action of the dynamics; the final result is then calculated by averaging over all samples. In most of our simulations we collected 1000 samples, although we increased the statistics to 10,000 in a few cases.

A detailed damage spreading analysis has recently been performed for KH with continuous opinions [47], for the case in which the agents sit on the sites of a BA network. The dynamics of the KH model are fixed by a single parameter, the confidence bound ϵ, which plays the role of temperature in the Ising model. As in the Ising model, it is interesting to analyze the damage propagation as a function of the control parameter ϵ; it turns out that there are three phases in the ϵ-space, corresponding to zero, partial and total damage, respectively. The existence of a phase in which the initial perturbation manages to affect the state (here the opinions) of all agents is new for damage spreading processes, and is essentially due to the fact that opinions are real-valued. In this case, in fact, the probability of a "damaged" opinion recovering its value in the unperturbed configuration is zero; on the other hand, to perturb the opinion of an agent it suffices that one of its compatible neighbours is affected, and the probability of having a compatible "disturbed" neighbour increases with the confidence bound ϵ. The only circumstance that can stop the propagation of the damage is when the perturbed agents are not compatible with any of their neighbours. The considerations above allow us to understand why the critical threshold $\epsilon_s = 1/2$ found in [47], above which damage spreads to all agents of the system, coincides with the threshold for complete consensus of the model, as in this case all agents share the same opinion and so they are all compatible with each other, which means that all agents are affected by each of their neighbours at some stage. Another interesting result from [47] is the fact that the two critical thresholds that separate the "damage" phases in the ϵ space do not seem to depend on the degree d_0 of the first node affected by the shock, although the Hamming distance at a given ϵ increases with d_0. This means that it is irrelevant whether the shock initially affected somebody who has many social contacts or somebody who is instead poorly connected: if damage spreads in one case, it will do so in the other too.

It is also important to study how damage spreads under D opinion dynamics. The hope is to be able to identify common features that would allow us to characterize the spreading process independently of the specific consensus model adopted. In Sect. 11.2.2 we stressed the analogies between the KH and the D model, so we expected to find similar results. For the D model we need to fix one more parameter to determine the dynamics: the convergence parameter μ. The value of μ only affects the time needed to reach the final configuration, so it has no influence on our results: we set $\mu = 0.3$. Figure 11.4 shows how the Hamming distance varies with the confidence bound ϵ for the D model on a BA network. The total number of agents is 1000. We remark that the damage is calculated here when the two replicas of the system have attained their final stable configurations. We have also plotted the corresponding curve for the KH model, as obtained in [47]. The two curves are

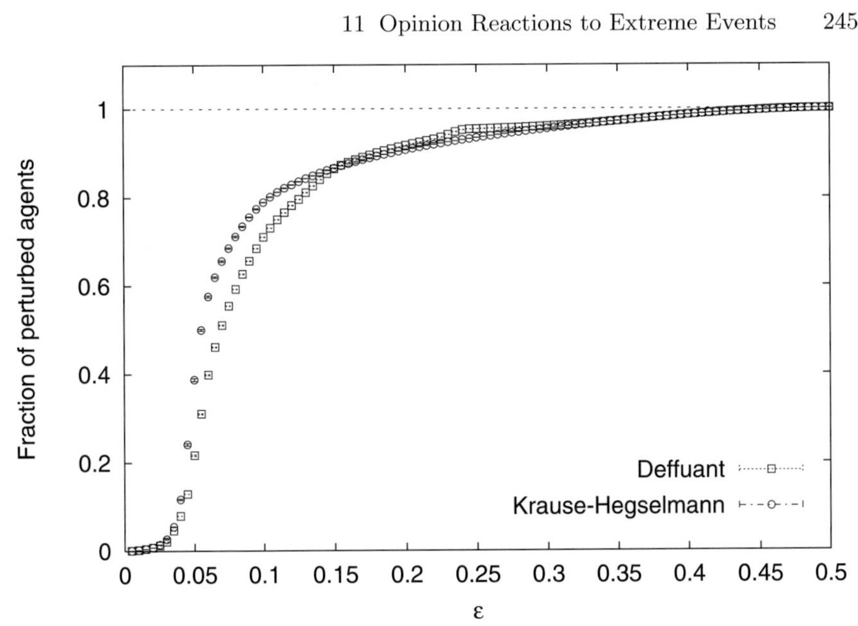

Fig. 11.4. Fraction of perturbed agents in the final configuration as a function of ϵ for the D and the KH model on a Barabási–Albert network

quite similar, as we expected, and the thresholds for the damage spreading transition are very close to each other. Again, for $\epsilon > \epsilon_s = 1/2$, all agents will be affected by the original perturbation.

As we explained in the *Introduction*, our main aim is to attempt a spatial characterization of the damage spreading process, which is impossible on a BA network. This is why from now on we shall focus on the lattice topology. Here we start by changing the opinion variable of the agent lying on the central site of the lattice; if the lattice side L is even, as in our case, the centre of the lattice is not a site, but the centre of a plaquette, so we "shocked" one of the four agents of the central plaquette. We refer to the initially shocked agent as to the origin. We will address the following issues:

- How far from the origin can the perturbation travel?
- What is the probability that an agent at a particular distance from the origin is affected?
- How does this probability $p(d,t)$ vary with the distance d and with the time t?

To discuss the first issue, we need to calculate the *range* r of the damage, in other words the maximum of the distances from the origin of the agents reached by the perturbation. The damage probability $p(d,t)$ is the probability that, at time t, a randomly chosen agent at distance d from the origin changed its mind, due to the initial shock. Here the time is represented, as usual, by the succession of opinion configurations created by the dynamics. The time unit we adopted is one sweep over all agents of the system. To calculate $p(d,t)$ we proceed as follows: after t iterations of the algorithm, we select all sites that are at a distance d from the origin, and which lie on-axis with respect to the origin, as in the scheme below:

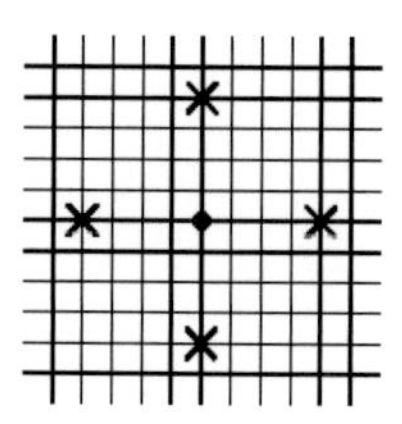

where the black dot in the middle represents the origin and the crosses mark the agents to be monitored. The damage probability is simply given by the fraction of those agents whose opinions differ from those of their counterparts in the unperturbed configuration (for example, if two of the four agents changed their mind, the probability is $2/4 = 1/2$). Note that by construction d must be a multiple of the lattice spacing (in our illustrated example $d = 4$). At variance with the evaluation of the damage range r, where we review all lattice sites, for the damage probability we neglect the off-axis sites because the lattice is not isotropic and the corresponding data would be affected by strong finite size effects due to the lack of rotational symmetry. To derive $p(d,t)$ from only four sites is of course difficult and we need to average over many samples for the data to have statistical meaning; we found that of the order of 10^3 samples are enough to obtain stable results. We calculated $p(d,t)$ for all distances from the centre to the edges of the lattice and for all intermediate states of the system from the initial random configuration to the final stable state.

We will mostly present results relative to the D model. The corresponding analysis for the KH model leads to essentially the same results. For the purpose of comparison with Fig. 11.4, we plot in Fig. 11.5 the Hamming distance as a function of ϵ, for the D and the KH model. The curves refer to a lattice with 40^2 agents: the two patterns are again alike. The damage spreading thresholds are close, but they lie quite a bit higher than the corresponding values relative to the BA network. This is basically due to the fact that in a BA network each vertex lies just a few steps away from any other vertex (small world property), and this makes spreading processes much easier and faster. Indeed, in the damage spreading phase, the time needed for the perturbation to invade the system is much longer for the lattice than for the BA network.

Since the amount of the damage is a function of ϵ, the range r of the damage is also a function of ϵ. It is interesting to analyze the histograms of the values of r for different values of the confidence bound. In Fig. 11.6 we show four such histograms, corresponding to $\epsilon = 0.10, 0.17, 0.18, 0.35$. Note that the values of r reported on the x-axis are expressed in units of $L/2$ (half

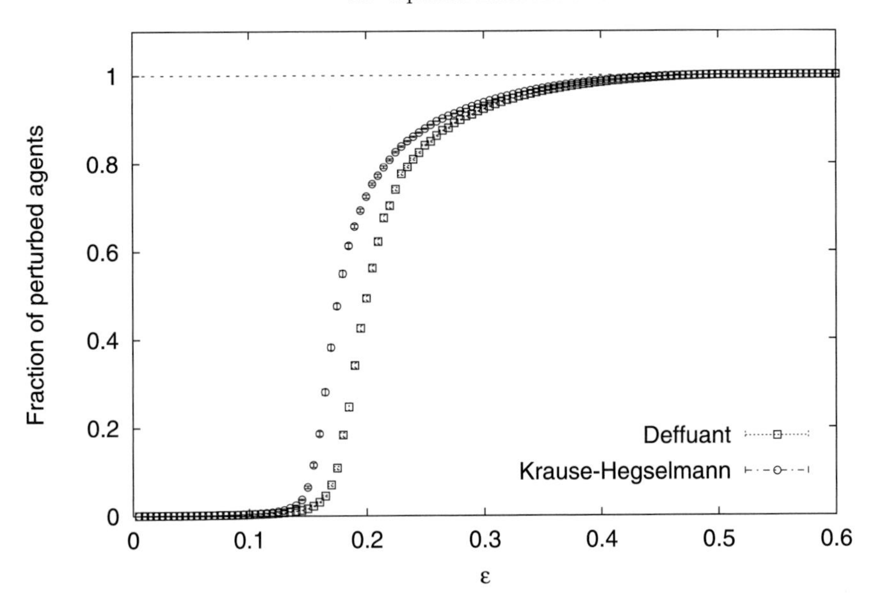

Fig. 11.5. As Fig. 11.4, but for agents sitting on the sites of a square lattice

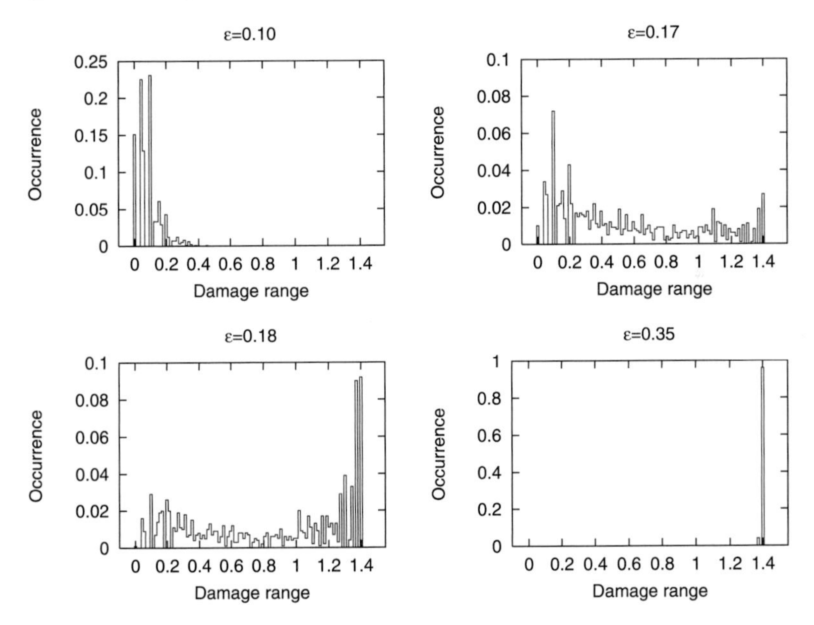

Fig. 11.6. D model, continuous opinions. Histograms of the damage range corresponding to four values of ϵ; the lattice size is 40^2

of the lattice side), which is the distance of the central site from the edges of the lattice; since the farthest points from the origin are the four vertices of the square, the maximal possible value of r is $L\sqrt{2}/2$ (which corresponds to $\sqrt{2} \sim 1.414$ in the figure). In the top left frame ($\epsilon = 0.10$) damage does not spread and in fact the histogram is concentrated at low values of r. In the other two frames, however, we are near the threshold for damage spreading, and we see that the damage often reaches the edge of the lattice ($r = 1$ in the plot) and even the farthest vertices ($r = \sqrt{2}$). The step from $\epsilon = 0.17$ to $\epsilon = 0.18$, despite the small difference in the value of the confidence bound, is quite dramatic, and signals the phase transition: in the first case (top right) it is more likely to have short ranges than long ones, in the other (bottom left) we have exactly the opposite. In the last frame, the range is almost always maximal; looking at Fig. 11.5, we can see that more than 90% of the agents are disturbed for $\epsilon = 0.35$, so it is very likely that the perturbation reaches one of the four vertices of the square.

The study of the damage probability $p(d, t)$ is more involved, as it is a function of two variables, the distance d and the time t. A good working strategy is to separately analyse the dependence of $p(d, t)$ on the two variables. We can fix the distance at some value d_0 and study how the damage probability at d_0 varies with time. We can also fix the time at t_0 and study how the probability at time t_0 varies with the distance from the origin. On top of that, we should not forget the dependence on ϵ, which determines the "damage" state of the system.

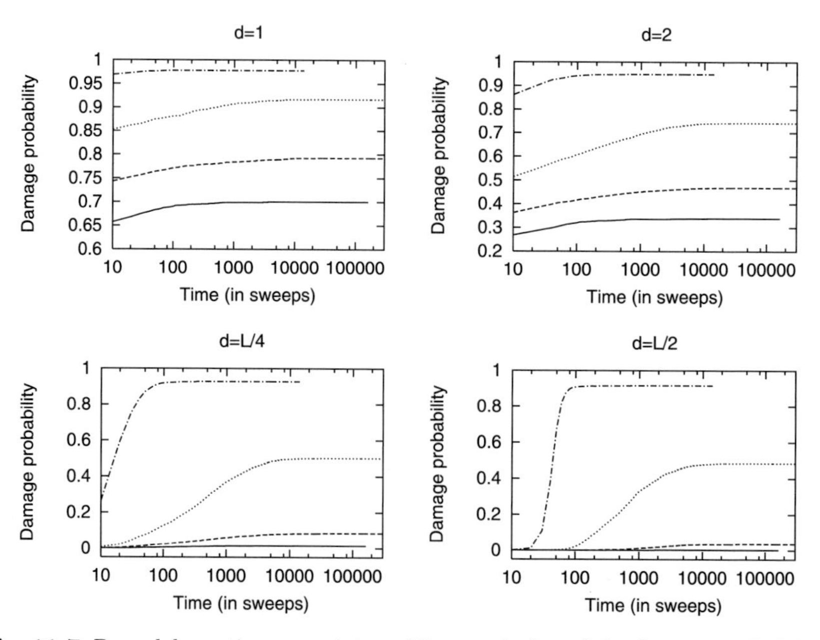

Fig. 11.7. D model, continuous opinions. Time evolution of the damage probability. Each frame refers to a fixed distance d from the origin, the curves are relative to different values of ϵ; the lattice size is 40^2

In Fig. 11.7 we explicitly plot the time dependence of the damage probability at four different distances from the origin, $d = 1, 2, L/4, L/2$. In each frame we have drawn four curves, corresponding (from bottom to top) to $\epsilon = 0.15, 0.17, 0.20, 0.30$. We remark that probability increases with ϵ, since this corresponds to a larger number of affected agents. All curves increase with time, which shows that the damage does not heal, and they reach a plateau long before the system attains the final opinion configuration. Note the rapid rise of the probability at the two largest distances ($L/4$ and $L/2$) for the two values of ϵ which fall in the damage spreading phase ($\epsilon = 0.20, 0.30$).

Figure 11.8 shows how the damage probability varies with the distance from the origin, at the end of the time evolution of the system. The distance values on the x-axis are renormalized to the maximal distance on-axis from the origin, $L/2$, as in Fig. 11.6. We again have four frames, one for each of the four values of ϵ we have considered in Fig. 11.7. We notice that for $\epsilon = 0.15$, which is slightly below the threshold, the damage probability at the edge (top left) is zero, whereas for $\epsilon = 0.17$, which is near the threshold, it is small but nonzero (top right) and it is about 1/2 for $\epsilon = 0.20$ (bottom left). We tried to fit the curves with simple exponential functions. We found that the tail off with distance is stronger than exponential: for low ϵ, $p(d, t)$ (at fixed t) is well approximated by $a \exp(-bd)/d$.

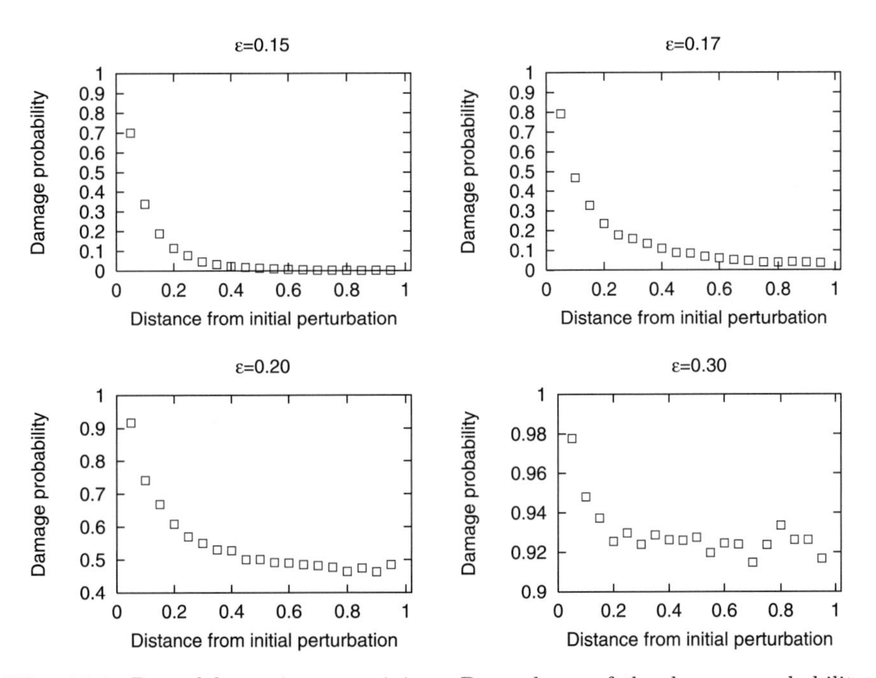

Fig. 11.8. D model, continuous opinions. Dependence of the damage probability on the distance d from the origin, when the system has reached the final stable configuration; the lattice size is 40^2

Note that we have chosen to introduce the shock into the system only at the beginning of the evolution. If one perturbs the system some time later instead, the amount of damage and the corresponding probabilities would decrease; however, the results of the analysis would be qualitatively the same.

11.3.2 Discrete Opinions

There is essentially one main reason for using real-valued opinions: the opinions of any two individuals are never exactly the same, although they can be arbitrarily close. This is what commonly happens in society, where no two persons have exactly the same idea or judgement about any issue. In fact, our opinion about somebody or a special event can fall anywhere between the two extremes "very bad" and "very good"; a situation analogous to the spectrum of visible light, where one can pass smoothly from red to violet.

On the other hand, for all practical purposes, this continuous spectrum of possible choices can be divided into a finite number of "bands" or "channels", where each channel represents groups of close opinions. This is actually what teachers do when they "mark" student essays. Electors also have to choose between a finite number of parties/candidates. Finally, for the case we are mostly interested in (the reaction of people to Xevents), the only quantitative investigation available to sociologists consists of performing polls where those interviewed must choose between a few options.

These examples show that it is more realistic to use integers rather than real numbers for the opinion variables of consensus models. Here we will repeat the damage spreading analysis of the previous section for the D model with integer opinions on a square lattice. We will see that the results are quite different to those we found before, due to the phenomenon of *damage healing*.

To start with, we must fix the total number Q of possible opinions/choices. Since we have performed simulations on systems with a few thousands agents, we decided to allow the number of choices to be of the same order of magnitude, so we set $Q = 1000$. The confidence bound must be an integer ℓ, but for consistency with the notation we have adopted so far, we will still use a real ϵ, again between 0 and 1, so that ℓ is the closest integer to $\epsilon\, Q$.

In Fig. 11.9 we show the variation of the Hamming distance with the confidence bound ϵ, for a lattice with 40^2 sites. We immediately notice that it is different from (the analogous) Fig. 11.5 for continuous opinions: after the rapid variation at threshold, the fraction of damaged sites reaches a peak, it then decreases and finally forms a plateau at large ϵ. Moving from real to integer opinions, we don't get any more overall damage – the perturbation can affect a fraction $f < 1$ of the total population (here $f \sim 0.6$) at the most – but it has no chance of affecting all agents. If we increase the number of agents N but keep Q fixed at the same value, the height of the final plateau decreases, going to zero when $N/Q \to \infty$.

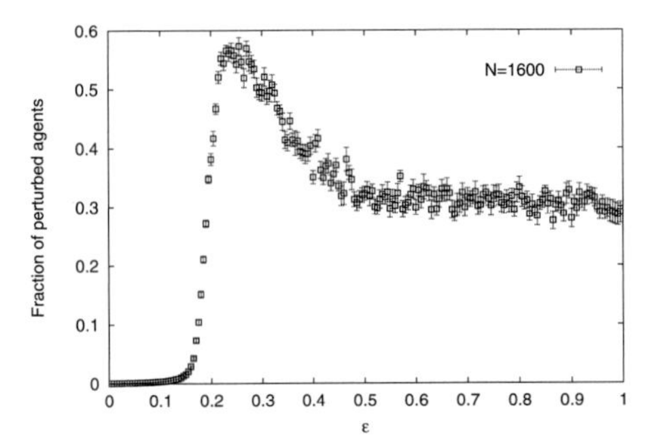

Fig. 11.9. D model, integer opinions. Fraction of perturbed agents in the final configuration as a function of ϵ for agents sitting on the sites of a square lattice

Why does this happen? Taking a look at Fig. 11.10 helps to clarify the situation. Here we see histograms of the damage range for $\epsilon = 0.18, 0.25, 0.35, 0.45$. If we compare the frame relative to $\epsilon = 0.18$ (top left) with its counterpart for continuous opinions (Fig. 11.6, bottom left), we see that they are basically the same. We are close to the transition so there is some finite probability of the damage reaching the edges and even the vertices of the square. We notice that the histogram is continuous, in the sense that any value of the range between the two extremes is possible. If we now look at the other three frames, the situation is very different: the range can be either

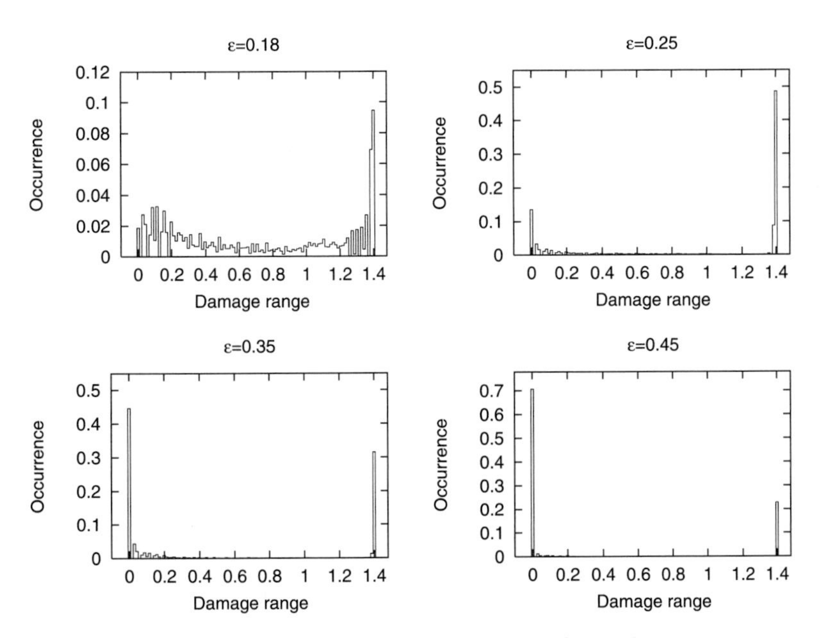

Fig. 11.10. D model, integer opinions. Histograms of the damage range corresponding to four values of ϵ; the lattice size is 50^2

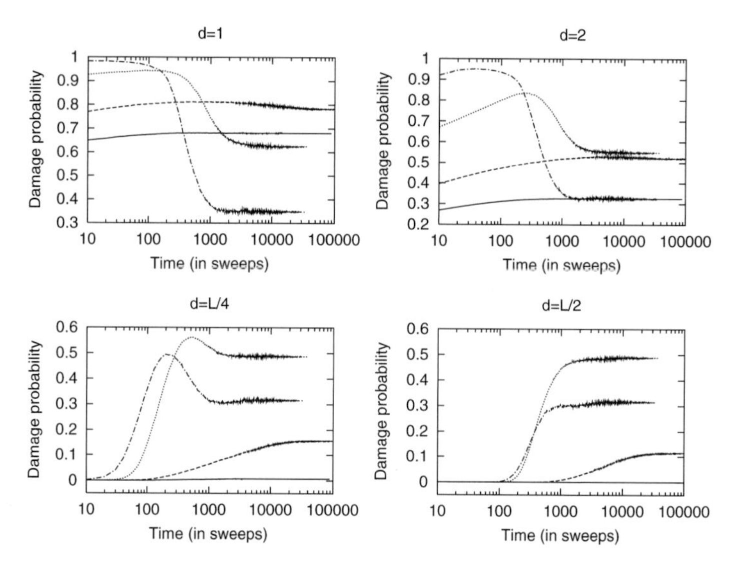

Fig. 11.11. D model, integer opinions. Time evolution of the damage probability. Each frame refers to a fixed distance d from the origin; the curves are relative to different values of ϵ; the lattice size is 50^2

very short or very long. In particular, when ϵ is very large (bottom right), the range is zero or maximal. That means that either the damage heals, or it spreads to all agents. In fact, for large ϵ ($> 1/2$), there is complete consensus in the final configuration (see Sect. 11.2.1), so all agents will end up with the same opinion. The question is then whether or not the final opinion in the perturbed configuration coincides with that of the unperturbed configuration; in the first case we have no damage, in the second total damage.

Now, real-valued opinions can be modified by arbitrarily small amounts, and this would still correspond to damage. On the other hand, the variations in integer opinions are discontinuous steps, and the latter are much more unlikely to occur. In this way, it is virtually impossible for a single agent to trigger a "jump" in the final opinions of all agents of the system to a different value. So, for large ϵ and many agents, the original perturbation will be healed by the dynamics[2] (no damage), whereas for continuous opinions even a small shock manages to shift the final opinion of the community a little bit (total damage).

The presence of damage healing is also clearly visible in Fig. 11.11, which is the counterpart of Fig. 11.7 for integer opinions. The four curves of each frame refer to $\epsilon = 0.15$ (continuous), 0.18 (dashed), 0.25 (dotted) and 0.35 (dot-dashed). The damage probability no longer increases monotonically as in Fig. 11.7; instead it displays various patterns depending on the confidence

[2] The non-vanishing probability for total damage in Fig. 11.10 is a finite size effect, as the total number Q of opinions is about the same as the population N.

bound and the distance from the origin. In particular, observe the behaviour of the curve for $\epsilon = 0.35$ and $d = 1$ (top left frame, dot-dashed line): here the probability is initially close to 1, because we are examining a neighbour of the shocked agent, but after few iterations it falls to about 0.3, due to healing. We also note the curious shape of the two upper curves for $d = L/4$, which recalls the pattern of the Hamming distance with ϵ of Fig. 11.9: the damage probability rapidly rises to a maximum and then it decreases to an approximately constant value.

Figure 11.12 shows the dependence of the damage probability on the distance in the final opinion configuration, for $\epsilon = 0.15, 0.18, 0.25, 0.35$. The curves look similar to those in Fig. 11.8. Again, the damage probability decreases faster than exponentially.

We conclude with some new results on damage spreading for the S model with two opinions on a square lattice [45], which complement the analyses from [43, 44]. Figure 11.13 shows the damage probability as a function of distance at various times. We see that the values of the probability are quite low; in fact, the system always evolves towards consensus, so the damage will heal over the long run, as with the D and KH models (with discrete opinions) when the confidence bound ϵ is above the threshold for complete consensus.

If damage spread like in a diffusion process, the distance covered by the propagation of the perturbation would scale as the square-root of the time t, and the probability of damaging a site at distance d would follow a scaling

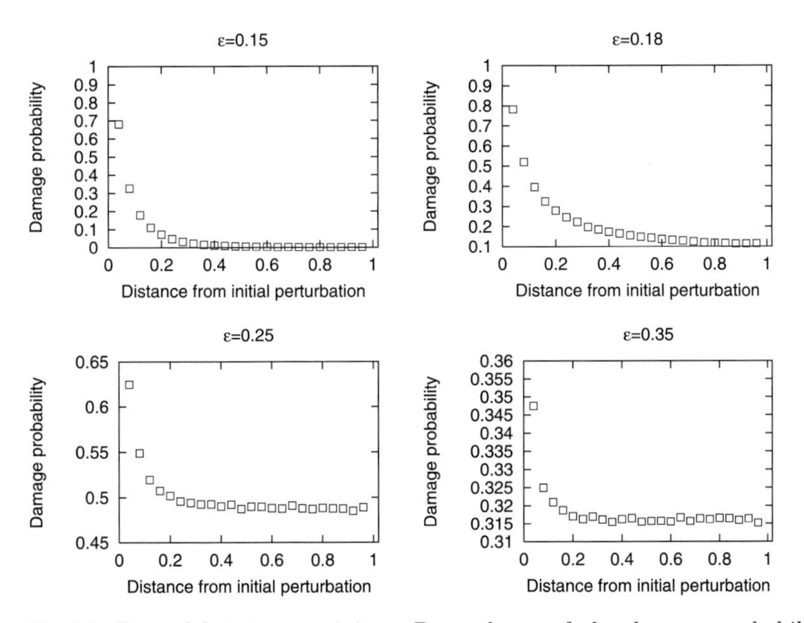

Fig. 11.12. D model, integer opinions. Dependence of the damage probability on the distance d from the origin when the system has reached the final stable configuration; the lattice size is 50^2

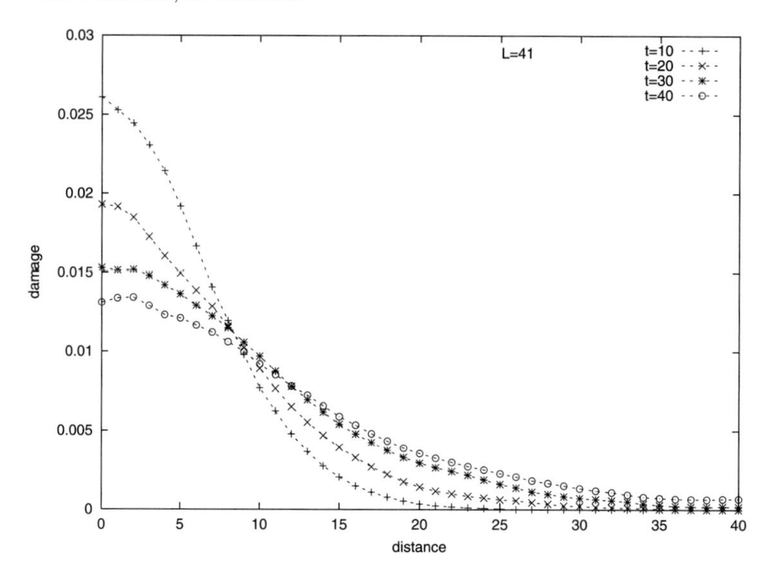

Fig. 11.13. S model, two opinions. Dependence of the damage probability on the distance d from the origin, for various times on a 41×41 lattice

function $f(d/\sqrt{t})$ over long timescales. Figure 11.14 shows that for $t \gg 1$

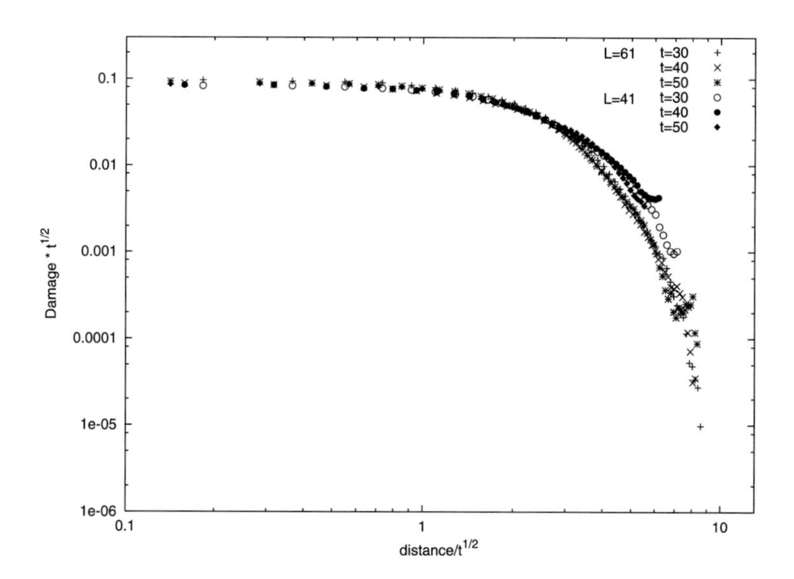

Fig. 11.14. S model, two opinions. Rescaling of the damage probability curves of Fig. 11.13 for larger lattices. Here we plot the damage probability time $\sqrt{t}$ versus $d/\sqrt{t}$ (t is the time). For $t > 20$, the curves for different times roughly overlap

this does indeed seem to be the case, even though damage spreading is not a random diffusion process.

Applications of these techniques to the case where people have opinions on several different themes are described elsewhere (see [48]; Jacobmeier, in preparation; Fortunato et al., in preparation).

11.4 Discussion

The three main models D, KH and S discussed in this chapter follow different rules but give similar results: they end up in a final state where opinions do not change any further. Depending on the confidence interval ϵ for continuous opinions, or ℓ for discrete opinions, this final state contains one opinion (consensus), two (polarization) or three and more (fragmentation) opinions. In the discrete case with Q different opinions, there is a maximum Q (2 for D, 3 for S, 7 for KH) for which a consensus is usually found. These numbers may correspond to the maximum number of political parties that may form a stable coalition government. The three rules differ in that S describes missionaries who don't care about the previous opinions of those whom they want to convince; KH describes opportunists who follow the average opinion of their discussion partners; while D describes negotiators who slowly move closer to the opinion of their discussion partner. Election results were successfully simulated by the model of S but not by those of D and KH, perhaps simply because nobody has tried to do so with those models yet.

The reaction of people to Xevents was investigated by performing a damage spreading analysis on the three consensus models we have introduced. The Xevent induces a change of opinion in one (or a few) agent(s); the dynamics propagates the shock to other agents. We represented the social relationships between people using a square lattice and a scale-free network, as for Barabási-Albert. In both cases we found that there is quite a wide range of values of the confidence interval ϵ (or ℓ) for which the original shock influences the opinions of a non-negligible fraction of the community. For very tolerant people and continuous opinions, the whole community will be affected by the event in the long run. By using integer-valued opinions, instead, we found that the perturbation does not affect more than a certain fraction of the population (which can be sizeable, however). Using the lattice we also studied how the influence of the Xevent on opinions varies with the distance in time and space from the event. The damage probability at a fixed distance from the original shock varies very rapidly with time; it increases up to a plateau for continuous opinions, and it follows more involved patterns for integer opinions. Our analysis also shows that the effect of the perturbation falls faster than exponentially with the distance from the place where the event took place.

What have we achieved with these simulations? We did *not* find a way to predict earthquakes or floods, nor have we provided a way to convince people to judge these dangers objectively, instead of being overly influenced by events that are close in time and space and forgetting the lessons from distant catastrophes that happened long ago. Our simulations give quantitative data on these space–time correlations of opinions and Xevents. Once sociology has delivered quality data on real people and their opinions [1], one can compare these results with the simulations and modify the simulations until they give a realistic description. Only then can the simulations be used to predict how danger perception will develop in space and time.

We thank J.S. Sá Martins for a critical reading of the manuscript. SF gratefully acknowledges the financial support of the DFG Forschergruppe under grant FOR 339/2-1.

References

1. R. Geipel, R. Härta, J. Pohl: *Risiken im Mittelrheinischen Becken.* In: *International Decade for Natural Disaster*, Reduction IDNDR, German IDNDR Series 4, ISBN 3-9805232-5-X, Deutsches IDNDR Komitee für Katastrophenvorbeugung, Bonn (1997)
2. G. Deffuant, D. Neau, F. Amblard, G. Weisbuch: Adv. Compl. Syst. **3**, 87 (2000)
3. G. Deffuant, F. Amblard, G. Weisbuch, T. Faure: J. Artif. Soc. Social Sim. **5**, issue 4, paper 1 (jasss.soc.surrey.ac.uk) (2002)
4. G. Weisbuch: Eur. Phys. J. B **38**, 339 (2004)
5. F. Amblard, G. Deffuant: Physica A **343**, 725 (2004)
6. R. Hegselmann, U. Krause: J. Artif. Soc. Social Sim. **5**, issue 3, paper 2 (jasss.soc.surrey.ac.uk) (2002)
7. U. Krause, in: *Modellierung und Simulation von Dynamiken mit vielen interagierenden Akteuren*, ed. by U. Krause, M. Stöckler, Bremen University (1997)
8. K. Sznajd-Weron, J. Sznajd: Int. J. Mod. Phys. C **11**, 1157 (2000)
9. P.I. Krapivsky, S. Redner: Phys. Rev. Lett. **90**, 238701 (2003)
10. L.R. Fontes, R.H. Schonmann, V. Sidoravicius: Comm. Math. Phys. **228**, 495 (2002)
11. R. Axelrod: J. Conflict Resolut. **41**, 203 (1997)
12. S. Galam: J. Stat. Phys. **61**, 943 (1990)
13. S. Galam: Physica A **336**, 56 (2004)
14. F. Wu, B.A. Huberman: *Social Structure and Opinion Formation*, cond-mat/0407252 at www.arXiv.org
15. D. Stauffer: AIP Conf. Proc. **690**, 147 (2003)
16. E. Ben-Naim, P. Krapivsky, S. Redner: Physica D **183**, 190 (2003)
17. G. Weisbuch, G. Deffuant, F. Amblard: Physica A 353, 555 (2005)
18. S. Galam, S. Moscovici: Eur. J. Soc. Psychol. **21**, 49 (1991)
19. T.C. Schelling: J. Math. Sociol. **1**, 143 (1971)
20. R. Albert, A.L. Barabási: Rev. Mod. Phys. **74**, 47 (2002)
21. D. Stauffer, H. Meyer-Ortmanns: Int. J. Mod. Phy. C **15**, 241 (2004)
22. D. Stauffer, A. Aharony: *Introduction to Percolation Theory*, Taylor and Francis, London (1994)

23. D. Stauffer, A.O. Sousa, C. Schulze: J. Artif. Soc. Social Sim. **7**, issue 3, paper 7 (jasss.soc.surrey.ac.uk)
24. S. Fortunato: Int. J. Mod. Phys. C **15**, 1301 (2004)
25. P. Assmann: Int. J. Mod. Phys. C **15**, 1439 (2004)
26. A.O. Sousa: cond-mat/0406766 at www.arXiv.org
27. J. Bonnekoh: Int. J. Mod. Phys. C **14**, 1231 (2003)
28. S. Fortunato: Int. J. Mod. Phys. C **15**, 1021 (2004)
29. D. Stauffer: Comput. Sci. Eng. **5**, 71 (2003)
30. S. Fortunato: Int. J. Mod. Phys. C **16**, 259 (2005)
31. R. Hegselmann, U. Krause: *Opinion Dynamics Driven by Various Ways of Averaging*, Universität Bayreuth (2004), Computational Economics **25**, 381 (2005); available at http://pe.uni-bayreuth.de/?coid=18
32. M. He, H. Xu, Q. Sun: Int. J. Mod. Phys. C **15**, 947 (2004)
33. A.T. Bernardes, D. Stauffer, J. Kertész: Eur. Phys. J. B **25**, 123 (2002)
34. M.C. Gonzalez, A.O. Sousa, H.J. Herrmann: Int. J. Mod. Phys. C **15**, 45 (2004)
35. C. Schulze: Int. J. Mod. Phys. C **15**, 569 and 867 (2004)
36. F. Slanina, H. Lavicka: Eur. Phys. J. B **35**, 279 (2003)
37. M. He, B. Li, L. Luo: Int. J. Mod. Phys. C **15**, 997 (2004)
38. J.J. Schneider: Int. J. Mod. Phys. C **15**, 659 (2004)
39. S. Fortunato: Int. J. Mod. Phys. C **16**, 17 (2005)
40. S.A. Kauffman: J. Theor. Biol. **22**, 437 (1969)
41. M. Creutz: Ann. Phys. **167**, 62 (1986)
42. H.E. Stanley, D. Stauffer, J. Kertész, H.J. Herrmann: Phys. Rev. Lett. **59**, 2326 (1987)
43. A.T. Bernardes, U.M.S. Costa, A.D. Araujo, D. Stauffer: Int. J. Mod. Phys. C **12**, 159 (2001)
44. B. Roehner, D. Sornette, J.V. Andersen, Int. J. Mod. Phys. C **15**, 809 (2004)
45. N. Klietsch: Int. J. Mod. Phys. C **16**, 577 (2005)
46. A.L. Barabási, R. Albert: Science **286**, 509 (1999)
47. S. Fortunato: Physica A **348**, 683 (2005)
48. D. Jacobmeier: Int. J. Mod. Phys. C **16**, 644 (2005)

For more information see Chap. 6 in:
D.Stauffer, S. Moss de Oliveira, P.M.C. de Oliveira, J.S. Sa Martins, Biology, Sociology, Geology by Computational Physicists, Elsevier, Amsterdam 2006 in press.

12 Networks of the Extreme: A Search for the Exceptional

Philippe Blanchard and Tyll Krüger

Summary. In this chapter, after a short survey of recent developments in the theory of complex networks, we discuss a class of random graph models for complex networks where the exceptional and Xevents play a crucial role in the formation of network structures. Indeed, some vertices – the "hubs" – have an extremely high number of connections to other vertices, whereas most vertices have just a few. These networks are generally "scale-free"; in other words, they exhibit architectural and statistical stability as the degree distribution grows. We also relate some extremal properties of the diameters of random graphs to the thresholds of epidemic processes, and we discuss robustness against system damage.

12.1 Extreme Events in Complex Systems and Our Perception of Them

Extreme events are the ones that grab our attention when we watch a news report, since they often are the events we are most afraid of (terrorist bombs on our doorstep or the bankruptcy of our bank, for example). What makes them so fascinating that they are the basis for thousands of scientific and a plethora of nonscientific books (almost every novel contains one or more Xevents), and how do they fit in to our visible and invisible world? The reasons for people's attraction to Xevents are psychological in nature, and therefore beyond the scope of this chapter, but we will claim (based on common sense) that the degree of extremality of an event is proportional to the inverse of its frequency of occurrence. The models we describe in Sect 12.3 are built on this principle, which we call "the Cameo principle".

What are Xevents? First of all, Xevents are rare (in time or in space); otherwise they wouldn't be "extreme". Put into a statistical frame, this means that the probability of observing them is very small (ignoring the precise meaning of "very small" for a moment). This is the usual way to characterize Xevents in natural sciences or mathematics. However, there is another way of looking at Xevents: they are usually of high relevance to the dynamics or structure of the system in which they appear. So their impact is somehow inversely proportional to the frequency of their occurrence. For complex networks – the subject of this chapter – we will show that our perception of the extreme, exceptional or rare means that the exceptional nodes in a network are also the functionally relevant ones.

Since extremality, in the sense described above, is the property of an event that occurs in the tail end of a probability distribution, one natural question to ask is whether there are typical tail distributions for rare events. It may seem a bit strange that there should be anything general to say about the distribution of rare events, except in the independent case (the classical approach, with characteristic normal or Poisson distributions). But scientific research – based on the large datasets that have become available over the last few years in a range of fields from social sciences to biochemistry – has provided strong evidence that Xevents are much more likely in complex real life systems than we originally thought. For decades, our thinking about the exceptional was influenced by the Bernoulli world picture, where the extreme is exponentially small and governed by laws like the central limit theorem (CLT) or Sanov's theorem in large deviation theory.

The some what surprising truth is that, for most (many) real world processes, the tail distribution – the statistics of the Xevents or large deviations – is of a power law type, indicating that real processes are far from independent; see [1, 2] for recent reviews. One of the most prominent examples is the Gutenberg-Richter law for earthquake distribution. Inspired by earthquake dynamics, a whole theory has emerged over the last 15 years called *self-organized criticality* (SOC). SOC mainly attempts to explain the frequency distributions with time for certain model processes. The basic idea, pitched by Per Bak ([3]) in 1990, is that complex systems stabilize themselves near a phase transition point (without parameter tuning) and so, as known from statistical physics, distributions are typically power law in nature (an example is the size distribution of finite clusters at the percolation threshold in lattice percolation). However, it is worth noting that this theory is largely heuristic and is only supported by results from computer simulations at present, even for the simplest models of SOC; see [4] and [5] for a more mathematical approach.

Almost ten years after the invention of SOC, a second class of power law distribution phenomena gained scientific attention, namely distributions characterizing the structure of complex networks like the Internet, the World-WideWeb (WWW), social networks and biochemical networks. Essentially every complex system has an underlying structure – usually encodable as a graph – providing an environment for dynamical evolution. Since this structure is also far from being a lattice in many applications, it is natural to study its basic graph properties. The simplest local property of a vertex is its degree: the total number of edges attached to a vertex, which is simply the number of the nearest neighbors of the vertex. Here it is mainly the degree distribution of the corresponding graph representation of the network that follows a power law. Since power law distributions have no characteristic size or "length", they are often called scale-free distributions (or graphs in the case of networks). In contrast to the SOC case, there is no clear link to phase transitions that could explain the emergence of power laws. But there are several building rules based on common sense psychology that do the job pretty well (see Sect. 12.2).

The theory of random graphs began with some studies of Erdös and Renyi [6] sixty years ago, in which they used probabilistic methods to show the existence of graphs with apparently contradictory properties. They discovered that there was a "typical" random graph among the random graphs they introduced: the random graph typically had certain sharply defined properties. The other great discovery was that all the standard properties of graphs appear rather suddenly. This phenomenon expresses the fact that there is a phase transition. The most spectacular example of a phase transition concerns the size of the largest component of a random graph. From this point of view, percolation theory is nothing else than the study of random subgraphs of different lattices. The classical spaces of random graphs have degree distributions that are Poissonian and are therefore not a good basis for embedding the high degree vertices (the hubs) that typically occur in real networks. One has to go beyond the independence assumptions in the classical theory.

12.2 A Short Survey of Scale-Free Networks

There is a vast amount of literature about complex networks. The best reference for an introduction to the subject is still the article by Albert and Barabási [7]. Other very readable specific surveys and books about complex networks include [8–15].

The first model of scale-free networks (or, more accurately, a random graph space with an asymptotic power law degree distribution) was also provided by Albert and Barabasi in a seminal paper [16]. The model they created is an evolutionary one, where one node is added at each time step, starting with a connected graph G_0 with at least two vertices. The new node added at time $t+1$ "generates" $m \geq 1$ independent edges identically distributed among the existing vertices in G_t such that vertex y is chosen with a probability $p(y,t) = \frac{d_t(y)}{\sum_{z \in G_t} d_t(z)}$ where $d_t(y)$ is the degree of $y \in G_t$. Multiple choices are allowed and therefore each node creates at most m edges (since the probability of multiple edge generation tends to zero, each vertex almost certainly creates exactly m edges). It can be proved (see Bollobas and Riordan [17]) that the asymptotic degree distribution is indeed a power law with an exponent of 3 independent of the choice of m, and that the diameter scales as $\frac{\log t}{\log \log t}$ for large t. The underlying rule for creating new edges with a probability proportional to the degree is called "preferential attachment" or "the richer you are the richer you get". This mechanism has its root in an old idea of Price [18]. Many variants of the above rule have been formulated that also allow for exponents other than 3. Although the rule is very appealing, it has certain drawbacks that limit its range of applicability. First, it is not robust with respect to small changes in the pairing rule – replacing the preferential term $d_t(y)$ by $d_t(y)^{\varepsilon}$ with $\varepsilon \neq 0$ will not give a scale-free graph (for $\varepsilon < 1$ one obtains a Poisson distribution and for $\varepsilon > 1$ a star-like graph emerges). Second, and this is especially true for social networks, the precise value of the actual degree of a vertex – for instance its number of friends – is hidden to the others. Motivated by these deficiencies, we searched for another psychological plausible rule of network formation that is more flexible in application and interpretation. Inspired by the view of rare events described in the *Introduction*, we believe that the main quantity that drives our view of rare events is the relative frequency of appearance. Formally speaking, if a family of discrete or continuous events $\{\omega\}$ is distributed with density φ, the attention we give to the event ω should be proportional to $\frac{1}{\varphi(\omega)^{\alpha}}$ where α is a positive individual parameter somehow describing a persons affinity to "sensations" or the "exceptional". Our personal view of rare events could therefore be summarized as "the exceptional is attractive". This is the basis of a building principle that we call "the Cameo principle" [3]. In the following section we describe a random graph model based on this principle.

12.3 Cameo Graphs

In this section we analyze some of the structural aspects of graphs built using the above-mentioned principle in detail. We believe that the Cameo principle is particularly well suited to the creation of social networks. According to this approach, the "the richer you are the richer you get" principle is replaced by "the rarer you are the more attractive you become". This can be applied to evolutionary as well as stationary graphs. The basic formal setting is the following: a positive, real-valued random variable ω with density φ is i.i.d. distributed on the vertex set $V_N : \{x_1\,; x_2\,; \ldots ; x_N\}$. The random variable ω is interpreted as the value of a property that has a enough of an attraction to others to prompt a contact; for instance richness, beauty, social influence, and so on. The crucial point is that the attraction and hence the probability of generating an edge is inversely proportional to a power of the frequency of appearance of the property ω:

$$\Pr\{x \to y \mid \omega(y) = \omega_0\} = const \cdot \frac{1}{N \cdot \varphi(\omega_0)^{\alpha}} \,; \alpha \geq 0 \,. \qquad (12.1)$$

Here $x \to y$ symbolizes the event where x generates an edge to y if x is about to make a new contact. In addition we should mention that the Cameo principle extends the concept of the random graph introduced by Erdös and Renyi obtained in the limit $\alpha \to 0$ [19]. For simplicity, in the following we fix the number of edges generated by each individual at $k_0 \geq 1$, but most of the statements will hold for more general situations where this number is itself a random variable. To avoid unnecessary mathematical complications, we first allow for multiple choices and later reduce this to the corresponding simple graph. Since the process of edge generation is a directional one – from the choice-making vertex x to the chosen vertex y – there is a natural notion of an out- and an indegree. Due to the possibility of multiple choices, the outdegree $d_{out}(x) \leq k_0$, but since the probability of a multiple choice scales at most like $\frac{const}{N^2}$ (double choice probability where $\alpha < 1$), then as $N \to \infty$ we almost certainly have $d_{out}(x) = k_0$. Behind the Cameo rule is the above-mentioned psychological experience that we are well adapted to recognizing relative differences but we find it very difficult to estimate absolute values of measures of rare events.

The interesting result is now the following: under very weak assumptions for the distribution $\varphi(\omega)$ ($\varphi(\omega) \underset{\omega\to\infty}{\to} 0$) we obtain a scale-free distribution for the degree that essentially only depends on the value of α. More precisely, we have the following theorem [20]:

Theorem 1. ***i)*** *Let* $\varphi(\omega) = \frac{1}{\omega^{\beta+o(1)}}$ *with* $\beta > \frac{1}{1-\alpha} > 0, \alpha \in (0,1)$, *then*

$$\lim_{N\to\infty} \Pr\{d(x) = k \mid x \in V_N\} = \frac{1}{k^{1+\frac{1}{\alpha}-\frac{1}{\alpha\beta}+o(1)}} .$$

ii) *Let* $\varphi(\omega) \in C^2([0,\infty))$ *and the second derivatives* $D^2(\varphi^\mu)$ *have no zeros for* $|\mu| \in (0,\mu_0)$ *and* $\omega > \omega_0(\mu)$ *(this is just an assumption of monotonicity in the tail of* φ*, and it implies that* φ *decays faster then any power law distribution); then*

$$\lim_{N\to\infty} \Pr\{d(x) = k \mid x \in V_N\} = \frac{1}{k^{1+\frac{1}{\alpha}+o(1)}} .$$

The emergence of a power law distribution independent of the choice of φ is not as surprising as it might first seem. The situation is best explained by an example. Let $\varphi(\omega) = e^{-\omega}; \omega \geq 0$ and define a new variable $\omega^* = \frac{1}{[\varphi(\omega)]^\alpha}$ $= e^{\omega\alpha}$. The new variable ω^* can be seen as the effective parameter to which the matching process applies. What is the induced distribution of ω^*? With $F(z) = \Pr\{\omega^* < z\}$ we obtain

$$F(z) = \int_0^{\frac{1}{\alpha}\ln\cdot z} \varphi(\omega)\, d\omega = 1 - z^{\frac{-1}{\alpha}} \tag{12.2}$$

and therefore the ω^*- distribution is given by $\phi(\omega^*) = (DF)(\omega^*) = \frac{1}{\alpha} \cdot \frac{1}{(\omega^*)^{1+\frac{1}{\alpha}}}$. This is a power law distribution where the exponent depends only on α. Proving the theorem in this special case is then simple, since the conditional indegree distribution for vertices x with a given ω- value ω_0 is a Poisson distribution, with expectation $const \cdot \varphi(\omega_0)^\alpha$, and the total indegree distribution is just the superposition of the conditional ones. The result for the exponential distribution can be generalized to hold for arbitrary distributions that satisfy the assumptions of Theorem 1. Because of its relevance, we formulate the main step as a theorem:

Theorem 2. *Let $\varphi(\omega) \in C^2([0,\infty))$ and let the second derivatives $D^2(\varphi^\mu)$ have no zeros for $|\mu| \in (0,\mu_0)$ and $\omega > \omega_0(\mu)$; then the distribution $\psi(y)$ of $y := [\varphi(\omega)]^\alpha := \varphi(\omega)^\alpha$ has density $\frac{1}{y^{1+\frac{1}{\alpha}+o_y(1)}}$.*

Proof: Let $G(x) := \int_0^x \varphi(z)\,dz$ be the cumulative distribution function of the random variable ω. The distribution function ψ of the r.v. $y = f(\omega)$ is then given by

$$\begin{aligned} D(\Pr\{f(\omega) \le y\}) &= D\left(\Pr\left\{\omega \le f^{-1}(y)\right\}\right) \\ &= D\left[G\left(f^{-1}(y)\right)\right] = DG\left(f^{-1}(y)\right) \cdot Df^{-1}(y) \ . \end{aligned} \tag{12.3}$$

For $f(\omega) := \varphi(\omega)^{-\alpha}$ we have $f^{-1}(y) = \varphi^{-1}\left(\frac{1}{y^{\frac{1}{\alpha}}}\right)$. With $Df^{-1}(y) = D\varphi^{-1}\left(\frac{1}{y^{\frac{1}{\alpha}}}\right) \cdot \frac{-1}{\alpha y^{1+\frac{1}{\alpha}}}$, and using the relations $D\varphi^{-1}(z) = \left[(D\varphi)\left(\varphi^{-1}(z)\right)\right]^{-1}$ and $DG = \varphi$ we therefore obtain

$$\psi(y) = \frac{1}{\alpha y^{1+\frac{1}{\alpha}}} \cdot \frac{-\varphi \circ \varphi^{-1}\left(y^{\frac{-1}{\alpha}}\right)}{(D\varphi)\left(\varphi^{-1}\left(y^{\frac{-1}{\alpha}}\right)\right)} \ . \tag{12.4}$$

The following technical lemma (the proof is given in the *Appendix*) states that the second term in the above formula is, for large values of y, smaller than any power of y, from which the proof follows.

Lemma 1. *Under the above assumptions, one has:*

$$\frac{-\varphi \circ \varphi^{-1}\left(y^{\frac{-1}{\alpha}}\right)}{(D\varphi)\left(\varphi^{-1}\left(y^{\frac{-1}{\alpha}}\right)\right)} = y^{o(1)} \tag{12.5}$$

With this result, it is now easy to prove the results claimed in Theorem 1. As already said, the conditional indegree distribution of vertices is asymptotically Poissonian with a given value of ω, with expectation $\frac{k_0 \cdot A}{[\varphi(\omega)]^\alpha}$ where $A^{-1} := \lim_{N\to\infty} \mathbb{E}\left[\frac{1}{N}\sum_{x\in V_N} \varphi(\omega(x))^{-\alpha}\right]$ is the normalization constant in the basic pairing probabilities of the Cameo principle. By Theorem 2, these expectation values are themselves distributed in the form of a power law with an exponent $1+\frac{1}{\alpha}$ for fast-decaying φ and with an exponent $1+\frac{1}{\alpha}-\frac{1}{\alpha\beta}$ if φ is itself a power law distribution with exponent β. The normalization constant is only well defined for $\alpha \in (0,1)$ and $\beta > \frac{1}{1-\alpha}$.

But what happens when $\alpha \geq 1$? The model is still well defined for each N, but there is no longer a uniform N-independent normalization. This means that, for a fixed (N-independent) ω value of a vertex y, the probability that $d_{in}(y) > 0$ goes to zero for $N \to \infty$. The indegree $d_{in}(y)$ stands for the number of directions ending in y. The reason for this is simple. Since the normalization constant $A = o(1)$ we almost certainly get $\Pr\{x \to y \mid \omega(y) = \omega_0\} = o\left(\frac{1}{N}\right)$. Further, the total number of edges is about $k_0 N$ and so $\mathbb{E}\left(d_{in}(y) \mid \omega(y) = \omega_0\right) = k_0 N \cdot . o\left(\frac{1}{N}\right) = o(1) \to 0$ as $N \to \infty$. Therefore, almost all of the edges are linked to a few number of vertices with extraordinarily large ω-values, and the graph becomes star-like. Instead of showing this result in its full generality, we will explain the situation for the case where φ is the exponential distribution $e^{-\omega}$, $\omega \geq 0$. First we derive a simple estimation for the largest ω value expected in the system for $\alpha > 1$. Since the expected number of vertices with ω values larger than ω_0 equals $N \cdot \int_{\omega_0}^{\infty} e^{-\omega} d\omega = N e^{-\omega_0}$, we get from the condition that this number should be about 1, giving the following estimation for the most likely maximal ω-value:

$$\omega_{\max}(N) \sim (1 + o(1)) \log N \ , \ \text{as } N \to \infty \ . \tag{12.6}$$

This implies that, with a probability converging to 1, the normalization constant A is of order

$$\left[\frac{N-1}{N} \int_0^{\log N} \frac{\varphi(\omega)}{\varphi(\omega)^{\alpha}} \left[\int_0^{\log N} \varphi(\omega) \, d\omega \right]^{-1} d\omega + \frac{1}{N \cdot \varphi(\log N)^{\alpha}} \right]^{-1}$$

$$= \left[\int_0^{\log N} e^{(\alpha-1)\omega} d\omega + N^{\alpha-1} \right]^{-1} \simeq \frac{1 - \frac{1}{\alpha}}{N^{\alpha-1}} \ , \ \text{as } N \to \infty \ . \tag{12.7}$$

Since the probability of an edge between x and y, with $\omega(y) = \omega$, is now

$$\Pr\{x \to y \mid \omega(y)\} = \frac{1 - \frac{1}{\alpha}}{N^{\alpha} \cdot \varphi(\omega(y))^{\alpha}} = \frac{1 - \frac{1}{\alpha}}{N^{\alpha}} e^{\alpha \omega(y)} \tag{12.8}$$

we only have a positive indegree for vertices y with $\omega(y) \geq \left(1 - \frac{1}{\alpha}\right) \log N$. Note that when the maximal ω value is larger than $\log N$, the same argument is still valid with an increased bound on the minimal ω value for which the indegree is still positive. Since the expected number of vertices with $\omega(y) \geq \left(1 - \frac{1}{\alpha}\right) \log N$ is given by

$$N \cdot \int_{\left(1-\frac{1}{\alpha}\right) \log N}^{\log N} e^{-\omega} d\omega \simeq N^{\frac{1}{\alpha}} \ , \ \text{as } N \to \infty \tag{12.9}$$

only an asymptotically vanishing fraction of vertices gains all of the (indegree) edges. Finally, we estimate the indegree expectation for the vertex $y_{\max}$ with the maximal ω value $\log N$. Since $\Pr\{x \to y_{\max} \mid \omega(y_{\max}) \sim \log N\} = 1 - \frac{1}{\alpha}$, one has $\mathbb{E}(d(y_{\max})) \sim \left(1 - \frac{1}{\alpha}\right) k_0 N$, so a positive fraction of all of the edges is linked to just one vertex – the superhub. In this sense, typical graphs for $\alpha > 1$ are star-like.

What does the resulting, N-dependent, degree distribution look like? Since Theorem 1 applies for all values of $\alpha > 0$, we can find out as follows. The random variable $k = Ce^{\alpha\omega(x)}$, the indegree for $C = \frac{k_0\left(1-\frac{1}{\alpha}\right)}{N^{\alpha-1}}$, is distributed like $\frac{1}{\alpha}\frac{C^{\frac{1}{\alpha}}}{k^{1+\frac{1}{\alpha}}}$ for $k > 0$, which implies that

$$\Pr\{d_{in}(x) = k; k > 0\} \simeq \frac{const}{N^{1-\frac{1}{\alpha}}k^{1+\frac{1}{\alpha}}} \tag{12.10}$$

for the degree distribution. Note that, by what was previously said, we have $\Pr\{d_{in}(x) = 0\} \underset{N\to\infty}{\to} 1$. It remains to say a few words about the degenerate case $\alpha = 1$. An analogous computation shows that a typical normalization constant is of order $\frac{1}{\log N}$. Therefore $\mathbb{E}(d(y_{\max})) \sim \frac{k_0 N}{\log N}$, and the smallest ω value that still gets a positive indegree is of order $\log\log N$, as $N \to \infty$.

An immediate consequence of the star-like structure is the finiteness (N-independence) of the diameter.

12.4 How Extremists Determine the Structures of Scale-Free Graphs

In real life, every parameter that would be used in a model would usually be a random variable varying from individual to individual. It is therefore natural to study the Cameo graphs of the previous section with a quenched disorder in the affinity parameter α [21]. As we will see, this generalization of the Cameo principle will radically alter fundamental features of the random graphs.

Let α be i.i.d. distributed with μ (here we do not require that the distribution is absolutely continuous) and let $\alpha_{\max} := \sup\{\alpha \mid \mu(\alpha) > 0\}$. The first surprising result is that, in case $\alpha_{\max} < 1$ and φ decaying faster than any polynomial, the exponent of the resulting power law degree distribution depends only on $\alpha_{\max}$. In other words, we have discovered that the vertices with the highest affinity α, that we call "extremists" for obvious reasons, are responsible for the exponent γ of the degree distribution. This result is easy to understand, although the detailed proof is somewhat involved. Therefore, we only give a short outline here. Assume that μ is supported on $I = [\alpha_{\min}, \alpha_{\max}]$, $1 > \alpha_{\max} > \alpha_{\min} > 0$. For fixed small $\varepsilon = \varepsilon(L) = \frac{\alpha_{\min} - \alpha_{\max}}{L} > 0$, we consider covering I with L subintervals $\left\{I_l^{(\varepsilon)}\right\}_{1 \le l \le L}$. For large N we almost certainly have $N \cdot \int\limits_{(l-1)\varepsilon}^{l\varepsilon} \mu(\alpha)\, d\alpha$ vertices with value α in I_l. Since the results from the previous section on Cameo graphs rely only on the fact that $const \cdot N$ edges are created according to the Cameo principle, for a distribution of the conditional indegree $d_{in}^{(I_l)}(x) := \#\{y \mid y \to x \wedge \alpha(y) \in I_l\}$ we obtain the following bounds (in the limit of large N):

$$\frac{1}{k^{1+\frac{1}{l\varepsilon}+o(1)}} \leq \Pr\left(d_{in}^{(I_l)}(x) = k\right) \leq \frac{1}{k^{1+\frac{1}{(l-1)\varepsilon}+o(1)}} \quad . \tag{12.11}$$

The rest of the argument relies on the dominating distribution principle: given a finite family of asymptotic power law distributions $\{\varphi_i\}$ with exponents $\{\beta_i\}$ then the distribution $\sum\limits_i \varphi_i$ is again an asymptotic power law distribution with exponent $\min\{\beta_i\}$. Applying this principle to the above situation, for each ε we get a bound that almost power law in nature, such that

$$\frac{1}{k^{1+\frac{1}{\alpha_{\max}}+o(1)}} \leq \Pr\left(d_{in}(x) = k\right) \leq \frac{1}{k^{1+\frac{1}{\alpha_{\max}-\varepsilon}+o(1)}} \quad . \tag{12.12}$$

Since this holds for any positive ϵ, and the constants involved stay bound as $N \to \infty$, we obtain the results claimed above.

Let us emphasize that the same result emerges if we randomize networks of the kind that Barabasi and Albert theorized. Let m be the positive integer describing the number of edges sent out by each new vertex. If m is a random variable, it is easy to show that the exponent of the power law describing the degree distribution only depends on $m_{\max}$, the maximum value that the random variable m can take.

The situation becomes more complicated for $\alpha_{\max} \geq 1$ (and more interesting). As explained in the previous section, there is no contribution from the set of vertices x with $\alpha(x) > 1$ to the indegree of vertices with fixed ω value. Therefore, the $k_0 \cdot N \int\limits_{1+\varepsilon}^{\alpha_{\max}} \mu(\alpha)\, d\alpha$ edges originating from that set of vertices with $\alpha(x) > 1+\varepsilon$ almost certainly link only to vertices with ω values bigger than $\left(1 - \frac{1}{1+\varepsilon}\right) \log N$ and generate a star-like substructure to the vertex with the biggest ω value. As an example, we estimate the expected indegree of a vertex y with $\omega(y) = \beta \ln N$, $\beta \in (0,1]$ arising from vertices x with $\alpha > 1$ where $\mu(\alpha) = 1/2$ for $\alpha \in [0;2]$. One has

$$\mathbb{E}\left(d_{in}(y) \mid \omega(y) = \beta \ln N\right) = \frac{1}{2} \cdot k_0 N \int\limits_1^2 \frac{1 - \frac{1}{\alpha}}{N^\alpha} e^{\alpha\beta \ln N} d\alpha \tag{12.13}$$

$$= \frac{1}{2} \cdot k_0 \int\limits_1^2 \left(1 - \frac{1}{\alpha}\right) N^{1+\alpha\beta-\alpha} d\alpha \,. \tag{12.14}$$

It is important to note that the indegree depends only on the value of $\alpha_{\max}$. Figure 12.1 gives a plot of this value for $N = 10^6$ and $k_0 = 2$ for $\beta \in (0.6; 1)$.

On the other side we can apply the result from the beginning of this section where $\mu(\alpha)$ was assumed to be supported in $(0,1)$ to the set of vertices x with $\alpha(x) < (1-\varepsilon)$. These vertices generate a scale-free indegree distribution with exponent $1 + \frac{1}{1-\varepsilon}$. Since ε can be taken to be arbitrary small, we get the following result in the mixed case where the α distribution μ has support in $(0,1)$ as well as in $(1,\infty)$. There is still a power law distribution for any fixed k:

$$\Pr\{d(x) = k\} \underset{N\to\infty}{\to} \frac{1}{k^{2+o(1)}}; k \text{ fixed} \tag{12.15}$$

but there is a single heavy tail of star-like form with a degree of the order of the total number of vertices. Let us emphasize that the very high degree values in Fig. 12.1 are of course out of reach in real social networks due to various reasons. For instance, capacity limitation causes some cut-off for the maximal degree. Also, in real life most vertices only have only potential access to a small fraction of the total population, so the Cameo principle applies in essence to small size substructure partitions of the society. Indeed, scaling the values in Fig. 12.1 down to village size populations of order 10^3 - 10^4 gives very reasonable contact structures.

We close this section with a numerical result showing that the asymptotic exponent $\gamma = 1 + \frac{1}{\alpha_{\max}}$ is achieved in networks of a reasonable size with $N = 10^6$ (Fig. 12.2). The distribution $\mu(\alpha)$ is taken to be the uniform one in $[0, \alpha_{\max}]; \alpha \le 1$. The finite size effect is accounted for by the factor 1.29 in front of the $\frac{1}{\alpha}$ term.

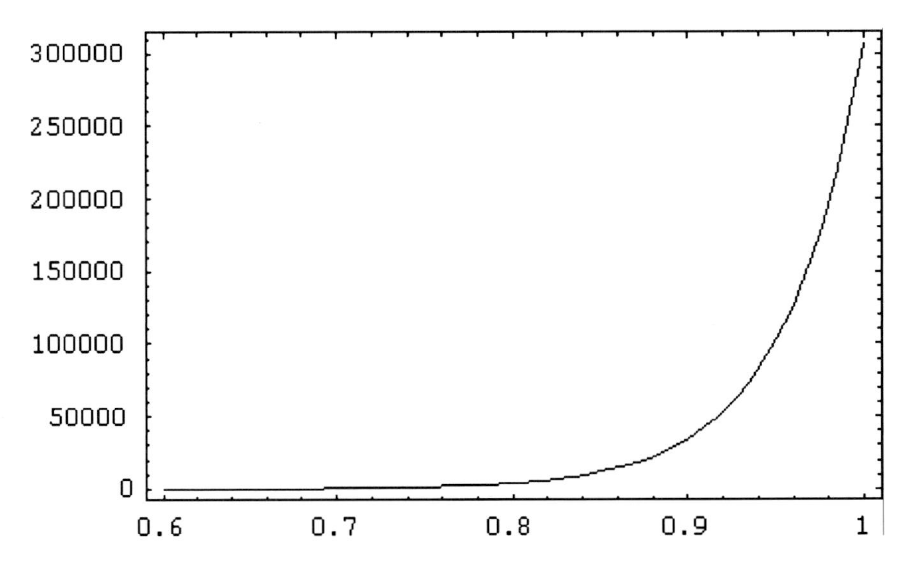

Fig. 12.1. Indegree of vertices with $\omega = \beta \log 10^6$ as a function of $\beta \in [0.6; 1]$

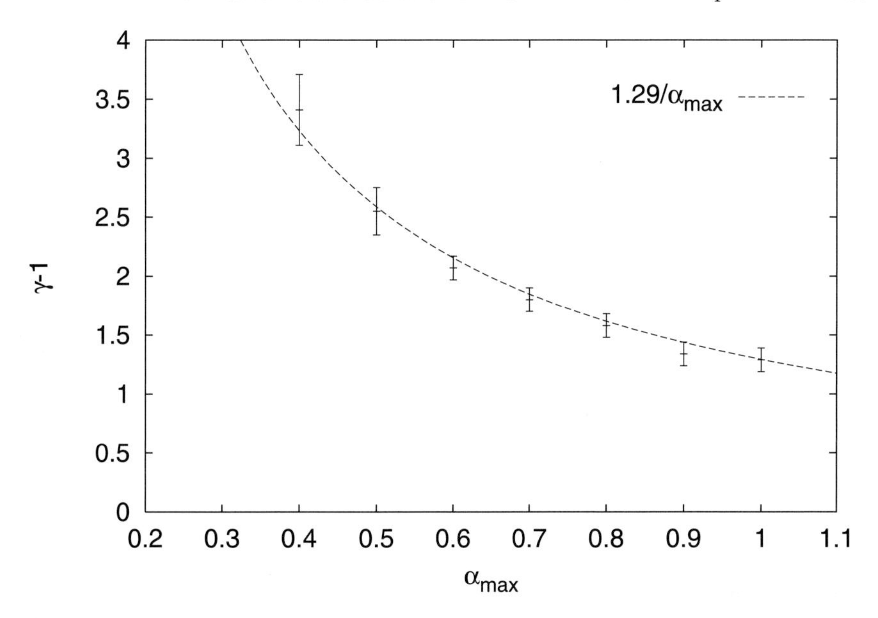

Fig. 12.2. Dependence of the exponent $\gamma - 1$ of the cumulative degree distribution on the upper limit $\alpha_{\max}$ of the affinity range. The data points can be fitted by the simple ansatz $1.29/\alpha_{\max}$

Summarizing the results of this section, one can say that the exponent of the power law is determined only by the vertices, which are most sensitive to the property ω. Acting on those vertices is an effective way to control the structures of evolving networks formed by Cameo-like rules.

12.5 Spreading of Epidemics in Scale-Free Networks and Robustness Under Random Attack

In this section we will study some aspects of propagation processes in scale-free graphs and study the influence of a heavy tail degree distribution on propagation properties. An epidemic is therefore any kind of stochastic transport phenomenon in social, biological or economical networks. Assume we have given a virus (or any other agent) that can be transmitted along edges with probability γ. Starting with a randomly chosen initial infected vertex, one may ask: what is the probability that a particular positive fraction of the whole population (vertex set) will eventually be infected (in the limit of large N)? For classical epidemic models that live on a lattice or a Erdös and Renyi random graph [19], and assuming that the number of edges is proportional to the number of vertices, a positive threshold value γ_c exists such that there will be no epidemic outbreak for $\gamma < \gamma_c$ but for $\gamma > \gamma_c$

a positive fraction of the total vertex set becomes infected. In other words, only sufficiently virulent diseases will cause epidemics. For scale-free graphs, the situation can be very different, as was first recognized by Pastor-Satorras and Vespignani [22] for the Albert and Barabasi model. Due to the presence of high degree vertices (hubs, the somewhat extreme vertices), the connectivity of a scale-free network can become so strong that a positive fraction of the total vertex set is within an certain N-independent constant distance. This happens for exponents less then 3 (in this case the second moment is no longer bounded) and it means that $\gamma_c \to 0$. The divergence of the second moment of the degree distribution is equivalent to the statement that the expected size of the 2-neighborhood (the size of the set of vertices at a distance of 2 from a randomly chosen vertex) diverges with $N \to \infty$.It is now easy to see that the random removal of, say, a fraction α of the edges ($\alpha \in (0;1)$) will not destroy this property, and so a cluster of finite diameter containing a positive fraction of all vertices still remains. For the above question about epidemics, this means that $\lim_{N\to\infty} \gamma_c N = 0$, since the eventual size of an epidemic starting in vertex x is just the size of the connected component containing x after deletion of $1 - \gamma\%$ of the edges. Therefore, the threshold γ_c needed to cause a widespread epidemic is 0. In other words, even the most inefficient infectious agent will spread widely. However, it is not clear whether this result has anything to say about current plagues in real world networks.

This result has another interpretation in terms of resistance to random attacks – the connectivity property of the net, on which its communicational skill relies, is essentially not affected by the random demolitions of either connections (edges) or transmission knots (vertices). The story becomes different if the attack is not random but focused on the hubs and/or the edges pointing to them. In that case it is easier to destroy the connectivity properties or communication function of a scale-free net than those of a classical random graph or lattice. This fact is rooted in the inhomogeneous topology of scale-free graphs. Random eliminations of vertices or edges mainly removes vertices with small degrees and will not significantly affect the connectivity properties of the graph. However, a reliance on high-degree vertices has very serious consequences; see [23, 24].

Of course, there are other ways to influence the structure of networks as was already mentioned at the end of the preceding section. Namely for evolving networks one could try to act directly on the parameters underlying the formation of the network. As an example let's take the so called terrorist networks causing so much attention nowadays. Instead of endless hunting the hidden hubs it might be much better to try to reduce the number of their most enthusiastic sympathizers (lowering so the $\alpha_{\max}$-value).

12.6 Conclusions and Outlook

In this article we have discussed some properties of scale-free random graphs, and a model where our perception of the rare produces a scale-free degree distribution irrespective of the distribution of the rare. Furthermore, only a small fraction of vertices (or individuals in the case of social networks) determine the overall structure of the network. The results presented are only a small fraction of what has been obtained for both Cameo graphs and general scale-free graphs. Many interesting and challenging mathematical, theoretical and practical questions remain for further research.

It is safe to say that scale-free networks have the potential to be of great theoretical use, since they could be used to answer a number of unanswered questions, particularly those associated with graphs that are not locally tree-like. There is not even a systematic program for characterizing the architecture of complex network. Work in this area has focused mainly on clustering properties (counting triangles) and the degree sequence, but there are many other quantities that are important to study.

The widespread presence of power laws has changed our point of view from regarding such distributions as exceptional to regarding them as the norm in complex systems. Since the assumptions made in the models that cause power laws are very weak, their appearance now appears to standard. Nevertheless, it has been difficult to relate these applications to phenomena in real life, since not every power law supposedly seen in nature is a "real" power law, and architectures based on different models (growth and preferential attachment, random graph methods like the Cameo principle, optimization performance under constraints) can give the same exponent for the degree distribution. It is therefore unclear as to which of the different processes or principles is really causing the scale freedom. Finding a compelling mechanism to explain a power law does not mean that there are not other, perhaps simpler explanations.

From a methodological point of view, we have significant hopes that relations between network architecture and concepts from statistical physics like phase transitions, criticality and self-organization can be exploited further, since they could give a much deeper understanding of many of these phenomena.

Acknowledgement. We are very grateful to Madeleine Sirugue-Collin, Santo Fortunato, Andreas Krüger and Andreas Ruschhaupt for interesting discussions, and would like to thank the support of the Volkswagen Foundation and DFG-Research Group 399 "Spectral Analysis, Asymptotic Distributions and Stochastic Dynamics".

12.7 Appendix

Here we prove the technical lemma used in the proof of Theorem 2, namely that the following holds under the assumptions of the theorem:

$$\frac{-\varphi \circ \varphi^{-1}\left(y^{\frac{-1}{\alpha}}\right)}{(D\varphi)\left(\varphi^{-1}\left(y^{\frac{-1}{\alpha}}\right)\right)} = y^{o(1)} \tag{12.16}$$

Since $\varphi(\omega)$ decays faster then any power law, we have

$$\varphi(\omega) < \frac{1}{\omega^l} \text{ for any } l \text{ and } \omega > \omega_0(l) \ . \tag{12.17}$$

Since $\varphi^{-1}\left(\frac{1}{y^{\frac{1}{\alpha}}}\right)$ goes to infinity for $y \to \infty$ we have to show

$$\frac{-\varphi(x)}{D\varphi(x)} = [\varphi(x)]^{o_x(1)} \ . \tag{12.18}$$

The last formula states that the negative logarithmic derivative of φ should not become too large or too small compared to φ and $\frac{1}{\varphi}$ respectively. For the following it is convenient to set $\varphi(x) = e^{-g(x)}$ with $g(x) \to \infty$ and rewrite (12.17) as

$$e^{-\mu g(x)} < \frac{1}{Dg(x)} < e^{\mu g(x)} \text{ for } \mu \in (0, \mu_0 > 0) \text{ and } x > x_0(\mu) \ . \tag{12.19}$$

Assume that (12.18) is not true with respect to the right hand side. Then, for a sequence of values $\{x_i\}$ and open intervals I_i around the x_i and some function $a(x)$, we have

$$\frac{1}{Dg(x)} = e^{\mu g(x)} a(x) \text{ and } a(x) > 1 \text{ for } x \in I_i \ . \tag{12.20}$$

Integrating the last equation gives

$$e^{\mu g(x)} = e^{\mu g(x_0)} + \mu \int_{x_0}^{x} \frac{1}{a(z)} dz \ . \tag{12.21}$$

Since our assumption that $D\left[\frac{1}{\varphi(\omega)}\right]^{\mu}$ is monotonous for $\mu > 0$ and $\omega > \omega_0(\mu)$ implies $a(x) > 1$, we eventually conclude that

$$e^{\mu g(x)} < e^{\mu g(x_0)} + \mu(x - x_0) \ . \tag{12.22}$$

However, the fast decay condition for $\varphi(x)$ expresses a growth condition for $g(x)$, namely for all k

$$g(x) > k \log x;\ x > x_0(k) \tag{12.23}$$

which clearly contradicts (12.21). Finally, we need to show that the left hand side of (12.18) also holds. Assuming the converse, we get

$$\frac{1}{Dg(x)} = e^{-\mu g(x)} \frac{1}{a(x)} \text{ and } a(x) > 1 \text{ for } x \in I_i \tag{12.24}$$

and after integration

$$e^{-\mu g(x)} = e^{-\mu g(x_0)} - \mu \int_{x_0}^{x} a(z)\, dz \ . \tag{12.25}$$

The monotonicity condition again implies that $a(x) > 1$ eventually, so

$$e^{-\mu g(x)} < e^{-\mu g(x_0)} - \mu (x - x_0) \tag{12.26}$$

which is a clear contradiction since the right hand side becomes negative for large values of x.

References

1. M.E.J. Newman: *Power laws, Pareto distributions and Zipf's law*, arXiv: cond-mat/0412004
2. M. Mitzenmacher: *A brief history of generative models for power law and log-normal distributions*, Internet Math., **I**, No. 2, 226–251 (2003)
3. P. Bak, C. Tang, K. Wiesenfeld: *Self organized criticality: An explanation of 1/f noise*, Phys. Rev. Lett., **49**, 4 (1987)
4. Ph. Blanchard, B. Cessac, T. Krüger: *What we can learn about self-organized criticality from dynamical systems theory*, J. Stat. Phys. **98**, 375–404 (2000)
5. B. Cessac, Ph. Blanchard, T. Krüger, J.L. Meunier: *Self-organized criticality and thermodynamic formalism*, J. Stat. Phys. **115**, 1283–1326 (2004)
6. P. Erdös, A. Renyi: *On random graphs*, Publ. Math. – Debrecen **6**, 290–297 (1959)
7. R. Albert, A.-L. Barabási: *Statistical mechanics of complex networks*, Rev. Mod. Phys., **74**, 47 (2002), arXiv:cond-mat/0106096
8. A.L. Barabasi: *Linked: new science of networks*, Perseus, New York (2002)
9. M.E.J. Newman: *Structure and function of complex networks*, SIAM Rev. **45**, 167–256 (2003)
10. S.N. Borogovtsev, J.F.F. Mendes: *Evolution of networks*, Adv. Phys. **51**, 1079–1187 (2002)
11. G. Caldarelli, A. Erzan, A. Vespignani: *Application of networks*, Eur. Phys. J. B **38**, Number 2, March II (2004)
12. S.N. Dorogovtsev, J.F.F. Mendes: *Evolution of networks: from biological nets to the Internet and WWW*, Oxford University Press, Oxford (2003)
13. S. Bornholdt, H.G. Schuster (eds.): *Handbook of graphs and networks. From the genome to the Internet*, Wiley, New York (2003)

14. R. Pastor-Satorras, A. Vespignani: *Evolution and structure of the Internet: a statistical physics approach*, Cambridge University Press, Cambridge (2004)
15. M.E.J. Newman, A.L. Barabasi, D. Watts: *The structure and dynamics of networks*, Princeton University Press, New York (2003)
16. R. Albert, A.-L. Barabási: *Emergence of scaling in random networks,* Science, **286**, 509 (1999)
17. B. Bollobás, O.M. Riordan: *Mathematical results on scale-free random graphs*, in; *Handbook of graphs and networks, from the genome to the internet*, eds. S. Bornholdt, H.G. Schuster, Wiley-VCH, Weinheim (2003)
18. D.J. de Price: *A general theory of bibliometric and other cumulative advantage processes*, J. Am. Soc. Inform. Sci. **27**, 292–306 (1976)
19. B. Bollobas: *Random graphs, 2nd edn.*, Cambridge University Press, Cambridge (2001)
20. Ph. Blanchard, T. Krüger: *The "Cameo principle" and the origin of scale-free graphs in social networks*, J. Stat. Phys., **114**, 5–6 (2004), arXiv: cond-mat/0302611
21. Ph. Blanchard, S. Fortunato, T. Krüger, Importance of extremists for the structure of social networks, Phys. Rev. E71 056114 (2005)
22. R. Pastor-Satorras, A. Vespignani: *Epidemic spreading in scale-free networks*, Phys. Rev. Lett. **86**, 3200 (2001)
23. D. Volchenkov, L. Volchenkova, Ph. Blanchard: *Epidemic spreading in a variety of scale free networks*, Phys. Rev. E **66** 046137 (2002)
24. Ph. Blanchard, C.H. Chang, T. Krüger: *Epidemic thresholds on scale-free graphs: the interplay between exponent and preferential choice*, Ann. Henri Poincaré Suppl. **2**, 957–970 (2003)

Part III

Prevention, Precaution, and Avoidance

13 Risk Management and Physical Modelling for Mountainous Natural Hazards

Michael Lehning and Christian Wilhelm

Summary. Population growth and climate change cause rapid changes in mountainous regions resulting in increased risks of floods, avalanches, debris flows and other natural hazards. Xevents are of particular concern, since attempts to protect against them result in exponentially growing costs. In this contribution, we suggest an integral risk management approach to dealing with natural hazards that occur in mountainous areas. Using the example of a mountain pass road, which can be protected from the danger of an avalanche by engineering (galleries) and/or organisational (road closure) measures, we show the advantage of an optimal combination of both versus the traditional approach, which is to rely solely on engineering structures. Organisational measures become especially important for Xevents because engineering structures cannot be designed for those events. However, organisational measures need a reliable and objective forecast of the hazard. Therefore, we further suggest that such forecasts should be developed using physical numerical modelling. We present the status of current approaches to using physical modelling to predict snow cover stability for avalanche warnings and peak runoff from mountain catchments for flood warnings. While detailed physical models can already predict peak runoff reliably, they are only used to support avalanche warnings. With increased process knowledge and computer power, current developments should lead to a enhanced role for detailed physical models in natural mountain hazard prediction.

13.1 Introduction

Mountainous areas tend to have an enhanced risk of natural hazards. This is because, in addition to general risks such as earthquakes and storms, extra risks are caused by the topography (such as avalanches, mud and debris flows). Because one cause of such hazards is the local terrain, local hazard mitigation is also often feasible. However, societies in rich countries have managed to create protection mechanisms for frequent local events. The greatest risk in this case often comes from extreme and rare events. Reducing these risks is a great economic challenge because cost – benefit estimations are based on weak statistics since data on Xevents is inherently scarce.

This contribution discusses risk management strategies using the example of a road over a mountain pass. The example illustrates that attempting to protect against Xevents results in exponentially growing costs. In this context, it is shown that organisational measures (road closures) based on a high quality forecast can be very cost-effective. In the second part of the contribution, we discuss the forecasting support that is available from physical modelling. An increased knowledge of the processes leading to slope failures, as well as improved meteorological forecasts, will also lead to better and more objective assessments of extreme natural hazards in alpine terrain over the long-term.

13.2 Risk Management Example for Mountain Roads

13.2.1 Integral Risk Management

Integral risk management for natural hazards means that protection measures involving forestry operations, land use planning, technical and organisational measures are coordinated and applied in an optimal manner. The optimal level of security is then reached with minimum cost and an optimal resource allocation is guaranteed. The combination of protective constructions, avalanche hazard maps, protective forests, systems for early warning, forecasting and alerting, closing off and securing areas, evacuations and artificially triggering avalanches is shown in Fig. 13.1. The optimization involves considering the duration times of the measures as well as the intervention strategies. Integral risk management in this sense has only recently been practiced. Integral risk management replaces protection strategies based purely on engineering structures. It makes use of existing (and new) structures but adds organisational measures. Increasing land-use, traffic (and thus economic value) in the mountain regions, together with increasing uncertainty about climate changes and limited resources available for prevention require a more flexible strategy [1].

13.2.2 Cost – Benefit Framework for Traffic Protection against Natural Hazards

When a variety of measures such as protection galleries or road closures need to be combined, their respective costs and benefits need to be assessed. We will show how this can be done in the example below. If we consider a traffic route that is unprotected from mountainous hazards, the society could face damage costs from material damage, injuries and deaths. The damage costs C_s can be reduced by protection measures, which in turn result in protection costs C_p.

Engineering structures only influence C_p and C_s. Organisational measures, on the other hand, also reduce the total benefit, B, of the road, because the traffic route will be temporarily closed at times. As a consequence, the social benefit loss from road closures ($-B_c$) needs to be assessed. Therefore, for integral risk management, the following minimisation must be made:

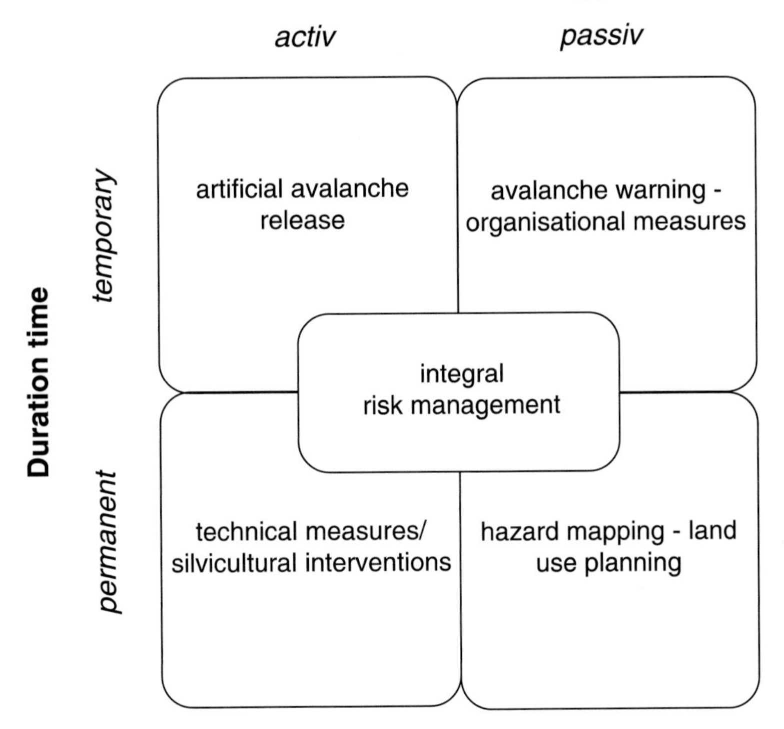

Fig. 13.1. Integral risk management via an optimal combination of various measures

$$C_s + C_p - B_c \rightarrow minimum \tag{13.1}$$

The minimisation process must be analysed step-by-step as new measures are added. For now, let us include the social benefit loss through road closures ($-B_c$) in the protection costs (C_p). In Fig. 13.2, the cost is shown as a function of risk reduction. Starting with an initial state, measures can be invoked and the remaining risk will decrease. The economic optimum is reached when the marginal costs of protection measures are equal to the marginal costs of risk reduction. Since measures are chosen in order of their cost-effectiveness, the protection cost increases exponentially as the risk is reduced still further. The higher the security attained, the higher the marginal cost for further risk reduction. Furthermore, as the security attained increases, the value of a further risk reduction drops (decreasing marginal benefit). The aim of minimising the total cost (C_{tmin}) causes the economic optimum (R_{opt}) to be where the marginal cost is equal to the marginal benefit (reduced damage cost). This point describes the economically optimal level of security. It also follows from this framework that Xevents are very difficult to deal with, and prevention of Xevents is only possible at very high cost.

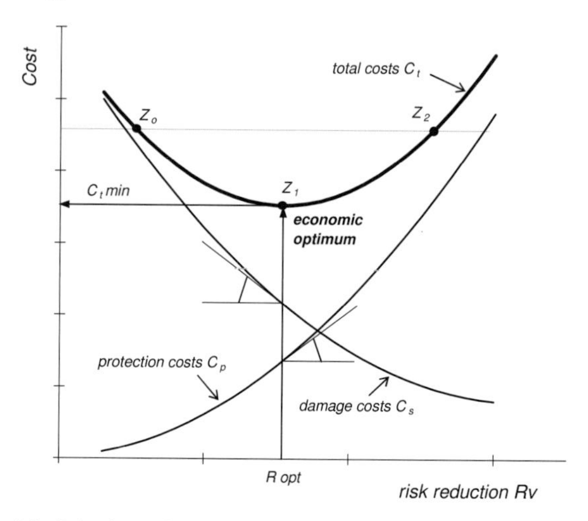

Fig. 13.2. Model of total cost minimization for natural hazard reduction

13.2.3 Case Study: Flüela Pass, Switzerland

System Definition and Introduction of Risk Approach

Until 1971, the Flüela Pass (2305 m ASL), which proceeds from Davos to the Engadin, was closed in winter for 156 days per year on average. Between 1971 and 1999, the pass road was also kept open during winter time. A protection plan consisting of temporary road closures and artificial avalanche release was applied.

Over a distance of 19.3 km, 47 avalanche paths cross the pass road for a total length of 10.1 km. On average, 38 natural and 27 artificially triggered avalanches hit the road each winter when the road is open. The maximum avalanche activity occurred in winter 1991/1992, when 117 events blocked the road. The average winter daily traffic (WDT) is 1000 vehicles/day. The cost to keep the pass road open in winter was, on average, 0.5 million CHF. 80% of this was spent on the removal of snow and avalanche deposits. Using the protection plan, the closure was reduced to 25 days.

In general, risk can be written as the product of probability and amount of damage [2,3]. Avalanche risk must be assessed by temporally and spatially overlapping the two independent processes of avalanche danger and land use of a certain area. The parameters given in Fig. 13.3 are important when recording risk situations [4].

The probability of occurrence of an avalanche can be calculated as the reciprocal of the mean return period T. The possible amount of damage is determined by the probability that objects (characterised by their monetary value) or people are present in the avalanche track. The probability of the presence of vehicles is given by the average winter daily traffic WDT, the mean width of the avalanche in the area of the road, g, and the speed of

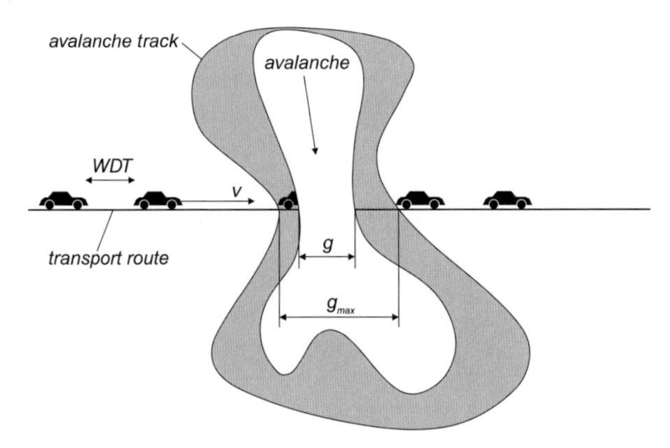

Fig. 13.3. Model used to assess the risk from avalanches along transport routes

the vehicles, v, in the path of the avalanche. The maximum width of the avalanche path in the area of the road, g_{max}, is then decisive when estimating protective measures (such as galleries) and the investment costs permitted. The probability of extent (the vulnerability), λ, is given as the probability that people are caught in the avalanche provided that the avalanche has occurred and that a car was in the avalanche track during that time.

The collective risk, R (fatalities/year), is then calculated as:

$$R = WDT\beta \sum_{i=1}^{n} \frac{g_i \lambda_i}{T_i \nu_i} \quad \text{(deaths per year)} \qquad (13.2)$$
$$(i = 1, \ldots, n \text{ avalanche paths})$$

The vulnerability for deaths in a vehicle ($\nu = 0.18$) and the mean number of occupants ($\beta = 1.61$; people/vehicle) are taken from statistical data. If a transport route is affected by more than one avalanche track, the risk per avalanche track can be approximated by simply adding the individual contributions under the assumption of small individual ratios of $g_i T_i^{-1}$. Risk peaks, such as those following an incident where a queue of vehicles is formed, require additional model assumptions [5, 6] and are therefore not discussed here.

The outset risk without any safety measures at the Flüela Pass of 0.7 fatalities per year only applies to moving vehicles. In order to make the pass road safer, controlled release of avalanches were used in combination with temporary road blocks between 1971 and 1999. In this way, an annual cost (K_a) of 0.22 million Swiss francs was incurred for artificial release, and on average 25 days of road closure per winter were necessary. The remaining risk was 20% of the initial risk. Despite the high number of closure days, this resulted in a very high (unacceptable) remaining risk of 0.14 deaths per year. We will show later that closure days are very expensive, so it can be concluded that this original risk management scheme was far from being optimised.

An evaluation of a large number of projects with galleries or permanent retaining structures along transport routes in Switzerland [6] suggested a social willingness to pay, WTP, an average cost of 10 million CHF and marginal costs of up to 40 million CHF in order to prevent one statistical fatality. The results can now be used for a cost-effectiveness analysis. In the following we try to quantitatively analyse the Flüela Pass situation.

Protection by Technical Measures Only

First we explore the situation where only technical measures of risk reduction are used. Figure 13.4 shows the curve for marginal costs up to complete risk reduction (remaining risk = 0) where risk is reduced by constructing galleries and dams [7]. The analysis is based on statistical avalanche occurrence data, traffic and current construction costs. Complete risk reduction of 0.7 deaths per year is reached with yearly costs of 10 million CHF, which translates into an average cost effectiveness of 14 million CHF per saved life (prevented fatality). According to Fig. 13.4, marginal costs (C_m) of less than five million CHF per saved life are to be expected if five individual measures are used (gallery constructions at certain locations). If the average accepted C_m value in Switzerland of 20 million is taken as the reference, 21 galleries need to be constructed. Employing the three most cost-effective measures alone reduces the outset risk by 25%.

Protection by Organisational Measures Only

In this section, the risk reduction seen when only organisational measures are taken is explored. Figure 13.5 presents the expected risk reduction as

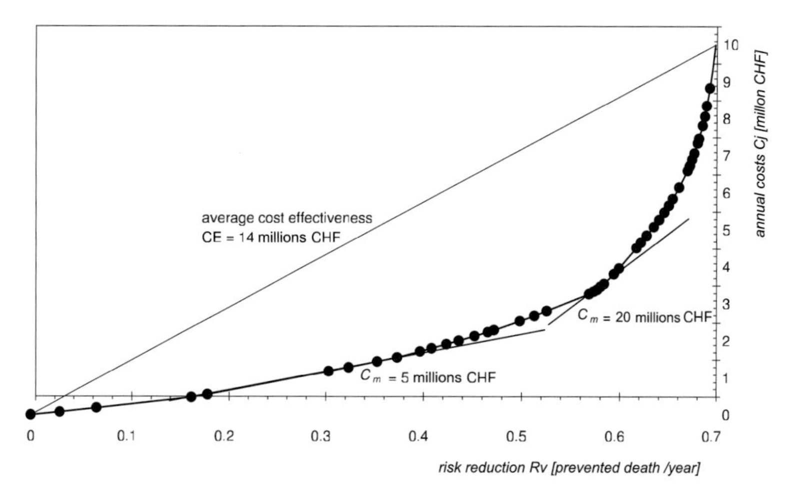

Fig. 13.4. Marginal costs and cost effectiveness of technical measures for a complete reduction of the outset risk in the Flüela pass example

a function of the closure days required. In order to achieve a full reduction of 0.7 fatalities per year (analogous to a full reduction by technical measures), 110 closure days per winter are necessary. These days have been arbitrarily combined into 25 closure units of 4.4 days each. A closure unit presents a unit organisational measure and is motivated by the fact that a dangerous avalanche situation usually persists for several days. As can be seen in Fig. 13.5, using the three most cost-effective closure units together reduces the outset risk by almost 50%.

Integral Protection by an Optimal Combination of Measures

According to Fig. 13.2, the aim is to find an optimal combination of measures to minimize the total cost, C_T. In order to do so, one additional difficulty needs to be overcome, namely to find a monetary representation of closure days. In this example, we work with the replacement cost approach: at the point of total risk reduction, 110 closure days correspond to a cost of 10 million CHF for technical measures. This results in a cost estimation of 0.09 million CHF per closure day. Note that an alternative method is discussed in the *Conclusions* and in [8]. The damage cost (C_S) only includes the fatality risk at a marginal cost of 20 million CHF and it neglects the costs from possible injuries or infrastructure damage.

Figure 13.6 shows the cost functions when the technical and organisational measures are combined in an optimal way. Because we now have now also defined monetary damage costs, we can calculate the optimal protection solution, which has a total cost C_T of approximately six million CHF per year and a risk reduction of 90% of the outset risk. This optimal level of

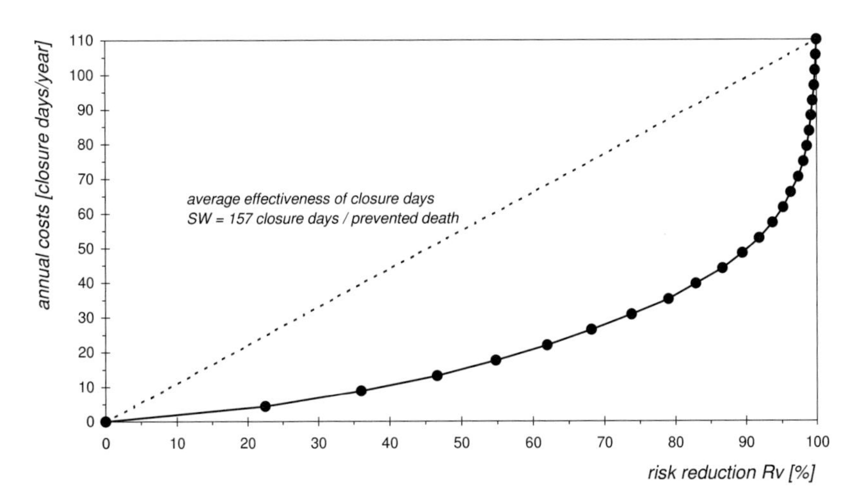

Fig. 13.5. Marginal costs (closure days) and cost effectiveness of organisational measures for a complete reduction of the outset risk in the Flüela pass example

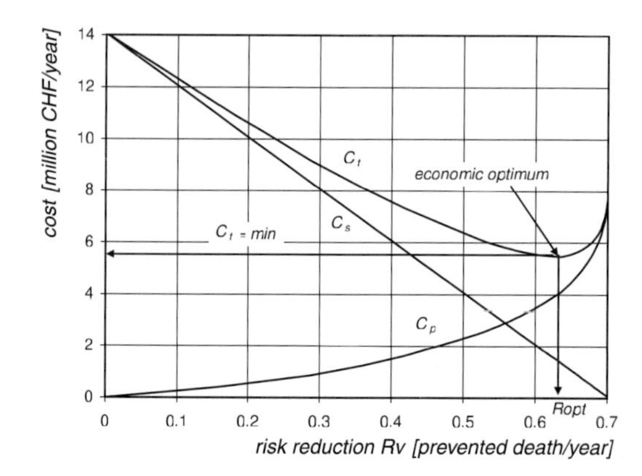

Fig. 13.6. Total cost minimisation with optimal combination of technical and organisational measures in the Flüela pass example

security only requires 13 closure days and 13 galleries. The yearly cost of the measures is 1.6 million CHF, which is significantly higher than the cost incurred during the years of opening. With the high cost and high impact of closure days, it is very important to have a reliable and objective method of invoking organisational methods.

13.3 Physical Modelling of Alpine Surface Processes to Support Natural Hazard Forecasting

The risk management approach discussed above shows that organisational measures are a very valuable addition to permanent prevention strategies such as engineering structures. Because they can be applied in a very flexible way, they are also the method of choice for Xevents. In the past, the full potential of organisational measures has not been exploited. This is partly because they require a reliable forecast of danger. In addition to reliability, objectivity is required because a potential failure to forecast a dangerous situation will certainly have legal consequences. Even with hypothetical perfect forecasting, accidents will happen from time to time since organisational measures work with an accepted and hopefully acceptable remaining risk.

In the reminder of this contribution we therefore explore the potential ability of physical modelling to provide an objective and reliable basis for forecasting natural hazards. This exploration is motivated by the fact that after many decades of development, numerical weather prediction models are now in a state that they can provide just such a reliable and objective method for weather prediction. Forecast models for alpine natural hazards have not reached that state yet. However, recent improvements in our understanding of natural hazard generation suggest that this will be possible in the future.

The primary cause of natural alpine hazards is meteorological forcing, especially precipitation. Abundant snow precipitation leads to avalanches and heavy rain leads to mud flows and flooding. However, knowledge of meteorological forcing alone is not sufficient to determine the danger. Many attempts to statistically link natural hazards to meteorology alone have shown only limited success [9]. The current condition of the snow or soil cannot be neglected. For example, an unstable snow cover will only need a small additional load to produce an avalanche, and soil and snow moisture are important factors in determining slope stability with respect to mud flows or the runoff response of creeks and rivers. Figure 13.7 is an impressive illustration that how deeply a weak layer is buried (and thus how big a potential avalanche might become) is also very important. Only if the complex interaction processes between the atmosphere and the surface are adequately represented is it possible to use physical modelling to support natural hazard forecasting. The task remains challenging because it requires modelling single snow grains in order to predict the avalanche danger of a whole area. In terms of runoff generation for alpine catchments, the complex and small-scale interaction processes between atmospheric forcing, snow cover, glaciers, vegetation and soil need to be considered. Comparing again to meteorological forecasting, an important observation can be made: while large-scale processes are now predictable with high accuracy, small-scale convective events are still not predicted satisfactorily. Forecasting the dangers from avalanches and floods requires the modelling of even smaller processes than those used in meteorology. In the following we will discuss the most important processes and present results from snow stability estimations and runoff predictions using a high resolution physical model.

Fig. 13.7. Photograph of the fracture line of a massive slab avalanche

13.3.1 Summary of Alpine Surface Processes

Figure 13.8 schematically shows the complex processes that occur in the atmosphere–snow–soil system. Precipitation can occur as snow, rain, graupel, hail or rime. With sufficient wind, snow will start to drift. In complex terrain, this results in irregular snow deposition with maximum snow depths of up to ten times the average snow depth. The wind will also influence the snow crystals and might form a hard crust at the snow surface. Vegetation (if present) will alter the surface water balance considerably through interception, unloading and evapo-transpiration. As soon as the snow is on the ground, the snow settles and water vapour fluxes cause the snow crystals to change. This change is also called metamorphism and is heavily influenced by the snow energy balance. From the ground, the snow cover receives a small but constant flux of energy. The exchange of energy with the atmosphere is much more intense. During daytime, the snow cover absorbs shortwave radiation. Energy is usually lost to the atmosphere by longwave radiation. The turbulent fluxes of heat and moisture can bring or remove energy and mass (latent heat only) from the snow cover. On clear nights, surface hoar is often formed by moisture sublimation on the snow surface, which then creates a dangerously weak layer. All of these processes change rapidly in space and time and interact with the topography. As soon as enough energy has entered the snow, it starts to melt and produce water. This will lead to a large change in the structure of the snow and to the possible formation of wet snow avalanches. A refreezing event will however then produce a stable snow cover. The melt water is first stored in the snow pores but then starts to percolate downwards, often along "preferential flow paths".

The complex processes continue in the soil. Depending on the soil grain size and soil history, more or less water can be stored in the pores. In springtime or after heavy precipitation, the soil may already be saturated and any

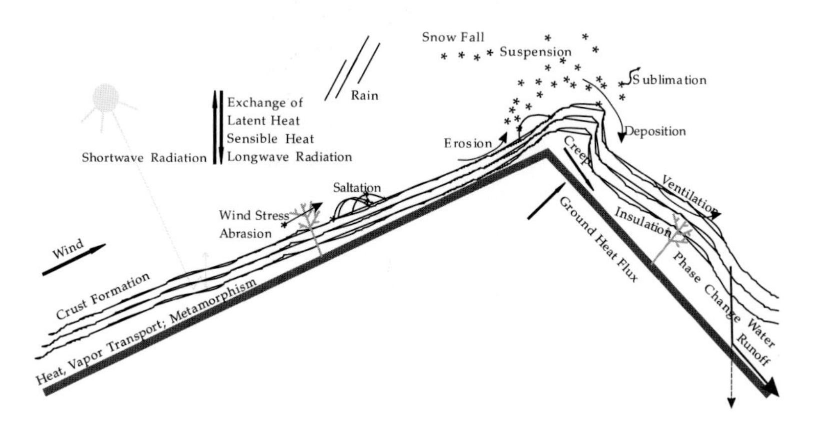

Fig. 13.8. Schematic representation of alpine surface processes as modelled in the ALPINE3D and SNOWPACK models

water entering the soil will immediately produce surface runoff. The dynamical water transport and storage behaviour is altered by the layered structure of soil and bedrock, which may contain impermeable layers such as clay or very permeable bedrock such as karstic limestone.

These processes are implemented in the alpine surface modelling system ALPINE3D. The model system uses a meteorological model to create atmospheric forcing fields such as air temperature, humidity, wind, radiation and precipitation at a very high spatial resolution of (currently) 25 m. The input is refined by including terrain effects such as shading, surface reflections and terrain emissions in the surface radiation balance. In a next step, a module calculates drifting snow and predicts snow redistribution. The combined atmospheric input is calculated on a full three-dimensional numerical grid. At every grid point, vegetation, snow and soil processes are then simulated. The runoff is processed via a combination of linear reservoirs.

13.3.2 Estimating Snow Cover Development and Snow Stability

The French SAFRAN-CROCUS-MEPRA (SCM) chain [10] led the way to operational snow cover modelling and combined the results from a snow cover simulation with the expert system MEPRA [11], giving an interpretation of the stabilities of the model profiles. MEPRA is based on a classical stability index approach [12], which is combined with a set of rules to evaluate the profile in terms of stability classes. Since the simulated snow covers are placed on hypothetical pyramids for a range of altitudes and expositions [10], this method has the potential to relate individual snow profiles to the local or regional danger from avalanches.

We discuss here a more direct approach, in which a stability index is derived directly from snow cover simulations driven by meteorological data from automatic stations. Our contribution explores the link between individual snow profiles and avalanche danger by using stability criteria applied to modelled snow profiles. The Swiss snow cover model SNOWPACK [13–15] is successfully used to assess new snow precipitation, drifting snow and snow cover development at (currently) approximately 100 automatic weather stations in the high alpine zone of the Swiss Alps. A good prediction of snow metamorphism and surface hoar formation [15] allows the simulation of weak layer development with reasonable accuracy. Figure 13.9 shows a comparison between a modelled and a simulated grain profile for the winter of 1999. The meteorological data from the Weissfluhjoch Versuchsfeld station has been used to predict the time development of the snow cover. A colour code is used for grain types. Many thin layers, which represent potential weak layers in the snow cover, are present in the simulation as well as in the observed profile. However, this only suggests that the model can reproduce the local snow cover development. Another question is whether knowledge of the local snow cover is useful when predicting the average slope stability in that region.

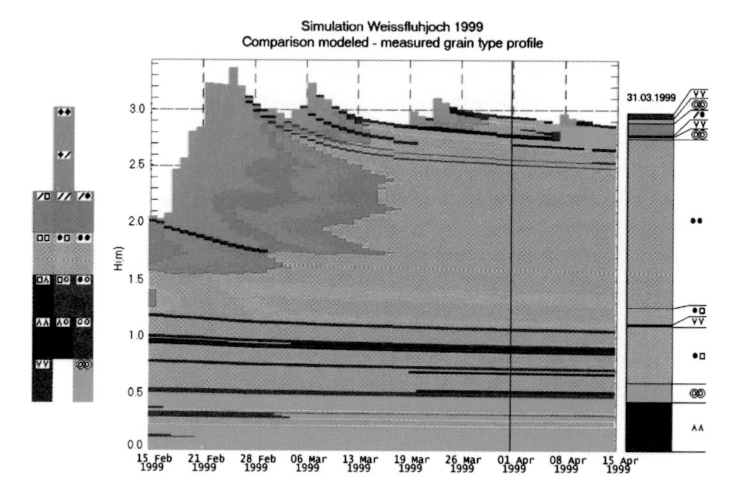

Fig. 13.9. Development of SNOWPACK-predicted grain types for the winter of 1999, and comparison with the profile observed on 31st March

To this end, we now try to calculate an estimate for the stability from the modelled snow profile. More detail on the stability formulations can be found in [16]. Here we briefly discuss the development of the avalanche danger compared to the development of the stability index for the winter of 2002/2003. As a first and simple measure of correlation between the stability index and the regional avalanche danger, we look at the coefficient of determination (r^2) from a linear regression. A first regression of the skier stability index at the station Klosters Gatschiefer yields an r^2 of 20%. This means that the stability index at a single (well selected) location already explains 20% of the total variation in estimated avalanche danger during three months of the winter of 2002/03. In [16] it is further hypothesized that the correlation between the regional avalanche danger and the SNOWPACK stability index can be improved by including more than just one location for the simulation. Therefore, we now consider two locations with automatic stations (Klosters Gatschiefer and Klosters Madrisa) and simply average the stability indices. This increases r^2 significantly and we can already explain 25% of the total variation in the hazard level. Note that Klosters Madrisa alone yields an r^2 of 20%. A slight improvement in r^2 to 27% can be obtained by including slope simulations for the two locations.

Figure 13.10 gives the time series representation of the average skier stability index for the two Klosters locations. The avalanche danger level valid for the area is plotted on the same graph. Note that the stability index is plotted on an inverse axis to allow a better visual comparison to the stability index. Ideally, changes in the stability index should predict changes in the hazard level. As discussed above using the linear coefficient of determination,

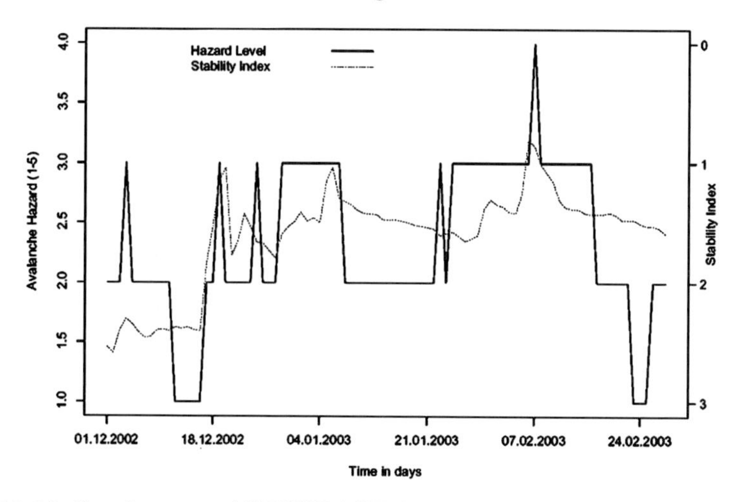

Fig. 13.10. Development of SNOWPACK skier stability index and avalanche hazard level. The stability index is the average of the two locations Klosters Gatschiefer and Klosters Madrisa and explains 27% of the variation in the danger level

this is not always the case. However, the most dangerous situation, on 7th February, is particularly well represented by the stability index.

13.3.3 Improvement in Extreme Runoff Forecasts from Alpine Catchments

Runoff forecasts based on simple and highly parameterised models have been very successful for gauged catchments. The simple models are often based on day-degree methods (see [17]) for snow-melt dynamics and only pass the combined input from precipitation and melting snow through a series of linear reservoirs [18]. The successful reproduction of runoff with those models is based on extensive calibration. Such a procedure is problematic if real Xevents need to be forecasted or if such a model is to be used for ungauged basins, for which no calibration is possible. We compare the performance of such a well calibrated model (PREVAH, [19]) with an uncalibrated version of ALPINE3D. Figure 13.11 shows a comparison between October 1990 and March 1992. The calibrated PREVAH model is better at predicting the base flow. This is reflected in the higher efficiency coefficient [19] $E2$ of 0.90 when compared to ALPINE3D's 0.82. Since our main interest is in Xevents, we now focus on the snowmelt month June/July 1991, when the spring runoff produces the highest runoff peaks. For this situation, Fig. 13.12 again compares PREVAH (Fig. 13.12a) and ALPINE3D (Fig. 13.12b) to the measured runoff. Since we mainly want to compare the dynamic response of the uncalibrated ALPINE3D model to the calibrated PREVAH model, we now again use the coefficient of determination instead of the efficiency coefficient, which

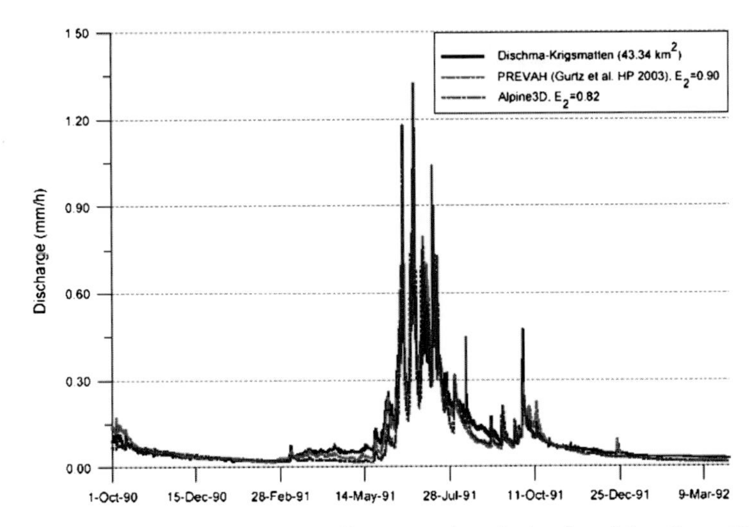

Fig. 13.11. Dischma seasonal runoff measured and simulated by the calibrated PREVAH model and the uncalibrated ALPINE3D model

overstates the influence of the base flow. For the critical spring melt situation, ALPINE3D already represents the runoff dynamics better, as indicated by the higher r^2 value of 90% versus 83%. That this result is really achieved by including the physical processes in the model in great detail can be shown by repeating the simulation without some of the process descriptions. If we switch off the modules for vegetation, radiation balance, interpolation of the meteorological variables in ALPINE3D and additionally only work with a uniform soil representation, the r^2 value goes down to 46%. The corresponding runoff curves are presented in Fig. 13.12c. We can see that the highly decreased correlation comes from the fact that the relative magnitude of the three main runoff events is not well represented now. While the first event is overestimated, the last event is severely underestimated.

13.4 Conclusions

By applying methods of integral risk management, we have shown that, for a mountain pass road (as an example), organisational measures based on hazard forecasts can be highly cost-effective when compared to more conventional protection measures. We have further pointed out that organisational measures are also flexible enough that they should be applied for Xevents. In this context, we have also defined a practical approach for how to use integral risk management for natural hazards in alpine surroundings. One reason why the potential of organisational measures is still not exploited sufficiently is the fact that organisational measures need accurate and objective hazard

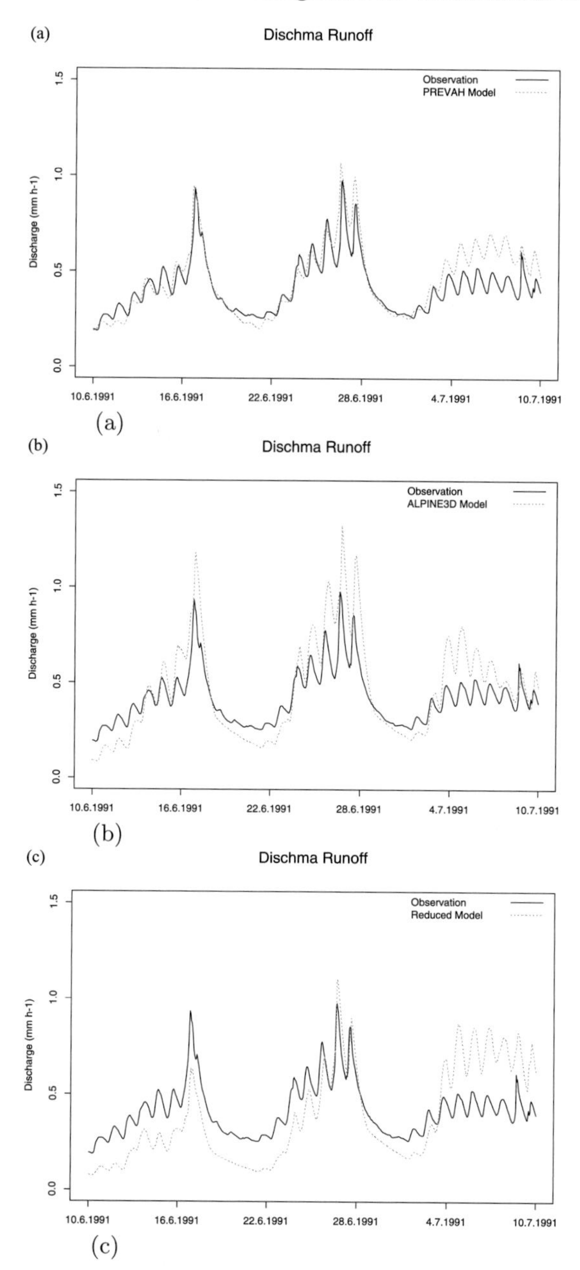

Fig. 13.12. Comparison of Dischma runoff simulation for the peak runoff time for the PREVAH model (**a**) and the ALPINE3D model (**b,c**). Using the full ALPINE3D model (**b**) results in a better agreement than that obtained for a reduced model configuration (**c**)

forecasts. Using the two examples of avalanche danger and runoff prediction, we tried to show that physical modelling has the potential to provide the basis for such high quality and objective hazard forecasts. While current forecasts of natural alpine hazards are mainly performed without significant model support, our increasing understanding of the physical processes involved and increasing computer power should open the door to the increased use of physical modelling within the framework of integral and sustainable risk management. It is also important to point out that much more work is required to reach that goal. The examples discussed above clearly show that, at present, forecasts of natural alpine hazards cannot be exclusively based on model simulations. While this may become possible in the future, human judgement and interpretation will remain crucial for the near future.

The basic approach and underlying principle of integral risk management, as presented here for the example of a mountain pass road, is compelling. More discussions are required on the details of the implementation, however. For example, there are several ways to determine the monetary cost of road closure days. Probably the most adequate would be to determine the willingness of potential users to pay. In our example we have worked with the replacement cost approach, which certainly gives other results. The replacement cost approach has been easy to implement for our example but suffers from the fact it is based on a linearisation of the full cost development of technical measures. In our example, we also had to assume monetary values for saved human lives or avoided injuries, as suggested by the Swiss PLANAT strategy (www.planat.ch). As methods of integral risk management become more and more widespread, broader discussion in society will lead to a broader basis for these values. The same applies to physical modelling. When the need for and the benefit of physical models become accepted in this field and more and more work is done on physical modelling, their contribution to natural hazard forecasting will increase.

Acknowledgement. This work has been accomplished with the help and support of many colleagues. We thank in particular Tuan Anh Nguyen, Massimiliano Zappa, Charles Fierz, Thomas Stucki, Manfred Sthli, Schagg Rhyner and Walter Ammann.

References

1. Ammann, W.: Schnee und Lawinen: Bestimmende Wirtschaftsfaktoren im Alpenraum. In: Le rôle de l'eau dans le développement socioéconomique des Alpes. Universit alpine d'été, Institut Kurt Boesch, Sion, 15 S, (1998)
2. Kaplan, S. and J.B. Garrick: On the quantitative definition of risk. Risk Anal. **1**(1), pp 11–27 (1981)
3. Starr, C.: Social benefit versus technological risk. Science **13**, pp 1232–1238 (1969)

4. Wilhelm, C.: Quantitative risk analysis for evaluation of avalanche protection projects. In: Proc. Int. Conf. for Snow Avalanche Res. Norwegian Geotechnical Institute, Oslo, Publ. **203**, pp 275–293 (1998)
5. Schärer, P: The avalanche-hazard index. Ann. Glaciol. **13**, pp 241–257 (1989)
6. Wilhelm, C.: Wirtschaftlichkeit im Lawinenschutz. Methodik und Erhebungen zur Beurteilung von Schutzmassnahmen mittels quantitativer Risikoanalyse und konomischer Bewertung. Mitt. Eidgenss. Inst. Schnee-Lawinenforsch. **54**, pp 309–309 (1997)
7. Wilhelm, C.: Kosten-Wirksamkeit von Lawinenschutzmassnahmen an Verkehrsachsen. Vollzug Umwelt, Praxishilfe. Bundesamt für Umwelt, Wald und Landschaft (BUWAL), Bern, pp 110–110 (1999)
8. Nöthiger, C.: Naturgefahren und Tourismus in den Alpen. Untersucht am Lawinenwinter 1999 in der Schweiz. WSL, Eidg. Institut für Schnee- und Lawinenforschung (SLF), Davos, p 245 (2003)
9. Brabec, B., R. Meister, U. Stöckli, A. Stoffel, T. Stucki: RAIFOS: Regional Avalanche Information and Forecasting System. Cold Reg. Sci. Technol. **33**, pp 303–311 (2001)
10. Durand, Y., G. Giraud, E. Brun, L. Mrindol, E. Martin: A computer-based system simulating snowpack structures as a tool for regional avalanche forecasting. J. Glaciol. **45**(151), pp 469–484 (1999)
11. Giraud, G., J. P. Navarre: MEPRA et le risque de dclenchement accidentel d'avalanches. In: Proc. Int. Symp.: Science and mountain – The contribution of scientific research to snow, ice and avalanche safety, ANENA, Chamonix, France, pp 145–150 (May 30–June 3, 1995)
12. Roch, A: Les variations de la résistance de la neige. AIHS Publ. **69**, pp 86–99 (1966)
13. Bartelt, P., M. Lehning: A physical SNOWPACK model for the Swiss avalanche warning. In: Handbook of sensory physiology, Part I: numerical model. Cold Reg. Sci. Technol. **35**(3), pp 123–145 (2002)
14. Lehning, M., P. Bartelt, B. Brown, C. Fierz: A physical SNOWPACK model for the Swiss avalanche warning; Part III: meteorological forcing, thin layer formation and evaluation. Cold Reg. Sci. Technol. **35**(3), pp 169–184 (2002)
15. Lehning, M., P. Bartelt, B. Brown, C. Fierz, P. Satyawali: A physical SNOWPACK model for the Swiss avalanche warning; Part II. Snow microstructure. Cold Reg. Sci. Technol. **35**(3), pp 147–167 (2002)
16. Lehning, M., C. Fierz, B. Brown, B. Jamieson: Modelling instability for the snow cover model SNOWPACK. Ann. Glaciol. **38**(1), pp 331–338 (2004)
17. Martinec, J., A. Rango: Parameter values for snowmelt runoff modelling. J. Hydrol. **84**, pp 197–219 (1986)
18. Gurtz, J., A. Baltensweiler, H. Lang: Spatially distributed hydrotope-based modelling of evapotranspiration and runoff in mountainous basins. Hydrol. Process. **13**, pp 2751–2768 (1999)
19. Zappa, M., F. Pos, U. Strasser, P. Warmerdam, J. Gurtz: Seasonal water balance of an Alpine catchment as evaluated by different methods for spatially distributed snowmelt modelling. Nordic Hydrol. **34**, pp 179–202 (2003)

14 Prevention of Surprise

Zuzana Chladná[1], Elena Moltchanova[2], and Michael Obersteiner[3]

Summary. Today there is common agreement that human actions are resulting in increasingly large-scale – even global – risks. Yet there seems to be a universal inability to stop these human, environmental and economic effects. In this chapter we consider the management of surprise in the framework of a wide spectrum of hazard levels. For instance, a reduction in greenhouse gases might reduce the probability of extreme climate changes. We have developed a general model for controlling extreme hazards. We first examine the dynamic behavior of a single global society and derive various optimal response strategies to counter the hazard. However, in real life such a global hazard management system does not exist due to a lack of international cooperation among nation states. A gaming model is constructed to elaborate the implications of hazard management when more nations are involved, and when expectations about the hazard are imperfect. While the models involved in this analysis are simple, the results from our numerical experiments are instructive and yield interesting insights into the economics of various institutions governing the interaction of societies and their capacity to mitigate risks. We discuss the outcome of the models in terms of its bearing on modern politics as well as what it might mean to the dangers that await us in the future.

14.1 Introduction

The large-scale disasters of the past few years – such as the recent hurricanes, unusually extensive flooding, devastating bushfires, violent ice storms in many parts of the world, as well as the emergence of previously unknown infectious diseases – have brought home to governments the realization that something new is happening to our global society. Such mega-risks have the potential to inflict considerable damage on the vital systems and infrastructures upon which our societies and economies depend, and create serious difficulties for traditional risk management and risk-sharing actors, such as the insurance industry. Preparing to deal effectively, in an anticipative and cooperative manner, with the hugely complex threats of the twenty-first century is a major challenge for decision makers in government and the private sector alike. In today's world, mankind is confronted with the following questions related to the disaster management process:

1. How do we deal with the large uncertainties and knowledge gaps about the hazards we are creating in a globalized world?

2. What are the best and/or sufficiently robust response strategies to control the hazards we create?
3. How can we design global institutions to implement these preventive measures?

However, we can only currently assess the hazard process imperfectly, and even the evolutions of the basic drivers of global hazards can only be poorly predicted. Today, we can observe three large-scale processes acting as drivers of global risks and opportunities: (1) deep global integration of mainly economic actor networks and intensification of interaction around the globe; (2) a transformation in humanity's relationship with its life-supporting systems due to new technological capabilities, and (3) the ever-increasing number of people occupying space and depleting resources due to increased consumption. These processes can be regarded as an exceptional confluence of three sets of powerful changes that give rise to self-reinforcing mechanisms of economic growth and concomitantly increasing hazard levels. It is precisely the reinforcing relationship of the basic drivers that are present during the Anthropocene that alters the nature of the risks that we have historically experienced. Risks can no longer be regarded as exogenous to human actions; on the contrary, the mega-risks of the twenty-first century are endogenously "produced" by human action. The central focus of this paper is that most of the systematic risks of the twenty-first century are new and socially produced. Yet, we may be unaware of some of these major hazards socially due to knowledge gaps (ignorance), uncertainty or social discounting of these endogenously produced risks. Discounting arises from the fact that humans do not communicate with their environment per se, but with a self-created image of that environment. This gives rise to imperfect assessments of risks and insufficient individual responses to danger.

Navigating and managing under deep uncertainty is the principal challenge to global governance. Differences in the urgency to react to global hazards as well as differences in the economic actions performed to lower risks are the main drivers of many international negotiations such as those around climate change. The question of whether to halt or delay measures to manage global risks is related to uncertainty management. Uncertainty is, in many cases, used to postpone actions. Thus, the biggest anxiety for conducting government operations is anticipating future conditions as perfectly as possible and raising the awareness of major hazards. It seems remarkable how infrequently the problem of managing under deep uncertainty is taken up in daily political discourse. One reason might be that negotiators prefer certainty in the prediction of outcomes and flexibility in terms of choices of outcomes. However, in the presence of continuous technological, social and environmental change, uncertainty about the global mega-risks we create is a defining feature of the "socially" absent character of most hazards. Our lack of knowledge about future outcomes is not a failure of due diligence on our part. Rather, it is an inherent outcome of the biophysical as well as social process of global change. Many worldwide exercises in foresight or forecasting studies, quantitative integrated assessment studies and scenario exercises either: (1) fail to acknowledge this inherent uncertainty, as the individual predictions are deterministic, or (2) are believed to be conducted to explore the uncertainty bounds and therefore are perceived as being exercises to reduce uncertainty. Investments in future studies seduce decision-makers with an illusion of control and rationality which undercut the effectiveness of hazard management under conditions of uncertain large-scale risks. Prediction models are constructed around consistency with historical events and fail to account for surprises that necessarily arise from systems that were perturbed in an unprecedented manner. For example, atmospheric greenhouse gas concentrations have increased to unprecedented levels at an unprecedented speed. Both climatologists and decision makers would be astonished to see the future actually unfold in the manner forecast by climate models. Thus, to support good decision-making, the issue is not one of being able to predict the unpredictable. Rather, the fundamental question is that, given that we cannot have reliable predictions of future outcomes, how can we prevent excessive hazard levels today and in the future in a cost-effective manner? A new decision-making framework is needed to address uncertain endogenous risks.

One of the prevailing approaches to uncertainty is adaptive management [7, 8]. Adaptive management is based upon the premise that the managed system is complex and inherently unpredictable. Adaptive management accepts the uncertainty that exists in the real world rather than ignoring it. Consequently adaptive management views management actions as experiments rather than solutions. That is, they craft plans through processes that are less deductive than inductive. Adaptive management is a structured process that reduces the costs of management experiments while increasing opportunities for social, technological and scientific learning. It is argued that in the face of deep uncertainty, policymakers (and humanity in general) should operate in just such an adaptive manner in order to benefit from the learning process. The theory advocates postponing actions if, for example, through technological advances, risk mitigation measures can be employed more effectively in the future. It is argued that adaptive solutions are likely to be robust across a wide range of alternative plausible outcomes. That is, the tactic is often not so much to maximize behavior conditioned for a particular set of circumstances, but to select one among a set of "good enough" actions that is most likely to remain good enough across a wide range of plausible outcomes. However, in situations where decision makers are confronted with mega-risks that result in unpredictable outcomes from perturbed complex systems, adaptive management might turn out to be a losing strategy, as learning is an endeavor that involves too much risk. In situations where "you only die once", a precautionary approach to risk is preferable. Thus, the approach to risk management crucially depends on the hazard's properties (incremental vs. abrupt, mega- vs. nano-scale, frequent vs. infrequent), our knowledge about the risk, how efficiently we can mitigate the risk, and how effectively the mitigation process is socially organized. In this chapter we will investigate these factors systematically and in more detail with the help of two related models.

The main point of this chapter is to elaborate on the framework conditions societies establish to prevent surprise rather than concentrating on the more technical and technological challenges of protective measures. In the following sections we will first present a dynamic model of endogenous risk. The model will help us to identify four strategy categories in a single global society setting of perfect knowledge about the endogenous risk. Then, using a static model we investigate the effects of different degrees of cooperation among societies and imperfect knowledge on optimally managing the endogenous risk. Finally, in the discussion we will put the two models in perspective and derive policy conclusions for managing large-scale endogenous risks that we face in the twenty-first century.

14.2 Dynamic Model

In order to examine the choice of the optimal decision strategy used under the threat of a catastrophe, we introduce a modified discrete version of the neoclassical macroeconomic model. Its key variation is not only due to the presence of an Xevent, but arises mainly from the endogenous probability that drives the event occurrence.

We consider an economy, which we call a *society*, solving a problem of optimal resource allocation. We assume that the initial value of the capital, K_0, is known. At the beginning of each period i a social planner has to distribute the current production Y_i between consumption C_i, capital investment I_i, and mitigation investment M_i:

$$Y_i = C_i + M_i + I_i \ . \qquad (14.1)$$

Naturally, the consumption, capital investment and mitigation investment must not be negative.

The choice of these three *decision variables* determines a *state variable* – the amount of capital in the next period, K_i. In the case of no Xevent, its value is only a sum of the capital from the previous period depreciated by the rate δ and the current capital investment. However, if catastrophe occurs in the given period, the level of the capital is reduced due to the losses caused by the event.

Consequently, at the end of the period, a new production Y_{i+1} is generated. Its level is determined by the simplified Cobb-Douglas production function with parameter γ assuming the capital to be the only input fed to the production process.

More formally, the amount of capital and the production in period i can be expressed as:

$$K_i = ((1-\delta)K_{i-1} + I_i)\, D_i$$
$$Y_{i+1} = K_i^{\gamma}\ .$$

Here D_i represents the proportional degradation of wealth: if a catastrophic event occurs, its value will be $1-d$, where d is the size of the damage. If no catastrophic event occurs, its value will be 1. In other words,

$$D_i = \begin{cases} 1-d & \text{with probability } p_i\ , \\ 1 & \text{with probability } 1-p_i\ . \end{cases}$$

We assume that the initial probability of the Xevent p_0 is known. Future evolution of the probability is endogenous, namely the probability of the catastrophe decreases as the mitigation investment M_i increases. Moreover, we assume that the extent of this reduction depends on two further parameters: φ, ω and on the previous probability state p_{i-1}. The parameter φ – *the mitigation investment efficiency* – determines the extent to which the mitigation investment influences the probability of the Xevent. The parameter ω expresses the natural deterioration induced by human activity.[1]

A social planner maximizes the expected utility over an infinite time horizon:

$$\max_{C_i, M_i} E_0 \left(\sum_{i=0}^{\infty} \frac{1}{(1+\rho)^i} U(C_i) \right) .$$

In the above expression, ρ is the discount factor and E_i is the conditional expectation subject to time i. For simplicity we employ a logarithmic utility: $U(C_i) = \ln C_i$.

The model that we have just proposed belongs to the category of dynamic programming problems. However, the endogenous probability makes the problem more complex. In order to keep the computational complexity at the lowest possible level, we focus on the three periods model only, instead of dealing with an infinite horizon.

One of the traditional approaches for dealing with a multi-period problem is to use Bellman's optimality principle and to solve the model backwards.[2] Basically, this means to first consider what the optimal solution for the last period will be and then work backwards to determine which decision for the initial period is optimal with respect to the coming periods.

---

[1] Here the function $p_i(M_i, p_i - 1) = \frac{\omega}{1+\varphi M_i} p_{i-1}$, where $\varphi > 0$, $\omega > 1$, has been used. However, other functions may be considered; see Sect. 14.3 for example.

[2] More about Bellman's principle and the computational methods can be found in [1].

Following this recursive algorithm, we solve the proposed three periods model in two steps. We start by analyzing the last period first and then move back to the first period.

In our model, in order to find the solution for the last period we need to solve the following subproblem:

$$\max_{C_2, M_2} E_1 \left(U(C_2) + \frac{1}{(1+\rho)} U(C_3) \right)$$

subject to

$$\begin{aligned} K_1^\gamma &= C_2 + M_2 + I_2 \\ K_2 &= ((1-\delta)K_1 + I_2)\, D_2 \\ C_3 &= f_3(K_2) \\ C_2 &\geq 0\,, \quad M_2 \geq 0\,, \quad I_2 \geq 0\,. \end{aligned}$$

The last equation is actually a terminal condition. Such a condition is necessary if we deal with a finite horizon problem. Here we set $f_3(K_2) = K_2^\gamma$. This means that in the last period the whole production is consumed. As a result we obtain the optimum values for C_2 and M_2 as functions of K_1 and p_1. Note that because of the boundary conditions, $C_2(p_1, K_1)$ and $M_2(p_1, K_1)$ are piecewise functions,[3] which increases the computational complexity.

After finding the optimal solution for the second period we proceed with the first period using the functions $C_2(p_1, K_1)$ and $M_2(p_1, K_1)$. In this way, we obtain the optimal values for consumption C_1, the capital investment I_1 and the mitigation investment M_1.

14.2.1 Possible Strategies

We focus on the optimal solution for the first period because the decision for the second period can be biased by the presence of the terminal condition.

For any set of initial parameters, the optimal strategy belongs to one of the following classes:

1. *Tactical approach:* $C_1 > 0$, $M_1 > 0$, $I_1 > 0$. As we have already mentioned in the *Introduction*, there are basically two ways in which the society can deal with catastrophic events:
   - A social planner decides to invest in capital in order to increase production in the future. Since the damage is proportional to the current state of the capital, even if a catastrophic event occurs, the rest of the wealth will still be higher than in the case of no current capital investment.
   - The social planner can decrease the risk of a loss by investing more in the mitigation (preventing society from suffering by decreasing the probability of the Xevent).

[3] To be more precise, there are five different cases which have to be considered.

Both ways apply within this approach.

2. *Ignorance approach:* $C_1 > 0$, $M_1 = 0$, $I_1 > 0$. A society ignores the presence of a catastrophe entirely. One reason for such an optimal strategy might be the relatively low mitigation efficiency in comparison with the higher effect of the capital investment on the overall expected utility.
3. *Panic behavior:* $C_1 > 0$, $M_1 > 0$, $I_1 = 0$. Such a behavior arises in the situation when the catastrophic event occurrence is perceived as highly probable – mainly because of the high chance of damage and the resulting high decrease in the consumption level. On the other hand, this strategy must be supported by the belief that there are enough resources and facilities to significantly change this unfavorable situation. As a result, the social planner decides to invest in mitigation at the expense of making capital investment. However, it is unlikely that this decision will lead to drastic decay in current consumption.
4. *"Enjoy life" approach:* $C_1 = K_0^\gamma$, $M_1 = 0$, $I_1 = 0$. A society recognizes that it does not have enough resources or capabilities to influence the outcome of its "game with Mother Nature". Neither capital nor mitigation investment (as described for the tactical approach) lead to a significant improvement. Therefore, the social planner decides to surrender and to "eat, drink and be merry" by spending all available resources today.

14.2.2 Numerical Results

If a doctor told you that there is a 10% probability of you catching flu, how much would this information bother you? Would your reaction be different if the doctor spoke of a heart attack instead? Would you change your lifestyle immediately? How drastic would the change be? We can ask very similar questions when discussing the field of catastrophic events.

Our main objective is to demonstrate how the anticipation of the catastrophic event influences the basic macroeconomic decisions. In this section we will use several examples to illustrate how and in which direction such anticipation drives the social planner's optimal strategy choice.

Parameters

The numerical computation requires us to set up the initial values for K_0, p_0 and for the model parameters: ρ, δ, γ, φ, ω and d. For the first three parameters (the discount, the depreciation and the production factor, respectively) we employ the usual macroeconomic values.

Recall that our attention is focused on the endogenous probability of the catastrophic event as a tool for managing the future. How effective the mitigation investment is depends on the *assumed* mitigation efficiency parameter φ of the society. Here we will consider three representative situations of very high, high and low mitigation efficiency. Under very high efficiency the society has the ability to decrease the probability of the catastrophic event from the initial level of say 50% to a subsequent level of 30% by investing only 10% of its current capital into mitigation. Under high mitigation efficiency, we mean the chance to decrease that probability to a level of 35% by the same mitigation investment, and finally a low efficiency means that we describe the assumption that the mitigation investment of 10% of the current capital will reduce the discussed probability to the level of 45%.

The formulation of the model allows us to deal with a wider class of catastrophic events; for example, we can consider discrepancies in the scale of the damage caused by the catastrophe. For the damage size parameter d, we use a similar categorization to that we used for the mitigation investment efficiency: *very high* damage corresponds to capital losses of 90%, *high* damage causes 50% losses, and *low* damage results in 10% of the capital being lost.

As the results show, one of the key determinants of the optimal strategy is the initial probability value. Usually, we will demonstrate how the distinct beliefs about the current probability state might lead to completely different decisions for an entire spectrum of probability values.

The last factor, which we took into account as a controller for the optimal strategy, is the initial capital value K_0. Mainly, we will deal with a society that possesses either a *high* or a *low* initial capital level.

Structure of the Results

Figure 14.1 illustrates how an optimal strategy might change if different initial probability values are employed. This particular example depicts a situation where a society faces very high damage and its expected mitigation efficiency is low. Two significant points can be observed in this figure: The first one

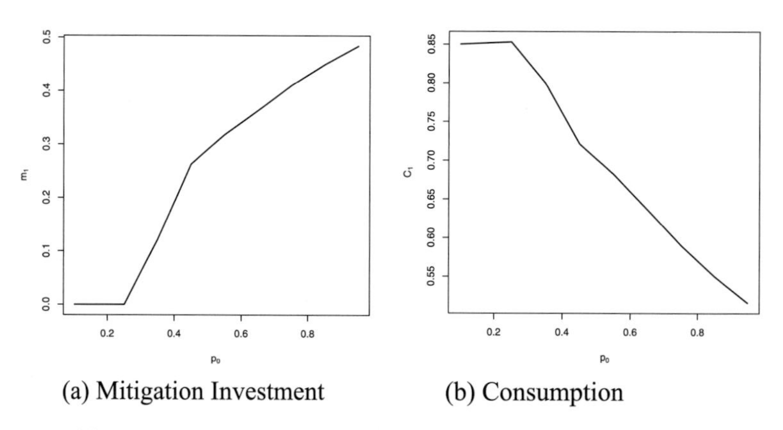

Fig. 14.1. Mitigation investment M_1 differs according to different beliefs about the initial catastrophic event probability (in this figure the initial capital level is "low", the damage size is "very high", and the value of mitigation investment efficiency is "low")

(for p_0 close to 25%) corresponds to the shift from "ignorance" to "tactical" behavior. The second one (near 50%) is induced by the switch from "tactical" to "panic" behavior.

We can interpret this as follows. In the case of a low disaster probability, the society ignores the option to mitigate despite the very high potential damage. The low mitigation efficiency naturally enhances this decision. However, if the initial probability increases, the perception of the potential large-scale damage becomes stronger, which results in a "tactical" approach. That is, the society's mitigation investment becomes positive at the expense of current consumption. The last part of the curves represents the response to the situation where a society faces both of the negative effects mentioned: very high damage together with high probability of a catastrophe. In this case panic begins: mitigation investment achieves higher priority than capital investment. Moreover, the society is willing to cut the current consumption in order to prevent a really apocalyptic situation.

Mitigation Investment Efficiency

Recall that a society is able to influence the probability of the catastrophic event by investing in mitigation. How powerful this effect will be depends on the assumed mitigation investment efficiency of the society. Figure 14.2 shows how the optimal mitigation investment and consumption vary depending on the society's assumed ability to mitigate.

Not surprisingly, when the ability to influence the probability of the event is small, the optimal strategy leads to "ignorance" behavior unless the occurrence of the catastrophe is almost certain. More interesting results can be

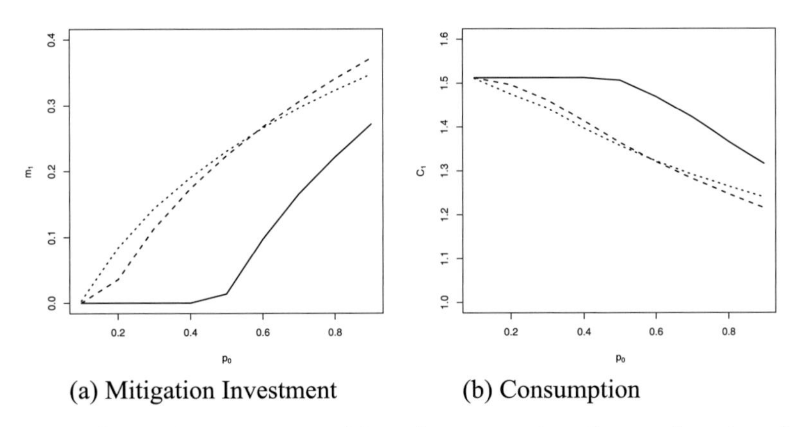

(a) Mitigation Investment (b) Consumption

Fig. 14.2. Mitigation investment M_1 and consumption C_1 as a function of the initial probability (in this figure the initial capital level is "high" and the damage size is "high"). The curves differ in the value of the mitigation investment efficiency used: the *dotted line* corresponds to a "very high" level, the *dashed line* stands for a "high" level, and the *solid line* for a "low" level

observed when comparing the curves to each other: if the initial probability is small, an increase in the mitigation investment efficiency will result in an increase of the actual mitigation investment M_1. However, the situation is somewhat different if the initial probability is high.

In order to explain this effect, it is necessary to consider the factors that influence the expected utility. The expected utility can be increased either by increasing current consumption or by lowering the probability of the catastrophic event. However, there is a trade-off between these two effects: the probability can be decreased only at the expense of current consumption. Then the problem of finding the optimal strategy is also about finding the solution to this trade-off. For a very low probability, a society with a low mitigation efficiency invests less in mitigation. This is probably because reducing the probability is costlier for a society with low efficiency than for a society with high efficiency. Thus, since the initial probability of the event is small they do not want to decrease their consumption because of a rather improbable event.

On the other hand, a society with high mitigation efficiency needs to invest less in order to produce the same effect. If the initial probability is high, the social planner may prefer certain consumption today to uncertain future gains due to mitigation investment.

Scale of the Damage

In this section we discuss the effect of the scale of the damage on the choice of the optimal strategy. More precisely, we would like to check whether the proposed model confirms our intuition: the higher the potential scale of the damage, the more we are willing to spend on prevention.

As Fig. 14.3 suggests, anxiety about a large catastrophic event almost immediately leads to a positive mitigation investment, irrespective of the probability of such an event. However, a social planner chooses the opposite strategy when a society anticipates a catastrophe producing small-scale damage. In this case a society ignores the possibility of mitigation even though the event is almost certain: society does not worry about such small damage, it prefers capital investment.

Surprise

The word *surprise* evokes in our minds the occurrence of an unexpected event (usually positive). Since we have been mainly discussing catastrophic events, a *surprise* in terms of our model is a negative notion and corresponds to the set-up where the social planner's beliefs about the probability of the event are low, but the possible scale of the damage might be high or even very high. From the viewpoint of the social planner, the occurrence of the catastrophe in this case appears to be a negative surprise. At the same time, the impact of the catastrophe is high, as the scale of the damage is high.

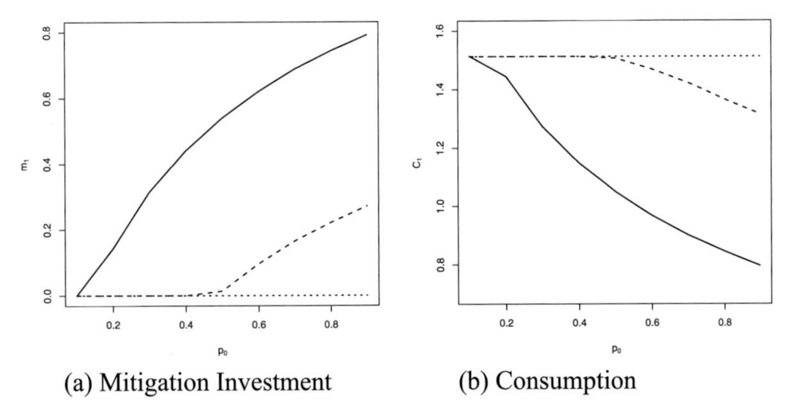

Fig. 14.3. Mitigation investment M_1 and consumption C_1 as a function of the initial probability (in this figure the initial capital level is "high" and the mitigation investment efficiency is "low"). The curves differ in the scale of the damage: the *solid line* corresponds to very high damage, the *dashed line* stands for high damage, and finally the *dotted line* for low damage

The social planner must be aware of this (improbable) possibility (as far as her assumptions about damage size of a catastrophe probability are correct) and take it into account. Therefore, what is the optimal investment strategy for such a situation?

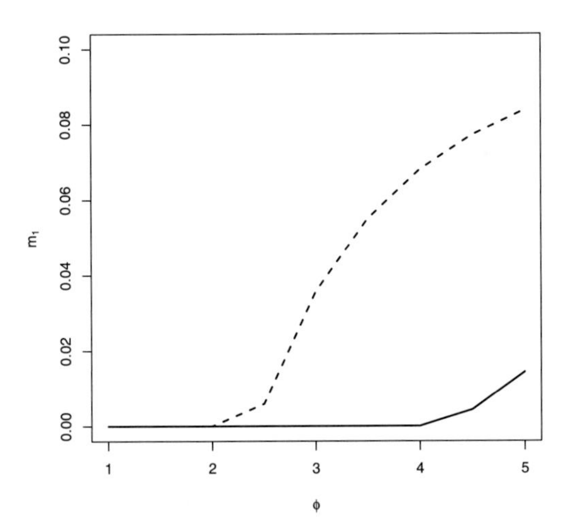

Fig. 14.4. Optimal strategies for an "unexpected" event. The *solid line* describes a society with higher initial capital, the *dashed line* is a society with a low level of initial capital (in this figure the initial probability is set at 20%, damage is "very high")

A major factor that influences the decision making in this case is the mitigation investment efficiency. This parameter describes how each additional unit of mitigation investment shifts the event probability. Figure 14.4 shows the impact of it on two societies that have different respective initial capitals.

14.2.3 Conclusions

Based on the above analysis we can claim that the following statements are true:

- No society invests in mitigation unless its perception of the probability of a catastrophic event is high; in other words either the damage size or the belief that the probability of a catastrophe (or both) are high.
- Even when there is evidence of high event probability or the potential for large-scale damage, some kinds of societies do not invest in mitigation at all. They either do not have enough resources or they are sceptical of their ability to improve the current situation by mitigation investment (their mitigation investment efficiency is small).
- In the case of a low probability event, the higher the mitigation investment efficiency, the smaller the probability of catastrophe needs to be in order to begin investing in mitigation. Societies with lower mitigation investment efficiencies are only willing to spend money on mitigation if they are almost certain that a catastrophe is going to happen.
- The optimal strategy for an event with large-scale damage and high event probability is "panic". On the other hand, the combination of small-scale damage and low event probability leads to the "ignorance" approach. However, the optimal strategy in mixed cases – large-scale damage and low event probability or small-scale damage and high event probability is not so straightforward. The capital available and the ability to mitigate are the main factors that determine the optimal strategy in this case.

14.3 Static Model

So far we have concentrated on the one-society world. However, in reality our world consists of many players, all of which differ widely culturally, scientifically and politically. Since a safe balanced environment is a common good, various externalities may arise. We now take a look at some situations which result from both cooperation and independent decision-making in order to better understand intersocietal interactions.

In order to examine both the behavior of societies in a simple game situation and the role of erroneous expectations we will now consider a more simplistic one-period model. Although this model is somewhat different from the dynamic model presented earlier, it possesses the same qualities and can legitimately replace the above model in a nondynamic setting. The main reason for this dual approach lies in the technical difficulties of dealing with the dynamic model.

14.3.1 Set-Up

Consider a one-period setting where society possesses capital K and may invest M in mitigation if it decides to do so. Also suppose that the catastrophic event of size d happens with probability $p(M)$. The expected utility function may then be written as:

$$\begin{aligned}\mathcal{U}(M) &= EU(M) = p(M)\ln\left((K-M)(1-d)\right) + (1-p(M))\ln(K-M)\\ &= \ln(K-M) + p(M)\ln(1-d)\ .\end{aligned}$$

Obviously the two-periods dynamic model may be transformed into this one by setting $\delta = 1$, $C_1 = 0$, $\gamma = 1$, and $\rho = 1$.

The probability of event $p(M)$ may depend on the mitigation investment M in various ways. The main property of such a dependence function, however, should be monotonicity: the higher investment should lead to a lower probability of disaster. A wide family of suitable functions can be described by two parameters: *the null mitigation probability* p_0 and *the mitigation efficiency*. The latter we will often describe by referring to *the* 10% *mitigation probability* $p_{.10}$, which is the probability of event when 10% of the capital K is invested in mitigation.

An example of the model's internal workings is shown in Fig. 14.5. The four cases displayed correspond to comparatively high and low initial probabilities of an Xevent and to comparatively high and low mitigation efficiencies. One important thing to notice is that, for low levels of potential damage, the optimal mitigation investment is 0. In fact, for each combination of parameters defining the probability function, a potential damage level exists at which non-zero investment becomes optimal (see Fig. 14.6). The higher the null-mitigation probability and the lower the efficiency of mitigation, the higher the potential damage needs to be to force a social planner to invest in mitigation.

14.3.2 Two Societies Game

By definition, the Nash equilibrium in a two-player game is a pair of strategies, each of which is a best response to the other. We may thus consider it to be a natural result of a non-cooperative game. The Pareto optimum is an allocation where one player cannot be made better off without hurting another player [3]. In general, Nash equilibria and Pareto optima do not necessarily coincide.

In the world of two societies, both players influence the common probability of disaster through their own mitigation investment. The relative influence

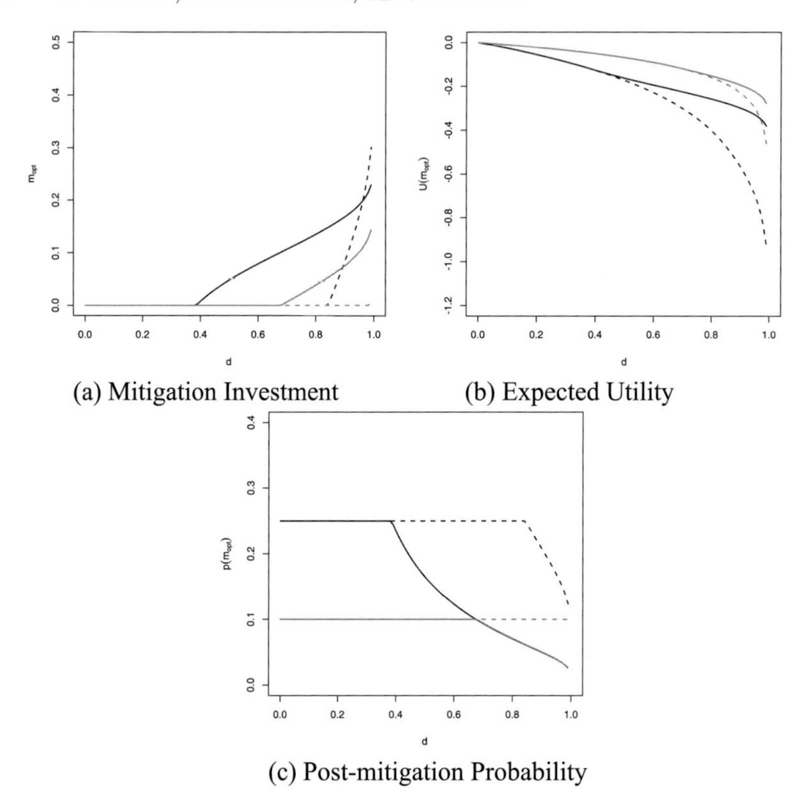

Fig. 14.5. Optimal mitigation investment (**a**) and the corresponding expected utility (**b**) and the post-mitigation probability (**c**) as a function of potential damage for $p_0 = .25$ and $p_{.10} = .10$ (*black solid curve*), $p_0 = .25$ and $p_{.10} = .20$ (*black dotted curve*), $p_0 = .10$ and $p_{.10} = .04$ (*gray solid curve*), and $p_0 = .10$ and $p_{.10} = .08$ (*gray dotted curve*)

each society will have on the environment depends not only on the relative size of this investment but also on the relative effectiveness of the mitigation of each society.

It should be noted that in this section we use upper indices to distinguish between societies because we used lower indices in the previous section to distinguish between time periods.

We assume that the mitigation probability depends upon mitigation investment through a logistic function:[4]

$$\text{logit}(p(M^{(1)}, M^{(2)})) = \text{logit}(p_0) + \beta^{(1)} M^{(1)} + \beta^{(2)} M^{(2)} ,$$

where $\beta^{(1)}$ and $\beta^{(2)}$ are the mitigation efficiency parameters for societies 1 and 2, respectively. Note that for negative $\beta^{(1)}$ and $\beta^{(2)}$, the above function

[4] $\text{logit}(x) = \ln(\frac{x}{1-x})$.

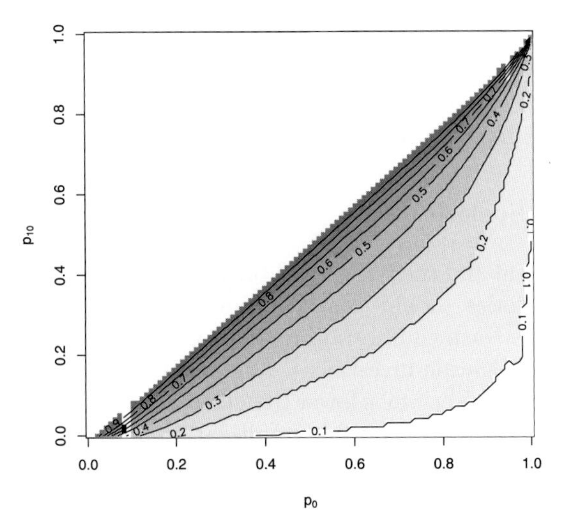

Fig. 14.6. Minimum vulnerability required to prompt non-zero mitigation investment

is a monotonously increasing function of both $M^{(1)}$ and $M^{(2)}$. The Nash equilibrium solution will then require:

$$\frac{\partial \mathcal{U}^{(1)}}{\partial M^{(1)}} = 0 \quad \text{and} \quad \frac{\partial \mathcal{U}^{(2)}}{\partial M^{(2)}} = 0$$

$$\Rightarrow \beta^{(1)} \ln(1 - d^{(1)}) \left(K^{(1)} - M_N^{(1)}\right) = \beta^{(2)} \ln(1 - d^{(2)}) \left(K^{(2)} - M_N^{(2)}\right) \quad (14.2)$$

in the case of a positive optimal mitigation investment for both societies.

Similarly, the Pareto optimum solution will require:

$$\frac{\partial \mathcal{U}^{(1)}}{\partial M^{(1)}} + \frac{\partial \mathcal{U}^{(2)}}{\partial M^{(1)}} = 0$$
$$\frac{\partial \mathcal{U}^{(1)}}{\partial M^{(2)}} + \frac{\partial \mathcal{U}^{(2)}}{\partial M^{(2)}} = 0$$

$$\Rightarrow \beta^{(1)} \left(K^{(1)} - M^{(1)}\right) = \beta^{(2)} \left(K^{(2)} - M_P^{(2)}\right) \quad (14.3)$$

in the case of a positive optimal mitigation investment for both societies.

It should be noted that unlike the case of the Nash equilibrium, the relative distribution of mitigation investment between the two societies under Pareto optimality does not depend on the difference in the potential damage levels. We will return to this peculiarity later.

We will now examine a special case of two identical societies and then take a look at the effect that discrepancies in material wealth (K), potential damage level (d) and mitigation efficiency (β) have on the comparative mitigation investments of the two societies.

Two Identical Societies

The Pareto optimal and the Nash equilibrium mitigation contributions of the two identical societies should clearly be identical as well. In this particular model setting, the Nash equilibrium mitigation investment never exceeds the Pareto optimal mitigation investment. As a result the post-mitigation probability at the Nash equilibrium is lower than that at the Pareto optimum. Thus, in the case where the Pareto optimum mitigation investment is non-zero, the failure to cooperate not only results in decreased economic prosperity of both societies, but it also hurts the environment. It may therefore be interpreted as a strong case for cooperation.

However, societies in the real world are often far from equal. Wide differences in wealth, vulnerability and scientific facilities exist. Although it would be impossible to investigate the whole spectrum of possibilities on these pages, we will now take a look at some special cases.

Wealth Discrepancy

If the only difference between the two societies lies in material wealth ($K^{(1)} \neq K^{(2)}$), both equations (14.2) and (14.3) become:

$$M^{(1)} = M^{(2)} + (K^{(1)} - K^{(2)}) \, ,$$

in the case of positive optimal mitigation investment for both societies. In general, therefore, a richer country should invest at least as much as a poorer one under both the Nash equilibrium and the Pareto optimum conditions.

Mitigation Efficiency Discrepancy

If the two societies differ only in their mitigation efficiency ($\beta^{(1)} \neq \beta^{(2)}$), then equations (14.2) and (14.3) become:

$$M^{(1)} = \frac{\beta^{(2)}}{\beta^{(1)}} M^{(2)}$$

in the case of positive optimal mitigation investment for both societies. Therefore, a country more efficient at mitigation will invest more under both the Nash equilibrium and the Pareto optimum conditions.

Discrepancy in the Scale of Potential Damage

Different scales of potential damage lead to a more interesting case than the previous two, since (other things being equal) under the Paretooptimum

$$M_P^{(1)} = M_P^{(2)} ,$$

whereas under the Nash equilibrium

$$\frac{K - M_N^{(1)}}{K - M_N^{(2)}} = \frac{\ln(1 - d^{(2)})}{\ln(1 - d^{(1)})}$$

in the case of positive optimal mitigation investment for both societies. Therefore, a society that is potentially more vulnerable will invest more than the more robust society under the Nash equilibrium, but the two societies should invest equally under the Pareto optimum. In some cases this leads to individual societies being better off in the absence of cooperation. An example is shown in Fig. 14.7.

Here we examine two societies; the only difference between them lies in their vulnerability to an abrupt event, $d_1 = 50\%$ versus $d_2 = 80\%$. Figure 14.3.2 displays the total expected utility function for different levels of investment, while Figs. 14.3.2 and 14.3.2 display cross-sectional profiles of this function for each society while the counterpart invests Nash or Pareto optimally. One can see that, although both societies invest equally under Pareto optimality, under the Nash equilibrium the more robust society does not invest at all! Numbers for this numerical example are presented in Table 14.1. Thus, although enforced cooperation would be more beneficial for the environment and for the world as a whole, it might not be the best choice for the individuals. Therefore, some redistribution of the benefits might be in order. A discussion of the nature and complexity of such a redistribution is unfortunately beyond the scope of this text.

Table 14.1. An example of the conflicts of interest seen with cooperation

Nash	M_i^N	U_i^N	Pareto	M_i^P	U_i^P
$i = 1$	0.0000	−0.0488	$i = 1$	0.0886	−0.1245
$i = 2$	0.1350	−0.2592	$i = 2$	0.0886	−0.1659
Both	0.1350	−0.3080	Both	0.1772	−0.2904

14.3.3 Erroneous Expectations

So far in our narrative we have implicitly assumed that the values of the parameters are known perfectly. Although this is reasonable with regard to

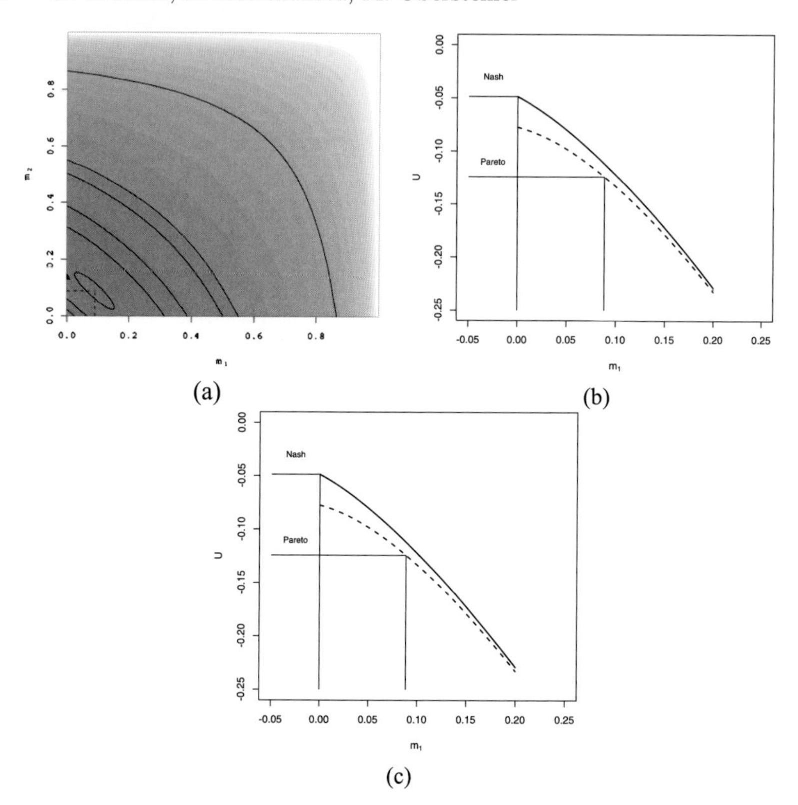

Fig. 14.7. An example of individual societies being better off without cooperation. The first graph (**a**) shows the utility surface ($K = 1$, $p_0 = 0.25$, $p_{.10} = 0.1$) for various combinations of mitigation investments. The *dotted segments* intersect at the Pareto optimum, the *black triangle* marks the Nash equilibrium. The curves on the other two graphs show the profiles of the utility function when the other society invests its Nash optimal (*dotted*) or Pareto optimal (*solid*) amount for the (**b**) vulnerable and (**c**) robust society

capital, it is hardly true when we turn to our knowledge of the environment. We do not really know how likely an abrupt climate event is, nor can we be sure about the effectiveness of our mitigation measures. Also, if something should happen, our estimates of the ensuing damage are very unlikely to turn out to be accurate. Does this mean that the previous discussion, dealing with an improbably omniscient social planner, has been completely in vain, or can we learn some useful lessons despite our imperfect knowledge? In this section we will once again return to a single society and take a look at the world of erroneous expectations.

First, it can be shown that a higher level of potential damage results in higher optimal investment. Therefore, higher expectations of potential damage – "a reverent fear of nature" – will lead to over-investment, which will in turn result in suboptimal utility. But the resulting post-mitigation probability of an abrupt climate event will be lower than the optimal post-mitigation probability. Underestimation of the danger, on the other hand, will naturally also lead to suboptimal utility, but will result in a higher than optimal post-mitigation probability of surprise, bringing the unknown disaster ever closer. It therefore seems prudent to err on the side of caution in this respect.

This case is not as clear-cut with respect to the initial probability of an Xevent and the mitigation efficiency. In general, the higher the initial probability, the higher the mitigation investment will be, but if such a probability is too high (over 80% in our model), society will prefer to invest less, perhaps hoarding the resources necessary for survival in the face of imminent disaster (see Sect. 14.3.3). The effect of erroneous expectations thus depends upon the actual state of the world and on the magnitude of our error. As mentioned at the outset, the models serve only to illustrate some aspects of abrupt climate phenomena and the values of the parameters should not be interpreted literally. However, it is probably safe to think that our world is not that close to destruction and, again, it is better to err on the cautious side.

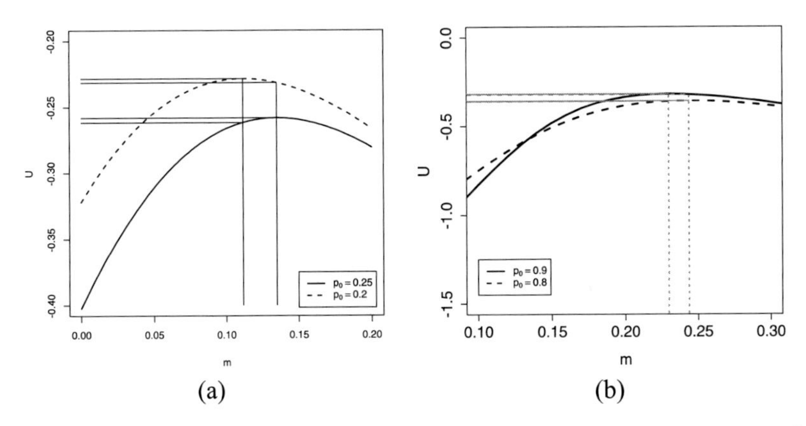

Fig. 14.8. Effect of erroneous expectations on the optimal investment in cases of (**a**) low and (**b**) high null-mitigation probabilities. In the first case, the exaggeration of risk is good for the environment, while in the second it is bad for both the economic welfare of the society and the environment

The influence of erroneous expectations regarding the mitigation effect is uncertain, as it depends on the confluence of other factors, such as the null-mitigation probability and the potential level of damage. Therefore, perhaps, theoretical research into this aspect would be most beneficial.

So far, the analysis of erroneous expectations has resulted in some general rules. It would certainly be beneficial to conduct research into the actual state of the world, in order to improve our understanding of the parameter values for d, p_0 and β. However, as has been demonstrated, it would be better to be cautious: to overestimate the danger and the potential damage and, at the same time, attempt to make the effect of current mitigation measures on nature clearer.

14.4 Discussion

Due to the lack of functioning global institutions controlling global risks, the pace of disaster development is currently undermining social systems (markets) and the safety nets of many countries, and is reducing their capacity to provide basic services for their people. It is paramount for the global community to prevent and manage catastrophic losses of human lives, livelihoods and natural and economic assets. The global common public good – resulting in a "less risky world" – is yet to be produced by a multitude of internationally coordinated actions. In this chapter we focused on societial performance and strategies in managing hazards, mainly focusing on the framework of conditions they establish to shape interactions among various decision-makers, rather than focusing on the specific prevention measures per se (see [4]). Collective efforts to shelter, protect and safely nourish the group have formed the backbone of human evolution from prehistoric times to modern civilization. Today, in a world of global integration, we are in the unique position of facing mega-risks that are global in nature (the climate change problem, the ozone hole, AIDS); the group has become the global community. However, nation states find it increasingly difficult to contribute to the public good due to political, cultural and scientific differences. Moreover, each nation state faces the temptation to take advantage of the public good without contributing to it: a phenomenon commonly referred to as the "Tragedies of the Commons", the "'Prisoners Dilemma" or the "moral hazard problem". The nation states have differing strategic judgments, differing assessments of risk and different opinions about the best way to address risk, as well as different methods of calculating interest, all of which lead to these social dilemmas. From our modeling efforts, we can predict that in the Kantian Pareto world, global welfare will increase, income disparities between countries will be minimized, and the resulting risk level will be low. If a Kantian global institution cannot be built and we continue to live in a non-cooperative and competitive Hobbesian Nash world, prevention of detrimental surprises is less plausible and the implementation of a risk management system that guarantees resilience is jeopardized. Furthermore, in the case of impaired resilience of the underlying system (when mitigation investment becomes ineffective once a threshold is exceeded) we might run into the danger of incurring irreversible damage. Thus, the challenge ahead for the global community is to build international institutions that have enough authority to enforce cooperative "Pareto Solutions" internationally. SARS, cybercrime and climate change are examples of issues that are too big for governments to handle by themselves [2].

Although we cannot offer solutions to bridging political and cultural gaps in global unity, we are able to take a closer look at the science behind dealing with Xevents and the "free rider problem".

Currently, there is little theoretical scientific work supporting the avoidance process of the social dilemma of free riding under endogenous mega-risks. We believe that the one major reason for this lack of theory is the difficulties involved with formally describing and solving the processes used to manage global mega-risks, even in highly stylized models. In this chapter we have made an attempt to work towards a better understanding of the key hurdles of such a process. The dynamic model allowed us to investigate investment strategies for mitigation of an endogenous risk using a dynamic formulation. Given optimality conditions, the resulting strategies range from panic behavior to ignorance of risks, depending on the initial conditions and the parametrization of the model. In the static model, we focused on cooperative behavior in the case of exogenous risk. The role of erroneous beliefs was also investigated.

We started our analysis with the global one-society world, possessing perfect information on the state of nature and the possible results of its own actions driving risk exposure. The probability of Xevents was assumed to be endogenous: to depend on the effort that society was willing to make (the mitigation investment). We then identified four different strategies that might be chosen, subject to the initial state of the world: (1) *tactical approach*, where the society still invests into both capital and mitigation, (2) *ignorance approach*, where the society invests in capital only, preferring not to deal with the threat of insignificant disaster, (3) *panic behavior*, where there is no capital investment and all resources are directed towards mitigation, and finally (4) the *"eat, drink, and be merry"* philosophy, where nothing is invested and everything is consumed in the last pre-apocalyptic grand party.

We then moved on to the interplay between two differing societies. We showed that, although cooperation would in principle be beneficial both economically and environmentally, the individual rational actors will have no incentive to invest or will jointly invest too little in risk mitigation measures. It is usually the more vulnerable society that is inclined to cooperate. However, in today's world, people from developing countries are particularly vulnerable to increasing global hazard levels as they lack insurance schemes and are usually poorly equipped to respond to large-scale catastrophes after the fact. These vulnerable countries, which would benefit most from cooperation to reduce hazard levels, lack the power to bring about a "Pareto solution". The challenge for long-term international cooperation is to establish "individual" responsibility for actions that increase global hazards (such as emission of greenhouse gases). Actions like the introduction of an international tort law on major global risks might turn out to be effective solutions in preventing surprise [5].

A major reason for the failure to cooperate is the fact that there is disagreement about the state of the world with respect to risk exposure. The adverse effects of increasing risk exposure are a matter of prognostication "Doomsday scenarios", such as global epidemics or a collapse of the Gulf Stream, and remain only hypotheses based on scientific projection. Inasmuch as they are triggered by unprecedented growth of population, natural resource exploitation and perturbation, and concentration of assets due to economic integration, contemporary natural disasters are different from any that have preceded them. Therefore, at the beginning of this chapter we turned to the notion of *imaginable surprise* [6], a situation in which perceived reality departs qualitatively from expectations. We therefore turned our attention to a world where natural catastrophes or Xevents are expected, but their probability or the extent of the potential damage is unknown. We have further assumed that the probability of such an event is endogenous and monotonously dependent on the mitigation investment, but that the potential damage is fixed. We have thus concentrated exclusively on *mitigation* without considering *adaptation* or vulnerability management. It would be of interest to consider a model augmented with a potential damage term. In this way the community would have to make a choice between investing towards global mitigation and/or local adaptation.

In the face of imaginable surprise, two options are appropriate: (1) reduction of the uncertainty, which is usually referred to as learning through data collection and research, and (2) management or integration of uncertainty directly into the decision-making or policy-making process [6]. Our analysis of the role of erroneous expectations has shown that while the uncertainty regarding the extent of potential damage can safely be integrated into decision making, the probability function of the event (the pre-mitigation probability and the mitigation effectiveness) needs to be investigated so that reasonably precise estimates can be arrived at. This creates a third possible avenue of investment, which we have tentatively coined *R&D investment*. Together with mitigation and adaptation, they constitute three possible objects of investment, each of which can be modeled endogenously. Such modeling is one of the goals of further investigation. A second direction is the extension of the current model to a multi-period, multi-societal framework. Among other things, this would allow investigation of the free-riding and redistribution mechanisms in more detail and to glance behind the potential political scene.

We believe that information management and knowledge building are necessary preconditions for implementation of the framework conditions, which allow countries to act jointly and effectively to manage the risks mankind faces in the twenty-first century. Disaster schemes still treat people as "clients" in disaster processes, where science and technology do things to them and for them rather than together with them [9]. In a world where centralized modes of risk management lose effectiveness, we also see tendencies towards increased cross-sectional complexity, increased effective participation by people, broadening liability on international scales and a move towards a "claim culture". In such a world, timely provision of information and application of knowledge to reduce exposure are paramount to establish democratization of the risk management process and to establish liability to endogenous catastrophic risks in order to guarantee intergenerational fairness. In this way, with the support of the scientific community, mankind could overcome the institutional obstacles to achieving an effective transition from nationally driven Pareto solutions to international strategies that are based on responsibility and liability.

References

1. J. Diaz-Gimenez: Linear-Quadratic Approximation: An Introduction. In: *Computational Methods for the Study of Dynamic Economies*, ed. by R. Marimon, A. Scott (Oxford University Press, New York, 1999), pp. 13–29
2. A. Etzioni: *Think Global, Act Global* (New Scientist, 11 September 2004), p. 16
3. H. Gintis: *Game Theory Evolving: A Problem-Centered Introduction to Modeling Strategic Behavior* (Princeton University Press, Princeton, NJ, 2000), pp. 12, 28
4. M. Obersteiner, C. Azar, P. Kauppi, K. Moellersten, J. Moreira, S. Nilsson, P. Read, K. Riahi, B. Schlamadinger, Y. Yamagata, J. Yan, J.P. van Ypersele: *Managing Climate Risks* (Science, 294(5543), 2001), pp. 786–787
5. OECD: *Emerging Risks in the Twenty-First Century: An Agenda for Action* (Organisation for Economic Co-operation and Development, Paris, 2003)
6. S.H. Schneider: Imaginable Surprise. In: *Handbook of Weather, Climate and Water: Atmospheric Chemistry, Hydrology, and Societal Impacts*, ed. by T.D. Potter and B.R. Colman (Wiley, Chichester, 2003), pp. 947–954
7. C.S. Holling, ed.: *Adaptive Environmental Assessment and Management* (Wiley, Chichester, 1978)
8. C.J. Walters: *Adaptive Management of Renewable Resources* (Macmillan, New York, 1986)
9. J. Weichselgartner, M. Obersteiner: *Knowing Sufficient and Applying More: Challenges in Hazards Management* (Environ. Hazards, 4, 2002), pp. 73–77

15 Disasters as Extreme Events and the Importance of Network Interactions for Disaster Response Management

Dirk Helbing, Hendrik Ammoser and Christian Kühnert

Summary. We discuss why disasters occur more frequently and are more serious than expected according to a normal distribution. Moreover, we investigate the interaction networks responsible for the cascade-like spreading of disasters. Such causality networks allow one to estimate the development of disasters with time, to give hints about when to take certain actions, to assess the suitability of alternative measures of emergency management, and to anticipate their side effects. Finally, we identify other fields where network theory could help to improve disaster response management.

15.1 Disasters as Extreme Events

Natural and man-made systems are usually robust to normal perturbations. They are constructed to handle them with variations of several standard deviations. However, preparation for Xevents is costly and often imcompatible with the requirements of everyday use. Therefore, it is often neglected. Moreover, Xevents [1,2] often do not obey common statistical distributions. Their distribution is instead characterized by "fat tails" [1–3], which implies a much higher frequency of occurrence than expected according to a normal distribution. These fat tails often follow a power law, which is characteristic of systems that reach a critical point and suffer from avalanche or cascade effects of a potentially arbitrary size. In some cases, it is even impossible to make statements about the mean value or the standard deviation of such events, as power-law distributions are not always normalizable. Typical examples of systems that exhibit power laws are

- avalanches of sand, debris or snow [4,5]
- earthquakes (see the Gutenberg-Richter law)
- crashes and bubbles at stock markets
- bankruptcies in banking networks
- disaster scenarios [6]

The detailed impact of rare events on a system is often unknown. Possible scenarios can, however, be anticipated using models that describe the interactions between different parts ("sectors") of the system. These interactions are mostly nonlinear and characterized by feedbacks. As a consequence,

Table 15.1. The 10 worst catastrophes in terms of victims between 1970 and 2003 [8]

No.	Victims	Date (start)	Event
1	300,000	11/14/1970	Storm and flood catastrophe, Bangladesh
2	250,000	7/28/1976	Earthquake in Tangshan, China (8.2 on the Richter scale)
3	>220,000	12/26/2004	Tsunami in the South Asian Sea
4	138,000	4/29/1991	Tropical cyclone Gorky, Bangladesh
6	60,000	5/31/1970	Earthquake in Peru (7.7 on the Richter scale)
7	50,000	6/21/1990	Earthquake in Gilan, Iran
8	41,000	12/26/2003	Earthquake in Bam, Iran (6.5 on the Richter scale)
9	25,000	9/16/1978	Earthquake in Tabas, Iran (7.7 on the Richter scale)
10	25,000	12/7/1988	Earthquake in Armenia, former USSR

small changes in the system state can have large effects when a certain critical threshold is exceeded. Such effects can be described by methods from systems theory and system dynamics, catastrophe theory [7], the theory of nonequilibrium phase transitions, nonlinear dynamics and the theory of complex, self-organizing systems. Insights from chaos theory and percolation theory are relevant as well. The same applies to the theory of networks.

Despite many reports on disasters [9,10], a scientific investigation of their general features and ways to fight them is still needed. Each year, about 250 million people are affected by natural disasters worldwide. Three billion

Table 15.2. The 10 greatest *insurance* losses due to disasters between 1970 and 2003 in millions of US dollars [8]

No.	Loss	Victims	Date (start)	Event
1	21,062	3,025	9/11/2001	Terrorist attack on WTC, Pentagon ..., USA
2	20,900	43	8/23/1992	Hurricane Andrew, USA & Bahamas
3	17,312	60	1/17/1994	Northridge earthquake, USA
4	7,598	51	9/27/1991	Typhoon Mireille, Japan
5	6,441	95	1/25/1990	Winterstorm Daria, France & UK et al.
6	6,382	110	12/25/1999	Winterstorm Lothar over Western Europe
7	6,203	71	9/15/1989	Hurricane Hugo, Puerto Rico & USA et al.
8	4,839	22	10/15/1987	Storm/floods in W. Europe, France, UK et al.
9	4,476	64	2/25/1990	Winterstorm Vivian, Western/Central Europe
10	4,445	26	9/22/1999	Typhoon Bart hits south of Japan

people live in endangered areas. The economic impact, and also the number and size of disasters seem to grow, potentially because of overpopulation and global warming due to CO_2 emissions and the greenhouse effect. In 2003, disasters took 60,000 victims and caused a damage worth 70 billion US dollars (see Table 15.1), while insurance schemes paid out 18.5 billion US dollars. Today, a single disaster can easily cost billions (see Table 15.2). For example, the losses due to the floods in Europe in August 2002 amounted to 21 billion Euros, the blackout in Northern America in 2003 to 6.7 billion US dollars, and the SARS outbreak in 2003/2004 to about 60 billion US dollars in China alone, not to mention the problems caused in Canada and other countries.

15.2 Examples of Causality Chains and Cascade Effects

The spreading of natural and man-made disasters can often be described by interconnected causality chains – a network reflecting how one factor or sector of a system affects others. In the following, we will give examples illustrating some of the complications that originate during disasters. For an event localized in time and space, it is often these cascade-like chain reactions that cause large-scale disasters that affect the whole system (in real terms, people in remote places around the world).

The tendency towards globalization of economic and other systems is likely to increase the frequency of large-scale disasters, as it reduces the diversity required to stop certain chain reactions and to adapt to changing economic and environmental conditions. Another danger is the ever-growing population and the trend to push social, economic, technological, and biological systems to their limits [11–14].

15.2.1 Earthquakes

Earthquakes (see Fig. 15.1) are caused by the relative movements of tectonic plates and continents. This builds up strain, which is reduced in sudden avalanche-like slides, giving rise to earthquakes. An earthquake can liberate energy equivalent to many atomic bombs. This causes strong vibrations, which are often enhanced by resonance effects. These vibrations can damage or destroy housing and facilities. Oscillating high-rise buildings can even damage each other, which may produce a domino effect. As a result of the tectonic activity, (infra)structures like bridges, tunnels, and streets are destroyed over wide areas. The same applies to electrical facilities, gas and water pipelines and the sewage disposal network, which causes serious supply and hygiene problems.

Some big earthquakes include those of San Francisco (1906), Guatemala City (1976), Mexico City (1985), and Bam (in Iran; 2003). An earthquake in Georgia (1991) caused a landslide that buried 85% of a village. Another earthquake in Southern Asia (2004) caused a tsunami with waves many meters high, which moved at a speed of 700–800 km/h and destroyed dozens of villages and hotels along the coastlines of India, Indonesia, Vietnam, Sri Lanka, The Maldives, Sumatra, Thailand, and even Africa. It killed more than 220,000 people and made approximately five million people hungry and homeless.

We will illustrate the earthquake-related problems in more detail using the disaster in Kobe (Japan, 1995). Kobe is not located in an area of major earthquake activity. Therefore, the earthquake came as a surprise, and so no particular preparation had been made for earthquakes. It took 12–18 hours for the official authorities to admit that they required international help to cope with the disaster.

About 6,400 people were killed, but initially, the official numbers were around 30. Nobody was able to make decisions. For example, Great Britain offered dog tracking units, but the legal regulations required one week of quarantine. Since nobody knew how to handle this problem, nobody dared to take responsibility. Interestingly, the Japan mafia (the "Yakuza") was better organized and it helped to distribute food and provisions, possibly in order to obtain more influence and to improve its reputation.

Massive destruction was inflicted upon the town and the highways, probably because Kobe was not constructed to withstand earthquakes. However, worse still were the hundreds of fires that broke out, which were caused by broken gas pipes in wooden houses between the skyscrapers. Widespread chaos was caused by the fact that the firefighters could not reach the fires because the street infrastructure was shattered and many water pipes were severed. Another problem came from the power supply lines hanging over the remaining streets, which seriously obstructed traffic, transport routes and supplies.

Thousands of people were made homeless, and people panicked during the aftershocks, which had the potential to cause damaged infrastructures and buildings to fall down.

Fires triggered by earthquakes can last for several days and can destroy the trading centers of a town, as in the San Francisco earthquake. There, the fires could only be stopped by evacuating and destroying a large number of villas in residential areas to produce a firebreak.

15.2.2 Power Blackouts

In recent years, electrical power outages ("blackouts") have affected larger and larger areas. This is because of

- the growing and highly fluctuating demands for power (due to, for example, an increased number of air conditioners),
- the increasing size and complexity of electrical power networks (often with power being exchanged across countries),

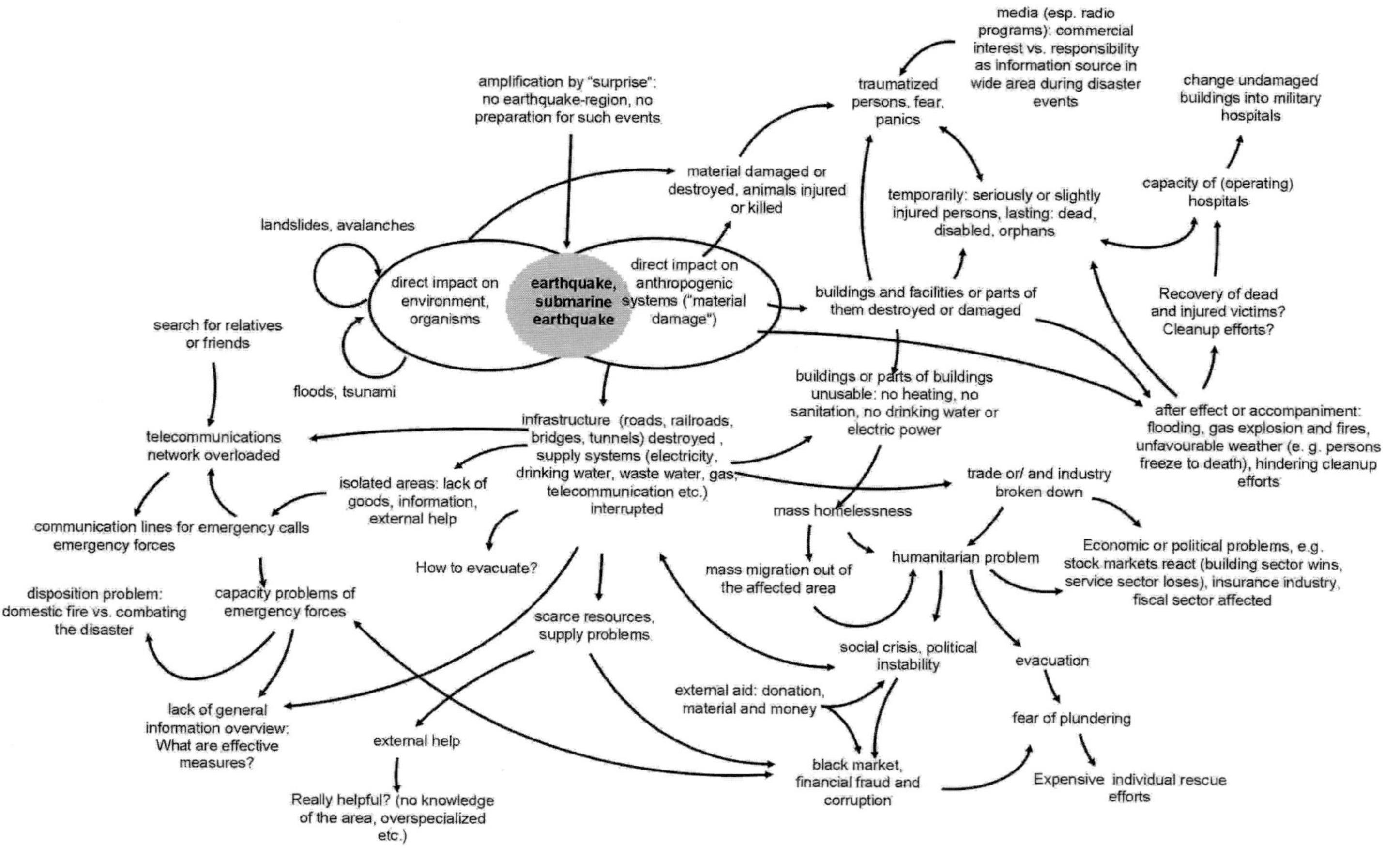

Fig. 15.1. Causality network of earthquakes

- the deregulation of the electricity market, which encourages profits with minimum investment

The largest blackout probably occurred in 2003 in north-eastern areas of North America (USA and Canada), which were followed by other major blackouts in Great Britain and Northern Italy in the same year.

The blackout in the USA and Canada left 50 million people without electricity for up to 48 hours. The sudden breakdown of one power station caused a cascade of shutdowns at other power stations in order to avoid overloading. The blackout affected the water supply as the water pumps stopped functioning so the water pressure dropped and contamination became more likely. The advice given, to boil water before use, was difficult to follow without electricity. Moreover, traffic systems stopped working, so thousands of people were imprisoned in elevators and subway trains, and many airports were closed. Traffic lights switched off, causing widespread traffic chaos. Petrol stations could not pump fuel due to the lack of electricity for their pumps. Although radio and TV stations did broadcast, most radios ran off mains power. The mobile phone network broke down due to overload. Only conventional telephones and laptops with internet connections remained functional so long as their batteries and accumulators had power. For this reason, the ability to inform the public about the situation was extremely limited. As gas pumps did not function, there was an explosion at one of the oil refineries, which meant that the population nearby had to be evacuated. Moreover, the use of candles caused several fires, which were hard to fight because the traffic chaos on the streets slowed down firefighter response. The blackout also had several long-term effects, among them reducing economic growth and delaying elections.

15.2.3 Hurricanes, Snowstorms, and Floods

(Thunder-)Storms are the most frequent cause of disasters, particularly in tropical areas. Hailstorms may produce hailstones of up to 1 kg in weight, as seen in Rostov (Soviet Union, 1923). However, much smaller hailstones than this can still injure people, damage cars and structures, trees, fields and plantations, which can, in turn, cause serious crop shortfalls and famines.

It is common to distinguish different kind of storms due to their geographic appearance or their meteorological character, such as hurricanes, tornados, cyclones, typhoons, monsoon rains, and others. In extreme cases, they have killed 300,000 people (Haiphong, Vietnam, 1881) and made 25 million people homeless (monsoon rains in Bangladesh, 1988).

Fifty million people may be forced to prepare for an evacuation when a full-scale hurricane is in sight. Panic-buying (hoarding) in advance of a forecasted storm is typical. The destruction caused by storms often interrupts air, train, and vehicle traffic due to high wind velocities and obstacles lying on streets and tracks. Strong rainfall may even make the operation of underground traffic impossible. Schools and many public activities are closed down. Broken electricity lines cause power blackouts. For example, during Hurricane Isabel, two million homes were without electricity.

Storms (see Fig. 15.2) often occur together with strong rainfall. This can cause serious floods, erosion, or landslides [4, 5], which can themselves be disastrous. Moreover, broken trees are often the source of insect plagues (bark beetle). During blizzards and snow storms, 60 cm of snow can easily fall per day.

This can stop public life, even in big cities such as Manhattan (1947) or Boston (1978), where 100,000 people were forced out of their homes. Moreover, the supply of coal to power stations, steel production and so on can be seriously endangered and a vast number of animals may die.

The floods in Central Europe in August 2002 [15] originated from extreme rainfall (up to 300 liters per square meter) and caused more than six billion Euro's worth of damage in Saxony (Germany) alone. Small streams had to cope with 100 times more water than usual, and flotsam reduced their flow capacity. As a consequence, rivers left their artificial river beds and flooded 15% of the Saxonian metropolis, including the center of Dresden and its disaster control center. Moreover, most hospitals had to be evacuated just when they were urgently needed. Tens of thousands of people also had to be evacuated, but the population often resisted official commands because it was afraid of plunder. This often necessitated expensive evacuations of single individuals by helicopters later on.

Evacuation, supply, and disaster response management was very difficult, as tunnels were full of water, many bridges were lost, and most of the remaining bridges could not be used for safety reasons. Electrical power supply was down in most areas of Dresden, for several weeks even in the center. The same applied to most telephones and faxes. The mobile phone network was overloaded and broke down as well. In some cases, information could only be communicated by messengers. Moreover, the ability to warn the population was seriously restricted, because church bells and sirens were not available or they required power.

All train connections to and from Dresden were interrupted for many months, with a single exception. Seven hundred kilometers of train tracks, 400 km of railroad embarkment, and 100 bridges were damaged or destroyed. Moreover, many electronic railway control centers stopped working. Water supply was a problem in some areas, as some waterworks supplying drinking water were flooded. Some clarification plants were flooded as well, which may have caused diseases. Additional health problems originated from the many drowned animals and the thousands of tons of mud and waste that the flood left behind. This caused one of the worst mosquito/insect plagues for decades.

The floods also endangered some of the most valuable cultural assets of Germany, affected radio and TV program, and damaged newspaper archives. Catastrophe tourism obstructed the recovery activities, as they generally obstructed many areas of disaster response management. However, it wasn't just

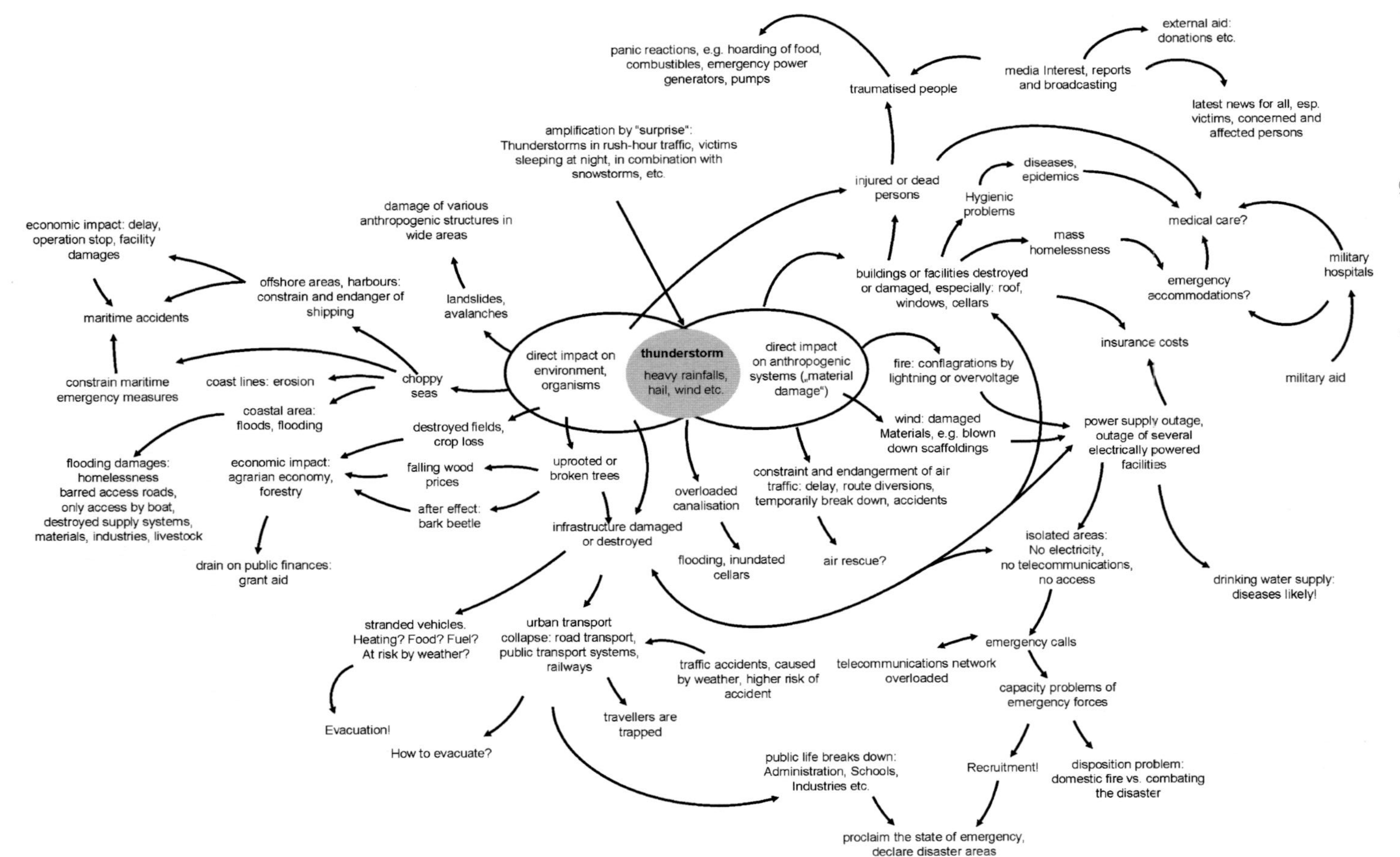

Fig. 15.2. Thunderstorms, their causality network and impact on environment

public infrastructure and facilities that were endangered. Thousands of cellars were flooded, but in many cases the water could not be removed/pumped out. Most buildings would not have resisted the high groundwater level.

The rumor of a broken dam almost caused panic in the city center of Dresden. Dams have broken several times in the past, for example in Fréjus (France, 1959), in Johnstown (USA, 1889), or along the Mississippi (USA, 1927). Fortunately, the rumor turned out to be false, otherwise tens of thousands people could have died in Saxony's metropole. However, let us finally mention that landslides can cause floods as well, as in the case of Vajont (Italy, 1963) [16].

15.2.4 Terrorist Attacks

Terrorist attacks [17, 18] have become an increasingly serious concern. In many cases, terrorists try to gain public awareness for certain religious or political interests or an ignored problem, for example a suppressed minority. In many cases, the ultimate goal is maximum damage. This is best illustrated by the terrorist attacks on 11th September 2001 in New York [19] and on 11th March 2004 in Madrid.

On 11th September 2001, four aircraft were hijacked. Two of them were flown into Manhattan's World Trade Center. Thousands of people had to be evacuated. According to the emergency plan, airports, tunnels and bridges in Manhatten were closed down. Together with panicked people, it produced a massive traffic problem. Even worse, the crashes caused large fires inside the Twin Towers, which weakened the steel framework of the buildings, so that the buildings finally collapsed. Many people, including a large number of fire fighters, were killed. Stock markets suffered; more than 1 trillion dollars were lost in a week.

As a consequence, many people cancelled their airplane tickets and reduced their number of trips. Together with other problems, several airlines filed a petition for bankruptcy (Swissair), while others had to merge. Moreover, international security laws were tightened and privacy of personal data has since been considerably restricted. An international fight against terror was started. This led to the wars on Afghanistan and Iraq, which in turn triggered many other terrorist attacks worldwide. The worst of them were in Djerba (Tunisia, 2002), Bali (Indonesia, 2002), Riad (Saudi Arabia, 2003), Casablanca (Morocco, 2003) and Madrid (Spain, 2004).

The attack in Madrid (Spain) on 11th March 2004, was characterized by successive explosions in several urban trains close to well-observed train stations. This strategy challenged the emergency measures in addition to the high number of injured and dead people. Hospitals were overwhelmed. There are signs that additional explosions should have killed the task forces trying to save the people, but these were avoided by jammer transmitters. As a consequence of this attack, the incumbent government lost the elections and the new government quickly withdrew Spanish soldiers from Iraq.

These events illustrate the truly global impact of some disasters. Other well-known examples of terrorist attacks are the Sarin gas attacks by the Aum sect in the Tokyo metro. One of the problems encountered in this case was that the victims were initially treated incorrectly, as the deadly chemical substance was not correctly identified. A similar problem occurred during a hostage rescue from a theater in Moscow (Russia, 2002), where the military used an secret anaesthetic gas.

15.2.5 Epidemics

The disasters capable of taking the most human lives are epidemic diseases (see Fig. 15.3), as they can easily spread across countries. Between 1500 and 1550, syphilis killed ten million people throughout Europe. Between 1735 and 1740, diptheria killed about 80% of all children under the age of ten. Malaria has killed several million people in the Soviet Union (1923) and India (1947). Other deadly epidemics include measles, pox, yellow fever, typhus and cholera. Some of these diseases occur if the drinking water has been contaminated, others when the general health of the population is lowered by hunger or cold. Many of them are transmitted by insects and animals, so that fighting epidemics often requires to destroying millions of animals (such as chickens).

Epidemics have sometimes determined the results of wars and the rise or fall of a nation or culture. One of the worst epidemics ever was the plague (pestilence), which killed about 75 million people. It reached Europe via the trade routes from Asia and was transmitted by rat flea, as well as by cough. Hundreds of people could die in one day in the same town, so much so that there was a scarcity of wood to burn the bodies. About 30% of the population of Europe died. Moral values decayed and criminal activity jumped up. Some social and racial minorities became the victims of pogroms. Economic activities broke down, as there was a lack of workers.

One of the worst epidemics of the last century was the Spanish influenza outbreak. It killed 20–40 million people between 1918 and 1920. Economic and social life was more seriously affected by the epidemic than by World War I. Banks, mines, and parliaments closed down. Trade and transport were interrupted. People tried to avoid infection by sealing their apartment windows, but many of them then died due to a lack of fresh air.

Influenza is still among the greatest danger of today, as the viruses responsible mutate quickly. A new influenza epidemic is expected every 10–20 years. It is most important to stop the spread of the disease as quickly as possible. Therefore, the World Health Organization (WHO) is monitoring the spread of diseases very carefully. Although the SARS outbreak in 2003 killed less than 900 people, it spread worldwide by air transport within weeks. Social cohesion was challenged, and pogroms occurred in some areas. Many public places like schools, theaters, restaurants, companies, and administrative offices temporarily closed down. Tourists avoided the region, and air traffic

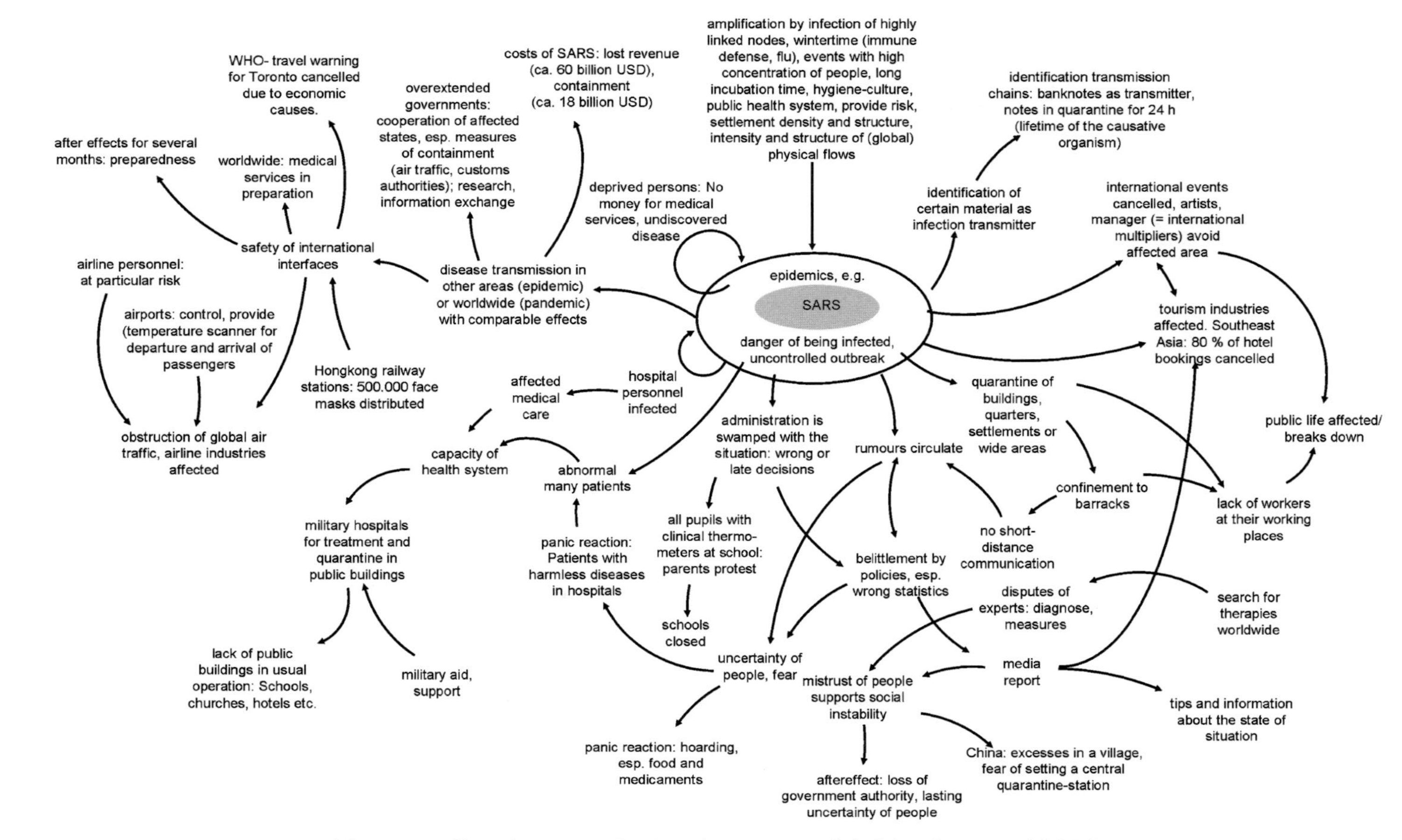

Fig. 15.3. Causality network of epidemics exemplified for the case of SARS

was restricted. The consequence was an overall economic loss of around ten billion US dollars worldwide. Correspondingly, stock prices went down.

Another serious disease is AIDS. Despite its relatively slow spread, it resisted effective treatment for many decades since it targets the immune system itself.

Although treatments are available today, many economies cannot afford the cost of them. In Africa, social structures have been already destroyed on a large scale by the high percentage of infections and the fact that many children have lost their parents. However, the economies in Eastern Europe and other countries in the world are also seriously affected.

Finally, we would like to mention the spread of computer viruses. For example, the virus "Slammer" caused an economic loss of 1.25 billion US dollars worldwide. However, the hazards go far beyond the direct economic damage due to computer downtimes and additional computer administration or software costs. Computer viruses seriously endanger the security and functioning of sensitive data systems and critical infrastructures, including communication systems.

15.2.6 Other Disasters

There are many other kinds of disasters we have not mentioned here. Among them are extreme aridity, locust plagues, meteorite impacts, overpopulation, disasters related to climate change, volcanic eruptions, bush and forest fires (see Fig. 15.4), inflation and economic crises. Moreover, we have not discussed man-made technological disasters. These include power plant accidents, such as nuclear radiation accidents of varying severity (including the level 7 major accident in the Chernobyl power plant, in the former USSR, 1986; the level 6 serious accident in Khystym, former USSR, 1957; and the level 6 accidents with off-site risk in Sellafield, UK, 1957, and Harrisburg, USA, 1979), large explosions (Enschede, The Netherlands, 2000; Toulouse, France, 2001), chemical disasters (Sandoz, Switzerland, 1986), and biological hazards or ecological disasters (including killer bees; ants endangering the red crab population of Christmas Island). Mine accidents, major train accidents (Eschede, Germany, 1998 [20]; London, UK, 1999; Hatfield, UK, 2001; Neishabur, Iran, 2004; Ryongchon, North Korea, 2004), aircraft crashes (New Dehli 1996; Paris, 2000; Bodensee 2002), and sunken ships (Estonia, Baltic Sea, 1994; Pallas, North Sea 1998; Tricolor, English Channel, 2002; Prestige, Atlantic Ocean, 2002) should also be mentioned. For obvious reasons, we will not discuss the issue of the vulnerability of critical infrastructures here. However, one can probably assume that the greatest threats in the future are potentially related to nuclear pollution, epidemic diseases, and disasters related to global warming (such as the melting of the polar icecaps, floods, and heavy storms).

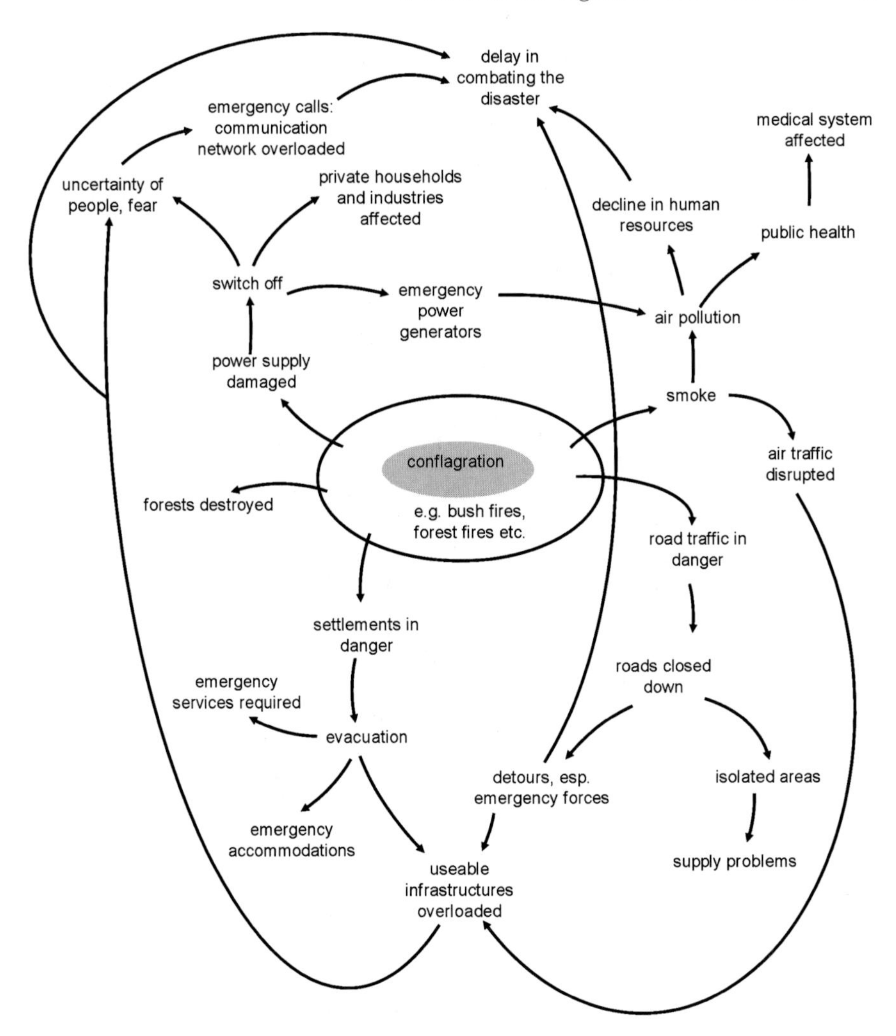

Fig. 15.4. Causality network of large-scale fires

15.2.7 Secondary and Tertiary Disasters

A disaster does not only spread in space and time and affect various sectors of a system. It may also trigger another kind of disaster. For example, an earthquake may cause power blackouts, a fire disaster, landslides, floods, or an interruption in the water supply. Thunderstorms may cause blackouts, fires, landslides, or floods. Floods may cause a lack of drinking water, blackouts, landslides, or epidemic diseases. Instead of adding more examples, we would like to refer the reader to Fig. 15.5.

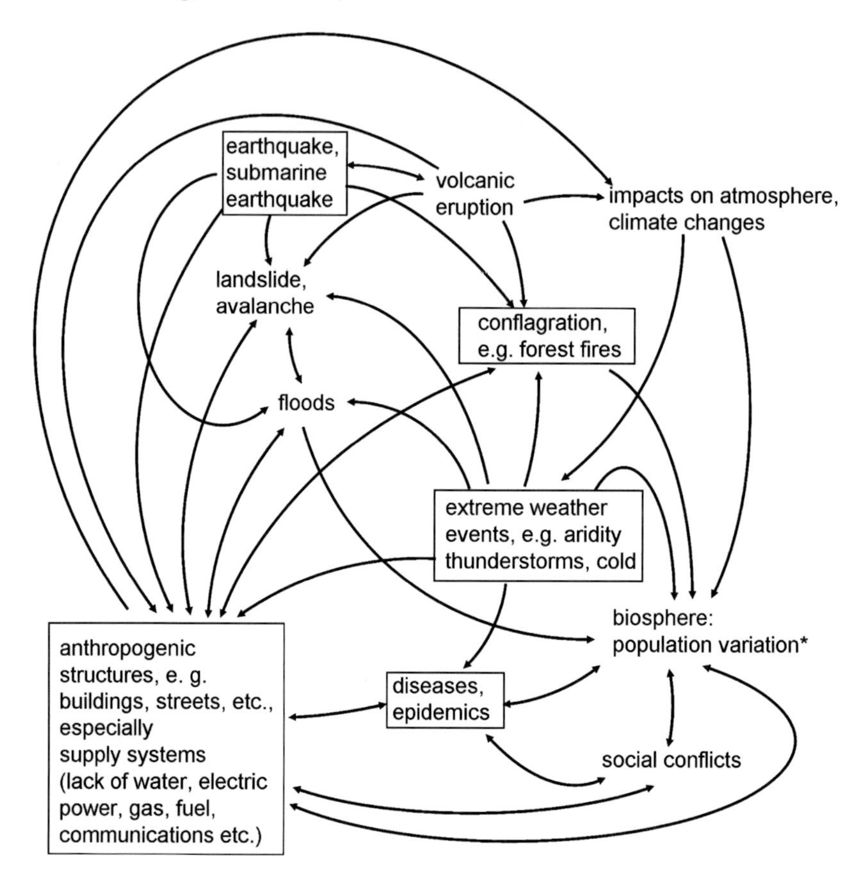

For more details see other figures.

)* Reaction of the biosphere: unusual population, all kind of organisms (incl. humans) can be affected in form of population increase or decrease, perturbing the balance of available and required resources including crop loss, shortage of food and famine, explosion of pathogens, pests, etc.

Fig. 15.5. Causality network illustrating how one kind of disaster may trigger another

15.2.8 Common Elements of Disasters

Despite the different origins of disasters, they share many common elements (see Fig. 15.6). We will summarize some of them here. Disasters often start with a large perturbation or disruption of some system component, and they spread via networks to other system components. Most disasters cause serious traffic, transportation and supply problems, and regular trade may break down. In the worst case, the disaster area is isolated from its environment and hardly reachable. For example, in 1970 a gigantic landslide on the highest mountain of Peru buried many villages and the city of Yungay after an earthquake. It took 24 hours for the total destruction of these towns to be recognized. One week later, two people arrived at the coast to inform the public that help was yet to arrive in the area. It took two months until NASA could identify the full scale of the disaster by air photographs, and after four and a half months, some villages had still not been reached by cars or planes. Another similar example was the Heta, the cyclone that devastated the South Pacific island of Niue in 2004.

During a disaster, a blackout of electricity is rather common. Note that this can have many serious implications (see also Sect. 15.2.2):

- Public transport is interrupted and streets are often congested (as long as fuel is available)
- Home heating systems stop working
- Water cannot be boiled, so a scarcity of drinking water may occur.
- Automatic teller machines and cashdesks in supermarkets do not work.
- Hospitals must be evacuated after a certain time period
- Communication breaks down

Even if power is available, information is a problem. There is often a lack of reliable information, and instead a flood of inconsistent data or rumors, and not enough time to evaluate them. Nevertheless, decisions must be made fast, in the right order, with the right priorities and under stress. Therefore, wrong decisions are likely. Apart from these problems, coordination is also a problem due to incompatibilities between communication systems, orientation problems in an unknown terrain (many road signs may have disappeared), administrative obstacles and legal responsibilities, which can reduce the flexibility of response when improvisation is needed.

Although the increased solidarity during disasters can be very helpful, it is hard to coordinate many people and different organizations that have not collaborated before and do not know each others' command structures. Such interaction must be exercised beforehand if fast and reliable actions are to be performed without the need for much discussion; in other words it should be based on certain codes and protocols.

When disasters strike, the surviving population tends to panic, particularly after events that may repeat, such as earthquakes. Moreover, panic buying (hoarding), if still possible, is typical. There are also people who use the opportunity to plunder shops and houses, particularly after the population has been evacuated. This often causes a resistance to evacuation measures from the population, so that expensive individual evacuation, by helicopter, may be needed later on. In any case, evacuation is a great burden on the population, as many thousands of people may become homeless. In the worst case, this can cause worldwide streams of migrants and refugees.

If resources are scarce, riots may break out, and a black market emerges. Criminal activity will go up, if the public authorities (police and military) lose control. Here, it must be considered that the task forces fighting the disaster will be exhausted after 72 hours at the most, which may cause a lack of manpower.

Pogroms may occur in the population if certain minorities are believed to be responsible for the disaster. This is particularly relevant to certain diseases, religious or racial affairs. Epidemics are a typical problem after disasters, either because water is contaminated, because the large number of corpses cannot be buried fast enough, or because the health of the population is poor anyway (due to hunger or cold). Finally, disasters have serious economic consequences, sometimes covering many years. Due to this and problems in disaster response management, the government's reputation may be tarnished and it may lose its power.

15.3 Modeling Causality Networks of Disaster Spreading

In this section, we will discuss a semi-quantitative method [21] that will allow us to:

- estimate the development of disasters over time
- get hints about when to take certain actions
- assess the suitability of alternative measures of emergency management
- anticipate the side effects measures of emergency management

To do this, it is necessary to take into account all of the factors that are relevant during the disaster and all direct and indirect interactions between them. This method follows the tradition of system dynamics [22].

We will start with a static analysis of interaction networks. For this, let us specify the approximate influence of different factors or sectors on each other. Such factors may, for example, be the energy supply, public transport, or medical support. In principle it is a long list of variables i, all of which may play a role in the problem under consideration. If we represent the influence of factor j on factor i by A_{ij}, we can summarize these (direct) influences using a matrix $\mathbf{A} = (A_{ij})$. However, in practical applications, one faces the following problems:

(i) The number of possible interactions grows quadratically with the number of variables or factors i. It is, therefore, difficult to measure or even estimate all of the influences A_{ij}.

(ii) While it appears feasible to determine the *direct* influence M_{ij} of one variable j on another one i, it is hard or almost impossible to estimate indirect influences on various nodes of the graph, which enter into A_{ij} as well. However, feedback loops may have an important effect and may neutralize or even overcompensate for the direct influences.

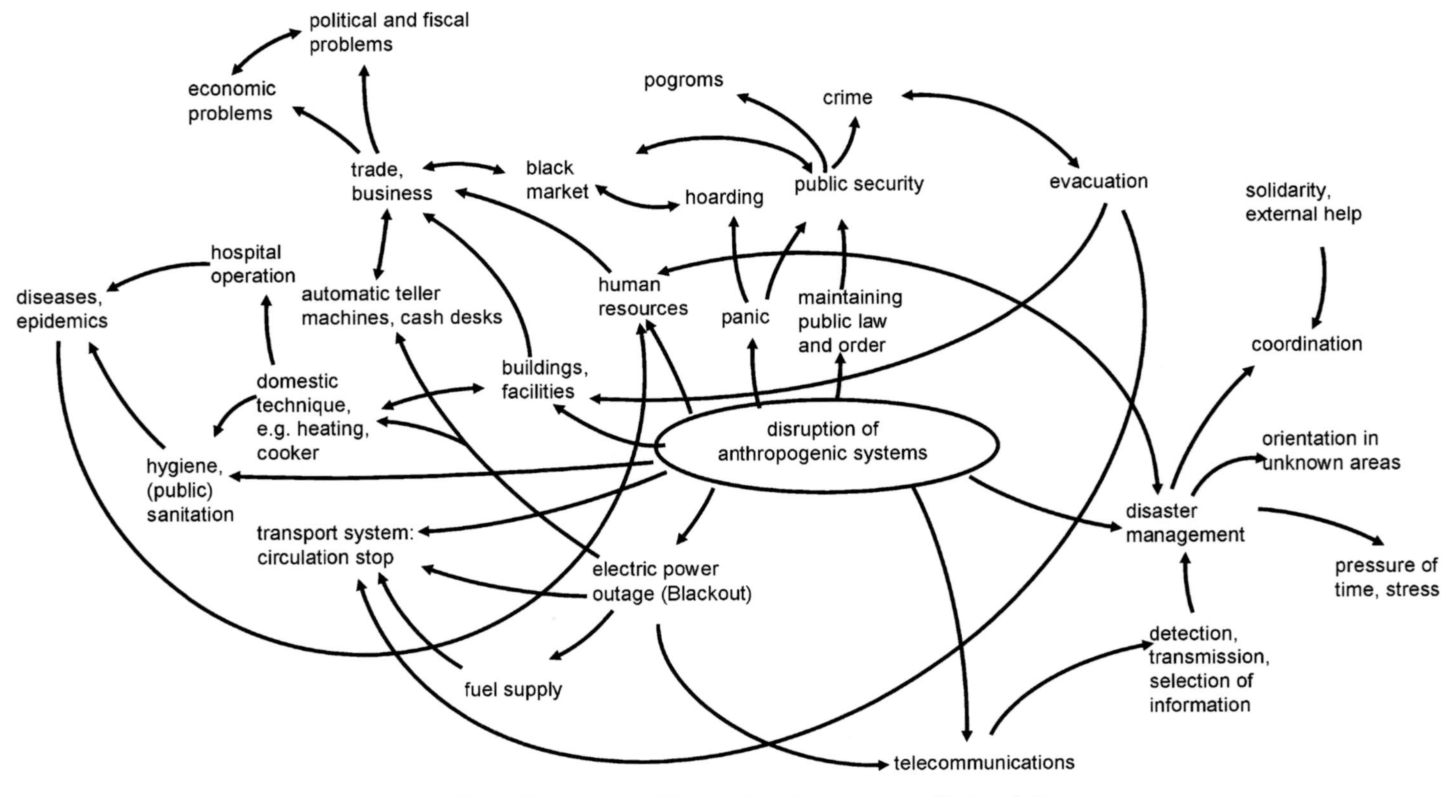

Fig. 15.6. Causality network illustrating the common effects of disasters

Problem (i) can be partially resolved by clustering similar variables and selecting a representative for each cluster of variables. The remaining set of variables should contain the main explanatory variables. Systematic statistical methods for such a procedure are available in principle, but intuition may be a good guide when the quantitative data required for the clustering of variables are missing.

Problem (ii) can be addressed by estimating the *indirect* influences due to feedback loops via the *direct* influences M_{ij}, which can be summarized using the matrix $\mathbf{M} = (M_{ij})$. We can use a formula such as

$$\mathbf{A}' = \mathbf{A}'_\tau = \frac{1}{\tau}\sum_{k=1}^{\infty}(\tau\mathbf{M})^k = \frac{1}{\tau}\sum_{k=1}^{\infty}\tau^k\mathbf{M}^k = \sum_{k=1}^{\infty}\tau^{k-1}\mathbf{M}^k \, , \qquad (15.1)$$

but as this only converges for small values of τ, we will instead use the formula

$$\mathbf{A} = \mathbf{A}_\tau = \frac{1}{\tau}\sum_{k=1}^{\infty}\frac{\tau^k\mathbf{M}^k}{k!} = \frac{1}{\tau}[\exp(\tau\mathbf{M}) - \mathbf{1}] \, , \qquad (15.2)$$

where $\mathbf{1}$ denotes the unity matrix. The expression $\mathbf{M}^k$ reflects all influences over $k-1$ nodes and k links, so $k = 1$ corresponds to direct influences, $k = 2$ to feedback loops with one intermediate node, $k = 3$ to feedback loops with two intermediate nodes, and so on. The prefactor τ^k is not only required for convergence, but with $\tau < 1$, it also allows us to take into account that indirect interactions often become weaker the more edges (nodes) there are in-between.

A further simplification can be achieved by restricting influences to a few characteristic discrete values. We may, for example, restrict ourselves to

$$M_{ij} \in \{-3, -2, -1, 0, 1, 2, 3\} \, , \qquad (15.3)$$

where $M_{ij} = \pm 3$ means an extremely positive or negative influence, $M_{ij} = \pm 2$ represents a strong influence, $M_{ij} = \pm 1$ a weak influence, and $M_{ij} = 0$ a negligible influence. Of course, a finer differentiation is possible wherever necessary. (For an investigation of stylized relationships, it can also make sense to choose $M_{ij} \in \{-1, 0, 1\}$, where $M_{ij} = \pm 1$ represents a strongly positive or negative influence.) The matrix $\mathbf{A} = (A_{ij})$ will be called the assessment matrix and it summarizes all direct influences ($\mathbf{M}$) and feedback effects ($\mathbf{A} - \mathbf{M}$) among the investigated factors. It allows conclusions about

- the resulting strengths of desireable and undesireable interactions, when feedback effects are included
- the effect of the failure of a specific sector (node)
- the suitability of possible measures for achieving specific goals or improvements
- the side effects of these measures on other factors

This will be illustrated in more detail by the example in Sect. 15.3.1.

One open problem is the choice of the parameter τ. It controls how strong the indirect effects are in comparison to the direct effects. A small value of τ corresponds to neglecting indirect effects, in other words

$$\lim_{\tau \to 0} \mathbf{A}_\tau = \mathbf{M} , \tag{15.4}$$

while increasing values of τ reflect the growing influence of indirect effects. This is often the case for disasters, as these are frequently related to avalanches or percolation effects. By varying τ, one can study different scenarios.

Note that τ may be interpreted as a time coordinate. Defining

$$\boldsymbol{X}(\tau) = \exp(\tau \mathbf{M}) \boldsymbol{X} \tag{15.5}$$

for an arbitrary vector $\boldsymbol{X}$, we find $\boldsymbol{X}(0) = \boldsymbol{X}$,

$$\frac{\boldsymbol{X}(\tau) - \boldsymbol{X}(0)}{\tau} = \frac{1}{\tau}[\exp(\tau \mathbf{M}) - \mathbf{1}]\boldsymbol{X}(0) = \mathbf{A}_\tau \boldsymbol{X}(0)$$

and

$$\frac{d\boldsymbol{X}}{d\tau} = \lim_{\tau \to 0} \frac{\boldsymbol{X}(\tau) - \boldsymbol{X}(0)}{\tau} = \mathbf{M}\boldsymbol{X}(0) .$$

From this point of view,

$$\boldsymbol{X}(\tau) = (\tau \mathbf{A}_\tau + \mathbf{1})\boldsymbol{X}(0) \tag{15.6}$$

describes the state of the system at time τ, and M_{ij} the changing rates. $\boldsymbol{X} = \mathbf{0}$ is a stationary solution and corresponds to the normal (everyday) state. An initial state $\boldsymbol{X}(0) \neq \mathbf{0}$ may be interpreted as a perturbation of the system by some (catastrophic) event. We should, however, note that the linear system of equations (15.6) is certainly a rough description of the system dynamics. It is expected to hold only for small perturbations of the system state, and it does not consider damping effects due to disaster response management. These aspects will be considered in Sect. 15.3.2.

15.3.1 Assessment of Disaster Management Methods

One advantage of our semi-quantitative approach to disasters is that it allows us to estimate the impact of certain actions on the whole range of factors [21]. As we have argued before, all direct and indirect effects are summarized by the matrix $\mathbf{A}$, which is determined from the matrix $\mathbf{M}$ of direct interactions. Different measures taken are reflected by the use of different matrices $\mathbf{M}$.

As an example, let us consider the spread of a disease. For illustrative reasons, we will restrict ourselves to a discussion of just five factors:

1. the number of infected persons
2. the quality of medical care

3. the public transport
4. the economic situation
5. the disposal of waste

These factors are not independent of each other, as illustrated by Fig. 15.7.

The corresponding matrix of the assumed direct influences among the different factors is

$$\mathbf{M} = \begin{pmatrix} 0 & -2 & +2 & 0 & -1 \\ -2 & 0 & +1 & +2 & +1 \\ -1 & 0 & 0 & +2 & 0 \\ -1 & 0 & +2 & 0 & +1 \\ -1 & 0 & +1 & +2 & 0 \end{pmatrix} \tag{15.7}$$

The correct choice for the sign of the direct influence M_{ij} of factor j on factor i is obtained as follows. We assume a positive sign if the factor i increases with an increase in factor j, while we assume a negative sign when factor i decreases with the growth in factor j. However, any determination of the absolute value of M_{ij} requires empirical data, expert knowledge, or experience. We have argued as follows:

- The growing number of infected persons affects all other factors in a negative way (see first column), as they will not be able to work. That is, economic problems will occur, as will problems with public transport and the disposal of waste. Health care is affected twice, since medical personnel may be infected and a higher number of patients will need to be treated, and capacities are limited. Therefore, we have chosen a value of -2 in this case, but -1 for the other factors.
- An effectively operating health system (second column) can reduce the number of infected persons efficiently, so we have chosen a value of -2 here. The health system was assumed to exert only an indirect effect on the economic situation and other factors (by reducing the number of ill persons).
- Public transport (third column) aids the spread of the infection assumed here (which could be, for example, SARS). Therefore, we have selected

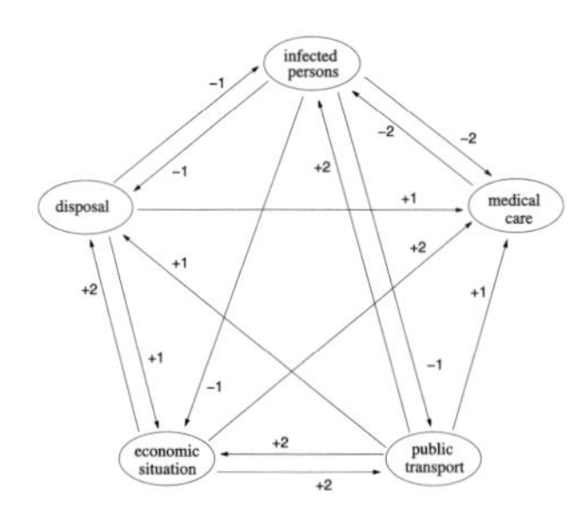

Fig. 15.7. Simplified interaction network for the example of the spread of a disease, as discussed in the text (after [21])

a value of 2. Transport is also an important factor for economic prosperity (leading to a value of 2 here), and transport is required to get medical personnel and workers in the disposal sector to their workplaces (which is reflected in the value of 1).

- The economic situation (fourth column) has a significant effect on the quality of the health system, public transport, and disposal, so we have chosen a value of 2 in each case.
- Waste may contribute to the spread of the disease if it is not properly removed. Therefore, a good disposal system (fifth column) may reduce the number of infections (giving a value of -1). It is also required for a functioning health system and steady economic production. This is why we have assumed a value of 1 here.

Depending on the respective situation, the concrete values of the direct influences M_{ij} may be somewhat different. When specifying them, it can be useful to check the values of A_{ij} for the direct and indirect influences for their plausibility, and to compare the sizes of the second-order or third-order interactions. For example, we see that the third-order feedback loop "number of infected persons→economic situation→quality of the health system→number of infected persons" is proportional to $(-1)\cdot(+2)\cdot(-2) = 4$. The same indirect influence is found for the feedback loop "number of infected persons→economic situation→public transport→number of infected persons". Moreover, according to our assumptions, the second-order autocatalytic increase in the number of infected persons due to its impact on the health system is four times as large as the one due to its impact on the waste disposal system. One surprising observation is that the number of infected persons drops due to its impact on public transport. In fact, once the number of buses drops (because the bus drivers are ill), the spread of the disease is slowed down. This suggests that in the event of a contagious disease, we should interrupt public transport; however, later on we will see that doing this has some serious side effects.or example Before we look at that, let us have a look at the resulting overall interaction matrix

$$\mathbf{A} = (A_{ij}) = \begin{pmatrix} 0.9 & -2.2 & 1.3 & -0.8 & -1.6 \\ -3.4 & 1.1 & 1.5 & 3.5 & 2.3 \\ -1.7 & 0.6 & 0.5 & 2.5 & 0.8 \\ -2.0 & 0.6 & 2.1 & 1.5 & 1.6 \\ -2.0 & 0.6 & 1.5 & 2.9 & 0.9 \end{pmatrix} \tag{15.8}$$

To calculate it, we have chosen the value $\tau = 0.4$, which will also be used later on to assess alternative actions for fighting the spread of the disease. In order to discuss a certain scenario, we will assume that X_j reflects the perturbation of factor j. Because of (15.6), the quantities

$$Y_i = \sum_j (\tau A_{ij} + \delta_{ij}) X_j \tag{15.9}$$

will be used to characterize the potential response of the system in the specific scenario described by the perturbations X_j (and without the damping effects resulting from the disaster response management discussed in Sect. 15.3.2). Here, δ_{ij} denotes the Kronecker function, which is 1 for $i = j$ and 0 otherwise. We will assume $X_1 = 1.0$, as the number of infected persons is higher than normal, and $X_2 = X_3 = X_4 = X_5 = -0.1$, as the other factors are reduced by the spread of the disease:

$$(X_1, X_2, X_3, X_4, X_5) = (1.0, -0.1, -0.1, -0.1, -0.1) . \qquad (15.10)$$

Moreover, if we attribute a weight of $w_1 = 0.5$ to the number of infected persons, a weight of $w_4 = 0.3$ to the economic situation, and weights of $w_2 = w_3 = 0.1$ to the quality of medical care and public transport, and ignore the issue of waste in our evaluation (so $w_5 = 0$), the resulting value of

$$F = F_\mathcal{T} = \left(\sum_i w_i Y_i^2 \right)^{1/2} \qquad (15.11)$$

will be used to assess the overall state of the system. In the stationary (normal) system state, F would be zero. Therefore, we want to find a strategy which brings F close to zero. For our basic scenario, we find

$$(Y_1, Y_2, Y_3, Y_4, Y_5) = (1.5, -1.8, -1.0, -1.1, -1.1) \quad \text{and} \quad F = 1.4 . \qquad (15.12)$$

These reference values will be compared with the values obtained for alternative scenarios which correspond to different actions taken to fight the disaster.

For example, let us assume that there are limited stocks of vaccine for immunization. Should we use these to immunize 1) the transport workers, 2) the medical staff, or 3) the disposal workers? In the first case, we have the modified matrix

$$\mathbf{M} = \begin{pmatrix} 0 & -2 & +2 & 0 & -1 \\ -2 & 0 & +1 & +2 & +1 \\ \underline{0} & 0 & 0 & +2 & 0 \\ -1 & 0 & +2 & 0 & +1 \\ -1 & 0 & +1 & +2 & 0 \end{pmatrix} , \qquad (15.13)$$

which implies

$$\mathbf{A} = \begin{pmatrix} 1.2 & -2.3 & 1.4 & -0.8 & -1.7 \\ -3.2 & 1.1 & 1.5 & 3.5 & 2.2 \\ -0.5 & 0.1 & 0.8 & 2.4 & 0.5 \\ -1.6 & 0.5 & 2.2 & 1.5 & 1.5 \\ -1.7 & 0.6 & 1.6 & 2.9 & 0.9 \end{pmatrix} , \qquad (15.14)$$

$$(Y_1, Y_2, Y_3, Y_4, Y_5) = (1.6, -1.7, -0.5, -1.0, -1.0) , \quad \text{and} \quad F = 1.4 . \qquad (15.15)$$

In the second case, when we immunize the medical staff, we find

$$\mathbf{M} = \begin{pmatrix} 0 & -2 & +2 & 0 & -1 \\ \underline{-1} & 0 & +1 & +2 & +1 \\ -1 & 0 & 0 & +2 & 0 \\ -1 & 0 & +2 & 0 & +1 \\ -1 & 0 & +1 & +2 & 0 \end{pmatrix}, \tag{15.16}$$

which implies

$$\mathbf{A} = \begin{pmatrix} 0.5 & -2.1 & 1.3 & -0.7 & -1.5 \\ -2.3 & 0.7 & 1.8 & 3.4 & 2.0 \\ -1.7 & 0.6 & 0.5 & 2.5 & 0.8 \\ -1.9 & 0.6 & 2.1 & 1.5 & 1.6 \\ -1.9 & 0.6 & 1.6 & 2.9 & 0.9 \end{pmatrix}, \tag{15.17}$$

$$(Y_1, Y_2, Y_3, Y_4, Y_5) = (1.3, -1.3, -0.9, -1.1, -1.1), \quad \text{and} \quad F = 1.2. \tag{15.18}$$

In the third case, when the disposal workers are immunized, we expect

$$\mathbf{M} = \begin{pmatrix} 0 & -2 & +2 & 0 & -1 \\ -2 & 0 & +1 & +2 & +1 \\ -1 & 0 & 0 & +2 & 0 \\ -1 & 0 & +2 & 0 & +1 \\ \underline{0} & 0 & +1 & +2 & 0 \end{pmatrix}, \tag{15.19}$$

which implies

$$\mathbf{A} = \begin{pmatrix} 0.6 & -2.1 & 1.3 & -0.7 & -1.6 \\ -3.1 & 1.1 & 1.6 & 3.5 & 2.2 \\ -1.6 & 0.6 & 0.5 & 2.5 & 0.8 \\ -1.7 & 0.6 & 2.1 & 1.5 & 1.5 \\ -0.8 & 0.2 & 1.9 & 2.8 & 0.6 \end{pmatrix}, \tag{15.20}$$

$$(Y_1, Y_2, Y_3, Y_4, Y_5) = (1.4, -1.7, -0.9, -1.0, -0.7), \quad \text{and} \quad F = 1.3. \tag{15.21}$$

While the immunization of the public transport staff has almost no effect on the overall state of the system, the last two measures can improve it. We see that it is more effective to immunize the medical staff than the disposal workers, although the best approach would be to immunize both groups. This

corresponds to

$$\mathbf{M} = \begin{pmatrix} 0 & -2 & +2 & 0 & -1 \\ \underline{-1} & 0 & +1 & +2 & +1 \\ -1 & 0 & 0 & +2 & 0 \\ -1 & 0 & +2 & 0 & +1 \\ \underline{0} & 0 & +1 & +2 & 0 \end{pmatrix}, \qquad (15.22)$$

and we obtain

$$\mathbf{A} = \begin{pmatrix} 0.2 & -2.0 & 1.2 & -0.7 & -1.5 \\ -1.9 & 0.6 & 1.9 & 3.4 & 1.9 \\ -1.6 & 0.5 & 0.5 & 2.5 & 0.8 \\ -1.7 & 0.6 & 2.1 & 1.5 & 1.5 \\ -0.8 & 0.2 & 1.9 & 2.8 & 0.6 \end{pmatrix}, \qquad (15.23)$$

$$(Y_1, Y_2, Y_3, Y_4, Y_5) = (1.2, -1.2, -0.9, -1.0, -0.6)\,, \quad \text{and} \quad F = 1.1\,. \qquad (15.24)$$

Other measures do not change the interactions in the system, but correspond to a change in the effective impact $\boldsymbol{X}$ of the disaster. For example, we may consider reducing public transport. With (15.8) and

$$(X_1, X_2, X_3, X_4, X_5) = (1.0, -0.1, \underline{-1.0}, -0.1, -0.1)\,, \qquad (15.25)$$

we find

$$(Y_1, Y_2, Y_3, Y_4, Y_5) = (1.0, -2.4, -2.0, -1.9, -1.7) \quad \text{and} \quad F = 1.6\,. \qquad (15.26)$$

We see that the number of infections can, in fact, be reduced. However, the overall situation of the system has deteriorated, as the economic situation and all of the other sectors were negatively affected by the reduction in public transport, because many people could not reach their workplace. Therefore, let us consider the option to increase the number of disposal workers. With (15.8) and

$$(X_1, X_2, X_3, X_4, X_5) = (1.0, -0.1, -0.1, -0.1, \underline{0.5})\,, \qquad (15.27)$$

we find

$$(Y_1, Y_2, Y_3, Y_4, Y_5) = (1.1, -1.3, -0.8, -0.8, -0.3) \quad \text{and} \quad F = 1.0\,. \qquad (15.28)$$

In conclusion, increasing the level of hygiene can be surprisingly effective.

Finally, let us assume that waste disposal is improved and that the medical staff and the disposal workers are both immunized. In that case, the interactions of the relevant factors are characterized by matrix (15.22), whereas the starting vector is again (15.27). The resulting response is

$$(Y_1, Y_2, Y_3, Y_4, Y_5) = (0.8, -0.7, -0.7, -0.6, 0.1) \quad \text{and} \quad F = 0.75\,. \qquad (15.29)$$

This is the only combination of measures that actually manages to reduce the number of infections compared to the initial state ($Y_1 < X_1$). However, we can also see that a negative impact on the economic situation and other factors is unavoidable. In any case, we can assess which measures are reasonable to use, what impact they will have on the system, and which of the measures need to be combined in order to control the spread of the disease (or other problems in different scenarios).

15.3.2 System Dynamics Treatment of the Spread of a Disaster

Before, we predominantly used the interaction network for a static assessment of the influence of different factors on each other. We will now try to extend this method in a way that allows us to perform a semi-quantitative analysis of the time-dependence of disasters for the purpose of anticipation, which helps to prepare for the next step in disaster response management or prevention [21]. We are especially interested in the domino or avalanche effects of particular events such as the failure of a particular factor or sector in the interaction network. We will assume that this failure spreads along, in the order of, the direct connections in the interaction network (causality graph). In terms of the example in Sect. 15.3.1, a failure of medical care would first affect the number of infected persons, and then the economic situation, public transport, and the disposal of waste.

For a description of the dynamics of the disaster, let us assume that $P_i(\tau)$ denotes the impact on factor i at time τ and W_{ji} the rate at which this impact spreads to factor j, while D_i is a damping rate describing the mitigation of the catastrophic impact on factor i by disaster response management. In this case, it is reasonable to assume the dynamics

$$\frac{d\boldsymbol{P}}{d\tau} = (\mathbf{W} - \mathbf{D})\boldsymbol{P}(\tau) = \mathbf{L}\boldsymbol{P}(\tau) \tag{15.30}$$

with $\mathbf{D} = (\delta_{ij} D_i)$, $\mathbf{L} = (L_{ij}) = (W_{ij} - \delta_{ij} D_i)$, and $\boldsymbol{P}(\tau) = (P_i(\tau))$. The symbol δ_{ij} again represents the Kronecker function, (1 for $i = j$ and 0 otherwise). When no better information is available, we may assume that the spreading rate W_{ij} is proportional to the strength $|M_{ij}|$ of the direct influence of factor j on factor i. With a constant proportionality factor c, this means

$$W_{ij} \approx c|M_{ij}| \, . \tag{15.31}$$

The formal solution of (15.30) for a time-independent matrix $\mathbf{L}$ is given by

$$\boldsymbol{P}(\tau) = \exp(\mathbf{L}\tau)\boldsymbol{P}(0) = \sum_{k=0}^{\infty} \frac{\tau^k}{k!} \mathbf{L}^k \boldsymbol{P}(0) = \mathbf{B}(\tau)\boldsymbol{P}(0) \, . \tag{15.32}$$

That is, $\mathbf{B}(\tau)$ describes the spread of an event in the causality network (interaction network) over the course of time τ, while $\boldsymbol{P}(0)$ reflects the initial impact of a catastrophic event.

When we assume

$$D_i = \sum_j W_{ji} , \tag{15.33}$$

(15.30) is related to the Liouville representation of the discrete master equation. In this case, we can apply all of the solution methods developed for it. This includes the so-called path integral solution [23], which allows one to calculate the probability of occurrence of specific spread paths. This has some interesting implications. For example, the danger that the impact on sector i_0 affects the sectors $i_1, i_2, \dots, i_n$ in the indicated order is quantified by

$$P(i_0 \to i_1 \to \dots \to i_n) = \frac{|P_{i_0}(0)|}{D_{i_n}} \prod_{l=0}^{n-1} \frac{W_{i_{l+1},i_l}}{D_{i_l}} \approx c^n \frac{|P_{i_0}(0)|}{D_{i_n}} \prod_{l=0}^{n-1} \frac{|M_{i_{l+1},i_l}|}{D_{i_l}} . \tag{15.34}$$

Moreover, the average time at which this series of events occurs can be calculated using

$$T(i_0 \to i_1 \to \dots \to i_n) = \sum_{l=0}^{n} \frac{1}{D_{i_l}} , \tag{15.35}$$

and the variance of this time is determined by

$$\Theta(i_0 \to i_1 \to \dots \to i_n) = \sum_{l=0}^{n} \frac{1}{(D_{i_l})^2} . \tag{15.36}$$

That is, (15.30) not only allows us to assess the likelihood of a certain series of events, but it also gives their approximate appearance times. In other words, we have a detailed picture of potential catastrophic scenarios and of their time evolutions, which facilitates specific preparation and disaster response management.

In the following, we do not want to restrict ourselves to case (15.33). If

$$D_i < \sum_j W_{ji} \tag{15.37}$$

for all i, the damping is weak and the solutions $P_i(\tau)$ are expected to grow more or less exponentially over the course of time, which describes a scenario where control is lost and the disaster spreads all over the system. In many cases, we will have

$$D_i > \sum_j W_{ji} \tag{15.38}$$

for all i; in other words, the impact of the disaster on the system decays over the course of time, and $\lim_{\tau \to 0} P_i(\tau) \to 0$. This determines how strong the damping effects need to be (or, in other words, the method of counteracting the disaster).

Finally, it may also happen that $D_i > \sum_j W_{ji}$ for some factors i, but $D_i < \sum_j W_{ji}$ for others. In such situations, everything depends on the initial impact $\boldsymbol{P}(0)$ and on the matrix $\mathbf{B}(\tau)$. However, in all of these cases, (15.34) to (15.36) remain valid.

15.4 Summary and Conclusions

In this contribution, we have discussed disasters as important examples of Xevents. They are often characterized by power laws, which is partly related to the tendency to drive a system to its critical threshold in order to increase its efficiency. Unfortunately, self-organized criticality is known to produce avalanche effects of a potentially arbitrary size. Such cascade effects can be observed in many different kinds of disasters.

Our modeling approach is based on identifying interactive causality chains, as has been illustrated for many different kinds of disasters. Moreover, we have suggested a semi-quantitative treatment that quantifies the strength of direct interactions in order to assess the relevance of indirect effects and feedback loops. This causality network approach allows one to assess not only the effectiveness of alternative measures of disaster response management and their side effects, but it also makes it possible to estimate the time at which certain events could happen via the spreading of perturbations within a causality network. We hope that this will help encourage anticipative rather than reactive disaster response management [24–32].

Network theory could certainly make further contributions to disaster response management. As disaster response management can be viewed as a problem of material, personal, and information logistics, models of supply networks [33] will be highly relevant. This includes issues of dynamic stability of disaster response management measures [21], as well as error and attack tolerances of networks [34,35]. The problem can be even viewed as a network of networks [36]. That is, it will not only be important to optimize the social, information, material, transportation and other networks involved [37], but also their mutual interactions. This means that both supply and coordination [38, 39] are crucial issues. In this respect, we hope to learn from biological systems, which have optimized network interactions over millions of years in an evolutionary way. Another promising issue is the development of new principles of disaster response management based on self-organization. It is potentially more effective to have autonomous units (task forces) with predefined interaction possibilities [34]. This could increase adaptiveness and flexibility [40–42] based on principles of decentralized control and collective intelligence.

Acknowledgement. The authors are grateful for partial financial support by the German Research Foundation (DFG project He 2789/6-1).

References

1. M. Falk: *Laws of Small Numbers: Extreme and Rare Events* (Birkhäuser, Basel, 1994)
2. J. Nott: *Extreme Events: Reconstruction from Natural Records and Hazard Risk Assessment* (Cambridge University, Cambridge, in press)
3. A. Bunde, J. Kropp, H.J. Schellnhuber (eds.): *The Science of Disasters: Climate Disruptions, Heart Attacks and Market Crashes* (Springer, Berlin Heidelberg New York, 2002)
4. A.K. Turner, R.L. Schuster (eds.): *Landslides: Investigation and Mitigation* (National Research Council, Transportation Research Board, Special Report 247, 1996)
5. R. Casale et al. (eds.): *Flood and Land Slides: Integrated Risk Assessment* (Springer, Berlin Heidelberg New York, 1999)
6. S.J. Guastello: *Chaos, Catastrophe, and Human Affairs: Applications of Nonlinear Dynamics to Work, Organizations, and Social Evolution* (Erlbaum, Mahwah, 1995)
7. G. Woo: *The Mathematics of Natural Catastrophes* (World Scientific, Singapore, 1999)
8. Swiss Reinsurance Company (ed.): Sigma **1** (2004), pp. 38–39, see http://www.swissre.com/
9. I. Asimov: *A Choice of Catastrophes: The Disasters That Threaten Our World* (Simon & Schuster, New York, 1979)
10. IFRC: *World Disasters Report (WDR) 2004* (International Federation of Red Cross and Red Crescent Societies, Geneva, 2004), see http://www.ifrc.org/publicat/wdr2004/index.asp
11. V. Linneweber: *Zukünftige Bedrohungen durch (anthropogene) Naturkatastrophen*, (Deutsches Komitee für Katastrophenvorsorge, Bonn, 2002)
12. G. Tetzlaff, T. Trautmann, K.S. Radtke (eds.): Extreme Naturereignisse. Folgen – Vorsorge – Werkzeuge, 2. Forum Katastrophenvorsorge, Conference Proceedings (Deutsches Komitee für Katastrophenvorsorge, Bonn, 2002)
13. Munich Re (ed.): Jahresrückblick Naturkatastrophen 2003, TOPICSgeo **2** (2004)
14. Munich Re (ed.): Schadenspiegel **2** (2004)
15. *Bericht der Unabhängigen Kommission der Sächsischen Staatsregierung Flutkatastrophe 2002*, see http://www.sachsen.de/de/bf/hochwasser/programme/download/Kirchbach_Bericht.pdf
16. M. Paolini, G. Vacis: *The Story of Vajont* (Bordighera, Boca Raton, FL, 2000)
17. P.M. Maniscalco, H.T. Christen: *Understanding Terrorism and Managing the Consequences* (Prentice Hall, Upper Saddle River, 2001)
18. P.V. Fellman, R. Wright: Modeling Terrorist Networks, Complex Systems at the Mid-Range. Paper presented at the *Joint Complexity Conference* (London School of Economics, 16th–18th September 2003)
19. National Commission on Terrorist Attacks upon the United States (ed.): *The 9/11 Commission Report* (9-11 Commission, Washington DC, 2004)
20. E. Hüls, H.-J. Oestern (eds.): *Die ICE-Katastrophe von Eschede. Erfahrungen und Lehren. Eine interdisziplinäre Analyse* (Springer, Berlin Heidelberg New York, 1999)

21. D. Helbing, C. Kühnert: Physica A **328** (2003), pp. 584–606 (and references therein)
22. J.J. Gonzalez (ed.): *From Modeling to Managing Security: A System Dynamics Approach* (Norwegian Academic Press, Kristiansand, 2003), and references therein
23. D. Helbing: Phys. Lett. A **212** (1994), pp. 130–137 (1994)
24. W. Zelinsky, L.A. Kosinski: *L.A. 1991. The Emergency Evacuation of Cities* (Rowman & Littlefield, Savage, 1991)
25. H.T. Christen, P.M. Maniscalco: *The EMS Incident Management System: Operations for Mass Casualty and High Impact Incidents* (Prentice Hall, Upper Saddle River, NJ, 1998)
26. K.N. Myers: *Contingency Planning for Disasters: Protecting Vital Facilities and Critical Operations* (Wiley, New York, 1999)
27. W.L. Waugh: *Living with Hazards, Dealing with Disasters: An Introduction to Emergency Management* (M.E. Sharpe, New York, 2000)
28. D. Alexander: *Principles of Emergency Planning and Management* (Oxford University Press, New York, 2002)
29. P.A. Erickson: *Emergency Response Planning: For Corporate and Municipal Managers* (Academic, New York, 1999)
30. G. El Mahdy: *Disaster Management in Telecommunications, Broadcasting and Computer Systems* (Wiley, Chichester, UK, 2001)
31. R. Shaw, L. Walley: *Disaster Management* (Butterworth-Heinemann, Amsterdam, 2002)
32. G.D. Haddow, J.A. Bullock: *Introduction to Emergency Management* (Butterworth-Heinemann, Amsterdam, 2004)
33. D. Helbing: Modeling and Optimization of Production Processes: Lessons From Traffic Dynamics. In: *Nonlinear Dynamics of Production Systems*, ed. by G. Radons and R. Neugebauer (Wiley, New York, 2003), pp. 85–105
34. K. Tierney, J. Trainor: Networks and Resilience in the World Trade Center Disaster. In: *Research Progress and Accomplishments 2003–2004* ed. by the Multidisciplinary Center for Earthquake Engineering Research (University at Buffalo, New York, 2004), pp. 157–172
35. L.K. Comfort, K. Ko, A. Zagorecki: *Modeling Fragility in Rapidly Evolving Disaster Response Systems* (Institute of Governmental Studies, Paper No. 2003'2, University of California, Berkeley, CA, 2003)
36. L. Dueñas-Osorio, J.I. Craig, B.J. Goodno: Probabilistic Response of Interdependent Infrastructure Networks. In: *2004 ANCER Annual Meeting: Networking of Young Earthquake Engineering Researchers and Professionals*, Proceedings (The Sheraton Princess Kaiulani, Honolulu, Hawaii, 28th–30th July 2004)
37. D. Braha, Y. Bar-Yam: Information flow structure in large-scale product development organizational networks. In: *Smart Business Networks*, ed. by P. Vervest et al. (Springer, Berlin Heidelberg New York, 2004)
38. J.E. Trainor: *Searching for a System: Multi-organizational Coordination in the September 11th World Trade Center Search and Rescue Response* (MA Thesis, University of Delaware, Newark, DE, 2004)
39. L.K. Comfort, K. Ko, A. Zagorecki: Am. Behav. Sci. **48**, No. 3 (2004), pp. 295–313

40. A.H.J. Oomes: Organization Awareness in Crisis Management. Dynamic Organigrams for More Effective Disaster Response. In: *ISCRAM 2004 Proceedings* (Brussels, 3rd–4th May 2004)
41. S. Duman, A.S. Petrescu: When What We Know Does Not Apply: Disaster Response, Complexity Theory and Preparing for Bioterrorist Threats. Paper presented at the *16th International Conference of the Public Administration Theory Network* (Anchorage, AK, 19th–22nd June 2003)
42. G.A. Koehler (ed.): *What Disaster Response Management Can Learn from Chaos Theory* (California Research Bureau, Proc. Conf. 18th–19th May 1995)

Index